The Essential HANDBOOK OF GROUND-WATER SAMPLING

The Essential HANDBOOK OF GROUND-WATER SAMPLING

DAVID M. NIELSEN
GILLIAN L. NIELSEN

CRC Press
Taylor & Francis Group
Boca Raton London New York

CRC Press is an imprint of the
Taylor & Francis Group, an informa business

This material was previously published in *Practical Handbook of Environmental Site Characterization and Ground-Water Monitoring, 2nd Edition* © CRC Press LLC, 2005

CRC Press
Taylor & Francis Group
6000 Broken Sound Parkway NW, Suite 300
Boca Raton, FL 33487-2742

Printed in the United States of America on acid-free paper
10 9 8 7 6 5 4 3 2 1

International Standard Book Number-10: 1-4200-4278-5 (Softcover)
International Standard Book Number-13: 978-1-4200-4278-8 (Softcover)

Library of Congress Cataloging-in-Publication Data

Nielsen, David.
The essential handbook of ground-water sampling/ David M. Nielsen.
p. cm.
Includes bibliographical references and index.
ISBN 1-4200-4278-5 (alk. paper)
1. Groundwater--Sampling. I. Title.

GB1001.72.S3N54 2006
628.1'61--dc22 2006045060

Visit the Taylor & Francis Web site at
http://www.taylorandfrancis.com

and the CRC Press Web site at
http://www.crcpress.com

Preface

Difficulty in accessing ground water without disturbing ground-water flow patterns, chemistry, microbiology, and the physical and chemical makeup of formation materials has made accurate characterization of *in situ* ground-water conditions a very challenging task. Ground-water monitoring and sampling methodologies have evolved significantly over the past few decades as we have learned more about the mechanics of subsurface processes (particularly contaminant fate and transport processes), the impacts that traditional sampling efforts have on these processes, and the detrimental effects that many commonly used sampling protocols have on sample integrity and data quality. Tremendous improvements in field and laboratory analytical technologies and methods have paralleled developments in the ground-water industry. New equipment and technologies used to evaluate ground-water samples have made it possible to reliably analyze samples at increasingly lower levels – in the sub-parts-per-billion range for many chemical constituents. These developments have the potential to advance our level of understanding of subsurface processes and conditions light-years ahead of what it was less than a decade ago.

However, despite these important advances in ground-water monitoring, sampling and analytical methodologies and technologies, there is a reluctance to change, on both the regulatory and practitioner levels. Many state and local regulatory agencies continue to *require* the use of outdated sampling practices (for example, purging multiple volumes of water from a well, or purging a low-yield well to dryness prior to sampling) that are only rarely capable of producing samples that satisfy data-quality objectives. Many practicing environmental professionals remain entrenched in older, less-efficient sampling practices that have consistently been proven in multiple research studies to be unreliable for providing representative samples – the primary objective of virtually every ground-water sampling program. Despite research that demonstrates convincingly that some traditional sampling practices provide neither accuracy nor precision in sample collection (for example, using devices that significantly agitate the water column in the well, resulting in high artifactual turbidity levels), these practices continue to dominate in the field. Reasons that are often offered for continuing the use of outdated sampling methods include: "it would cost too much for us to change the methods we use now"; "we have years worth of data created using our current methods, and changing methods might cause changes in the data, which would be difficult and expensive to explain"; "we'll have to purchase new field equipment, which will cost more"; "we'll have to re-train all of our field staff and that would cost too much"; and "changing our regulations (or our standard operating procedures) will take a lot of time and cost a lot of money." Continuing the use of inappropriate sampling methods for the sake of maintaining consistency (whether in terms of cost or quality of data) is not a valid argument; if questionable or bad data were being collected before, collection of good data becomes even more important. The arguments related to economics are also invalid, as described below.

The underlying theme behind the failure of the industry to keep pace with advances in technology is cost. However, many practicing professionals fail to recognize that the arguments they make for keeping costs low are often based on false economics, both in terms of short-term (i.e., meeting data-quality objectives) and long-term (i.e., lowering overall project cost) project goals. The justification for continuing to use outdated sampling practices to save money is shortsighted, both technically and economically. Field and analytical data produced by ground-water sampling programs are often relied upon as the basis for making potentially far-reaching and expensive decisions. It is therefore critical that the data generated by ground-water sampling programs be of the highest quality possible so that the numbers accurately reflect formation water-quality characteristics and that they do so consistently from one sampling event to the next. The old adage "garbage in equals garbage out" applies here. Poor ground-water samples submitted for analysis will yield poor quality (inaccurate) and inconsistent (imprecise) data that are then used to make potentially very expensive decisions on a number of critical issues such as:

- Whether or not a site is in regulatory compliance with permit requirements;
- Whether or not a site can be closed, and/or monitoring discontinued;
- Whether or not a site requires active remediation;
- Whether or not a site is a candidate for risk-based corrective action or natural attenuation; and/or
- Whether contaminant levels at a site are sufficiently high to pose a health risk to nearby receptors of ground water.

Producing poor quality data can ultimately cost all of the stakeholders involved in a sampling program much more than the cost of changing to sampling methodologies that produce higher quality data. For example, the cost of delayed (or never initiated) response to contaminant detection caused by sample dilution due to overpurging a well might result in a delay in implementation of (or a failure to implement) necessary remediation which, in turn, results in more widespread contamination and, ultimately, higher cleanup costs. Alternately, apparent detection of contaminants due to false positives caused by using sampling methods that generate turbidity in samples could result in either a determination to implement an expensive ground-water remediation program where no actual ground-water contamination exists or an expensive effort to explain and document the reason for the false positive.

The cost of making incorrect decisions based on poor quality samples is clearly much greater than any cost savings that might be envisioned by shortsighted practitioners who insist on using outdated sampling procedures. It must be recognized that all decisions based on sampling data of questionable value will themselves be questionable. All efforts in sampling ground water should therefore be directed toward collecting the highest quality samples possible to ensure that decisions based on sampling results are scientifically and legally defensible.

The primary objective of this book is to allow readers to develop an understanding of why the continued use of outdated sampling methods will ultimately call into question the representative nature of the ground-water samples being collected and the quality of data generated by many current sampling programs, and to encourage regulators to specify and practitioners to use improved methods. A secondary objective is to help users understand why using outdated methods can cost significantly more than improving sampling techniques. As is explained in great detail through the text, figures, tables and references to recent research that appear herein, updating sampling protocols will significantly improve the quality of data generated by any ground-water sampling program. This, in turn, will provide all of the stakeholders involved in the program with a more accurate and precise picture of subsurface conditions, and ultimately result in lower overall project costs.

Ground-water sampling is a key component of any effective ground-water monitoring program and nearly all environmental site characterization programs, from those conducted at large Superfund sites; Department of Defense and Department of Energy facilities; petrochemical facilities; solid- and hazardous-waste landfills; and open-pit and underground mines, to those done at the corner service station or dry cleaning shop. Developing a scientifically valid and cost-effective ground-water sampling program requires an understanding of a variety of factors that can affect the integrity of samples collected from ground-water sampling points and, ultimately, the quality of analytical results and other data generated from those samples. These factors include formation and well hydraulics; sampling point placement, design, installation and maintenance; purging and sampling device selection and operation; water-level measurement methods; field equipment cleaning methods; purging and sample collection practices; and sample pretreatment, handling and shipment procedures.

The first few chapters of this book provide the reader with a detailed discussion of these factors and the processes that occur in sampling points between sampling events and during sample collection activities. The many important developments in ground-water sampling that have occurred in the past few decades are described, explanations of how new sampling methodologies and technologies can be effectively used in the field are provided, and many new ASTM standards that have been produced over the past decade to document and provide technical information on newly developed sampling protocols are introduced. Using all of this information, readers should be able to make sound technical and economic decisions regarding the choice of sampling equipment, methodologies and procedures to meet the site-specific objectives of a ground-water sampling program, to develop a sound and cost-effective sampling and analysis plan for field practitioners to follow, and to successfully carry out ground-water sampling events in the field.

However, while optimizing sampling methodologies is critical to preserving the representative nature of a sample, the journey of a sample is not complete at the end of sample collection. It is only finished when the sample is analyzed in the laboratory, and when laboratory analytical results are processed and the resulting analytical data interpreted and presented in a way that will allow hydrogeologists, regulators, risk assessors, remediation specialists and other data users to make sense of it. The final few chapters of this book are devoted to describing in detail these important aspects of a ground-water sampling program.

We hope this book will alert readers to the many key components that comprise a well-conceived and executed ground-water sampling program. We also hope that the book provides sufficient guidance to help practitioners produce meaningful data of the highest possible quality in a cost-effective manner, and assists them in using those data to accurately depict subsurface conditions to ensure prudent decision making.

Happy Sampling!

David M. Nielsen/Gillian L. Nielsen
The Nielsen Environmental Field School
Galena, Ohio, USA

Contributors

David M. Nielsen, C.P.G., C.G.W.P., P.Hg.
Nielsen Ground-Water Science, Inc.
The Nielsen Environmental Field School
Galena, OH

Gillian L. Nielsen
Nielsen Ground-Water Science, Inc.
The Nielsen Environmental Field School
Galena, OH

Olin C. Braids, Ph.D
O.C. Braids and Associates, LLC
Tampa, FL

Robert D. Gibbons
Director, Center for Health Statistics
Professor of Biostatistics and Psychiatry
University of Illinois at Chicago
Chicago, IL

Martin N. Sara
ERM, Inc.
Vernon Hills, IL

Rock J. Vitale, C.E.A.C., C.P.C.
Technical Director of Chemistry/Principal
Environmental Standards, Inc.
Valley Forge, PA

Matthew G. Dalton
Dalton, Olmsted and Fuglevand, Inc.
Kirkland, WA

Brent E. Huntsman
Terran Corp.
Beavercreek, OH

Ken Bradbury
Wisconsin Geological and Natural History Survey
University of Wisconsin – Extension
Madison, WI

Contents

Dedication

This book is dedicated to the memory of a dear friend, colleague, and contributor to this publication, Martin Sara, who passed away as this book was being assembled. Marty worked for many years in the environmental industry, where his tireless efforts helped advance the science of ground-water monitoring and sampling. His energy, laughter, enthusiasm and keen wit will be missed by all who knew and loved him.

Dave & Gillian Nielsen

1

The Science Behind Ground-Water Sampling

David M. Nielsen

CONTENTS

Objectives of Ground-Water Sampling

The overall objective of most ground-water sampling programs is to collect samples that are "representative," that is, samples that accurately reflect *in situ* ground-water conditions in the formation of interest at the site under investigation. Ground-water sampling programs are implemented at a variety of locations where they are commonly, although not exclusively designed to characterize or monitor ground-water contamination using traditional monitoring wells. At other sites, single-event or "snapshot" evaluations of ground-water chemistry are increasingly being made using direct-push sampling tools, which permit rapid characterization of ground-water conditions without using traditional monitoring wells. This approach to ground-water characterization is especially viable for properties undergoing real-estate transfers, but it can be used at any site to rapidly collect samples representative of formation conditions.

Regulatory Compliance Monitoring

At many sites, ground-water monitoring programs are required to be implemented under one or more U.S. EPA (or state-equivalent) regulatory programs such as the Resource Conservation and Recovery Act (RCRA), the Comprehensive Environmental Response, Compensation, and Liability Act (CERCLA or "Superfund"), the Clean Water Act (CWA), the Toxic Substances Control Act (TSCA), and the Safe Drinking Water Act (SDWA). In most cases, the objective of such monitoring programs is to provide information on ground-water chemistry, which will help to determine whether a regulated facility is in compliance. These different regulatory programs, discussed in more detail in Makeig and Nielsen (2006), regulate a wide variety of sites at which ground-water contamination has a potential to occur, including industrial and municipal solid-waste landfills, chemical and petroleum production facilities, industrial manufacturing facilities, U.S. Department of Defense sites, U.S. Department of Energy sites, uncontrolled hazardous waste sites, and underground storage tank sites. Under structured regulatory programs, either the U.S. EPA or the equivalent state regulatory agency requires that ground-water sampling programs be established as part of a ground-water monitoring program to satisfy a wide range of secondary objectives including determining: (1) whether the operation of a facility has had an effect on ground-water quality (i.e., has resulted in ground-water contamination); (2) the physical and chemical nature of the contamination; (3) the three-dimensional extent of the contamination; (4) the rate and direction of movement of that contamination with respect to other properties or receptors (i.e., water supplies) in the area; (5) what the most appropriate methods of remediating the ground-water contamination may be; (6) the effectiveness of remediation methods implemented at a site; or (7) changes in long-term ground-water quality during or following site remediation or closure.

Non-Regulatory Monitoring

In addition to contaminant characterization and monitoring, ambient monitoring programs are conducted by a variety of government agencies including the U.S. Geological Survey and various state agencies (regulatory and non-regulatory) and tribal governments, as well as regional, county, and municipal government agencies. These programs are not concerned with contaminant fate and transport issues at a particular site, but are focused on large-scale hydrogeologic and geochemical characterization of aquifers to determine their suitability for specific uses or sensitivity to development. In these situations, ground-water sampling programs are implemented to determine the ambient quality of ground water available for public drinking water, agricultural, or industrial uses. For example, the tribal government in the Owens Valley area of California closely monitors water levels and water quality in regional wells to ensure that over-pumping does not occur within the valley. Over-pumping would have a significant and detrimental impact on both the water quality and water supply necessary for local municipal and agricultural use.

Collecting "Representative" Samples

The primary objective of most ground-water sampling programs is to collect samples that are representative of ground water in its *in situ* condition. A representative ground-water sample must accurately reflect the physical and chemical properties of the ground water in that portion of the formation open to the well to be sampled. While this objective is common to most programs, the working definition of "representative" is not. As a consequence, a clearly defined statement of objectives, which includes a site-specific definition of a representative sample, is a critical component of any site-specific sampling and analysis plan (SAP). The term representative may mean different things to different investigators, due mainly to differing project objectives. For example, those interested in characterizing ground-water quality from a water-supply well for the purpose of water-supply evaluation or risk characterization for receptors are most interested in volume-averaged concentrations of chemical constituents in the target water-bearing zone rather than "worst-case" conditions in the heart of a contaminant plume. Samples collected after pumping a significant volume of water from the well may be considered representative for these purposes, whereas samples collected using methods designed to focus on a specific, discrete portion of the formation (i.e., the center of a contaminant plume) may not be representative for that particular application. If contaminants were found in the latter type of sample, the concentrations present would not accurately represent the concentrations to which users of this water-supply well would be exposed; they could be much higher because of the lack of dilution.

To be considered representative for the purpose of contaminant plume characterization, samples should be collected in a manner that maintains the depth-discrete nature of the zone targeted for monitoring, as defined by the site-characterization program. Extensive pumping of contaminant monitoring wells, particularly those in low hydraulic conductivity materials (silts and clays, weakly fractured or solution-channeled rock), can result in volume-averaged samples, which greatly complicate interpretation of data concerning the concentration and spatial distribution of contaminants. If the open portion of the well is long relative to the thickness of the contaminant plume, a negatively biased (underestimated) concentration value would be produced as a result of dilution caused during the sampling process by mixing of water from within the plume with water from uncontaminated portions of the formation. Similarly, in fractured rock, because fracture porosity is typically very small (compared with intergranular porosity), extensive pumping during the sampling process may draw in water from a very large volume of the formation. In an example cited by McCarthy and Shevenell (1998), in a rock with a fracture porosity of 10^{-4}, pumping only 10 l (2.6 gal) of water from a well could conceivably draw in water from a volume of 100 m^3 (3531 ft^3) of rock. Mixing of water from different locations in the formation could greatly dilute the level of contaminants in the sample and the concentrations may not be in equilibrium with the adjoining matrix. Thus, if the well was installed with the intent of monitoring a discrete zone, the sampling process would have the effect of undoing all of the work that went into properly locating the well and the open or screened zone.

Many investigators have acknowledged the difficulty of obtaining samples that are truly representative of subsurface conditions (Grisak et al., 1978; Gibb et al., 1981; Schuller et al., 1981; Claassen, 1982; U.S. EPA, 1982, 1991; Gillham et al., 1983; Barcelona et al., 1987). There has always been almost universal agreement that to obtain representative samples and prevent sample alteration, subsurface disturbance and sample handling must be kept to a minimum (Puls and Powell, 1992a). However, the procedures that are currently used to access the subsurface to collect ground-water samples make some level of disturbance of subsurface conditions unavoidable. The goal in sample collection should be to use methods that result in the least disturbance or change in the chemical and physical properties of the water collected as the sample and that, therefore, produce the most representative sample possible.

A significant level of effort has been invested over the past few decades in establishing uniform laboratory methods for analyzing samples, while comparatively little effort has been invested in the development of uniform field sampling methodologies (Puls and Powell, 1992a). This deficiency has resulted in the production of large volumes of highly variable and sometimes meaningless data, and a significant waste of time and money. Although ASTM International has produced a number of standards related to sampling environmental media, such standards often take many years to be adopted by regulatory agencies and practitioners. Some groups, such as State regulatory agencies and professional trade associations, have considered developing performance-based certification programs for field sampling personnel, but they have recognized what an enormous undertaking developing and administering such programs would be, and none are yet available. Consequently, the discrepancy in standardization of procedures and quality assurance/quality control (QA/QC) measures between the laboratory and the field remains.

Ground-Water Sampling and Data Quality

Because the field and analytical data produced by ground-water sampling programs are often relied upon as the basis for making potentially far-reaching and expensive decisions, it is imperative that these data be of the highest quality possible. Some of the important decisions affected by ground-water sample data analysis from a ground-water monitoring program include:

- Whether a site is in compliance with permit requirements
- Whether a site can be closed and monitoring discontinued
- Whether a site requires active remediation
- Whether a site is a candidate for risk-based corrective action or natural attenuation

To ensure that these decisions are defensible, they must be supported by valid data of suitable quality that are both accurate and reproducible. The chemical analytical data must be representative of formation water quality within an acceptable level of certainty (accuracy), which is defined by data quality objectives (DQOs). An important element of obtaining a representative sample is that the sample collection procedures should be reproducible to ensure precision in samples collected for analysis during different sampling events. This aspect of sampling has gained increased emphasis with the requirement to follow Quality Assurance Project Plans (QAPPs) as part of ground-water sampling protocols (Gibs and Imbrigiotta, 1990). The goal of the DQO process and the QAPP is to document protocol and procedures for a given site that will ensure that the data generated are of sufficient quality and quantity to facilitate making sound technical decisions. Furthermore, it is important that the data from well to well and from sampling event to sampling event are reproducible with a minimum amount of bias and with no unexplainable outliers, to permit trend analysis. The key to successfully attaining these goals is to use sampling methodologies that generate data that are both accurate and precise.

Inconsistent and biased ground-water data collection is a serious issue for the interpretation of data for trend analysis and evaluation of remedial action performance (Barcelona et al., 1994). One of the main reasons for this is that the traditional methods used by many investigators for the collection of ground-water samples are not well suited to producing high-quality samples. The accuracy and precision of sample analytical data are only as good as the quality of the samples submitted for analysis, which is strongly controlled by sampling method.

Meeting DQOs: A Superfund Project as an Example

Ground-water sampling project objectives, in the context of a Superfund site investigation, include four main goals: contaminant detection, contaminant assessment, resource evaluation, and corrective action selection and evaluation. Specifically, ground-water samples are first used to determine whether hazardous substances are present in ground water underlying a site under investigation, and at what concentrations. Identifying contamination serves as an important factor in determining locations that require further investigation. If contamination is found to exist, ground-water samples then serve to assess the nature and three-dimensional extent of contamination. This is necessary for ascertaining whether the contamination can be attributed to the site under investigation and for determining the number and types of affected or potentially affected receptors. This information helps to establish the site priority for future remedial response. Whether or not contamination is present, ground-water investigations can establish subsurface characteristics such as aquifer boundaries, ground-water potability, aquifer interconnections, and ground-water flow rate and direction. These parameters are necessary to evaluate the relative risk associated with documented or potential releases (Thornton et al., 1997). If the contamination present at the site poses sufficient risk to receptors to require remediation, data from ground-water samples may then be used to help determine the optimum remedial approach and, after the approach is implemented, to evaluate remedial performance and effectiveness.

Owing to the "screening" nature of the early phase of a Superfund site assessment, data collection focuses on sampling "worst case" concentrations and locations (e.g., searching for highest concentrations within contaminant plumes) rather than determining average or overall site conditions. Because listing of sites on the National Priorities List (NPL) is a process that is subject to legal challenges, the data used to evaluate sites must be scientifically and legally defensible and of known and documented quality. This requires that DQOs be established that the field sample collection, handling, storage, and laboratory analysis procedures be documented as following previously established procedures and that a QC procedure be implemented (Thornton et al., 1997).

Subsequent phases of sampling typically focus on spatial and temporal contaminant plume definition, and finally, on selecting an appropriate remedial strategy and evaluating its effectiveness. This requires samples of a more discrete nature, which represent specific and usually relatively thin target zones within the formation. As some investigators (e.g., Ronen et al., 1987; Cherry, 1992; Puls and McCarthy, 1995) have recognized subsurface heterogeneities often result in distributions of many chemicals of concern (particularly those associated with dense non-aqueous phase liquids [DNAPLs]) that may vary by several orders of magnitude over only a few feet, both vertically and laterally. Ideally, sampling points are constructed so that they are open to and focused on zones that are known through site-characterization activities to be contaminated or that are determined to be the most likely pathways for ground-water and contaminant movement. Examples of preferential flow pathways include sand and gravel lenses within a fine-grained matrix or fractures or solution channels within otherwise competent rock.

A common objective of all phases of Superfund site investigations is the collection of high-quality reproducible data to support decision making at several different stages of the investigation. High-quality data collection implies data of sufficient accuracy, precision, and completeness (i.e., the ratio of valid analytical results to the minimum sample number called for by the sampling program design) to meet program objectives. Accuracy depends on the correct choice of sampling tools and procedures to minimize sample and subsurface disturbance for collection to analysis. Precision depends on the repeatability of sampling and analytical protocols (Puls and Barcelona, 1996).

Factors Affecting the Representative Nature of Ground-Water Samples

A number of factors influence the ability of samplers to collect representative ground-water samples. Table 1.1 provides a summary of the factors that must be evaluated for each site undergoing ground-water sampling to determine how each might affect the representative nature of samples to be collected and the sampling strategy and methods to be used.

Formation and Well Hydraulics

Developing an understanding of site-specific formation hydraulics is critical to the proper design and installation of sampling points used for site characterization and wells used for long-term monitoring and remediation applications, to the ability of investigators to collect the most accurate samples of ground water for chemical analysis, and to the accurate interpretation of geochemical data. Formation hydraulics directly affect a variety of activities, from three-dimensional well placement relative to known or suspected

TABLE 1.1

Factors Influencing the Representative Nature of Ground-Water Samples

Factors related to formation and well hydraulics
Ground-water flow paths and flow through wells
Hydraulics within a well between sampling events
Factors related to sampling point placement, design, installation, and maintenance
Placement of the sampling point with respect to the source(s) of contamination being monitored
Placement of the sampling point intake in the preferential flow pathway (the zone of highest hydraulic conductivity) within the formation(s) of concern
Installation method used for sampling point construction (i.e., drilling method or direct-push installation method)
Suitability of sampling point design with regard to material selection, diameter, depth of the sampling point intake, screen length, and screen slot size for the hydrogeologic and geochemical environment being monitored
Methods used during sampling point construction, including placement of annular seal materials and care in placement of filter pack materials
Method, timing, and duration of sampling point development
Long-term maintenance of the sampling point to ensure that the sampling point can continue to provide suitable samples for analysis with regard to representative chemistry that is not impacted by compromises in well integrity (e.g., cracked surface seals)
Factors related to geochemical changes associated with sample collection
Pressure changes in the sample
Temperature changes in the sample
Entrainment of artifactual particulate matter in the sample
Agitation or aeration of the water column in the well or the sample during sample collection

source areas of contamination, to incorporating wells that may not have been designed specifically for the purpose of monitoring contamination (e.g., residential water-supply wells or agricultural irrigation wells) into a monitoring network, to monitoring well intake design (screen slot size selection, screen length, and screen placement).

Understanding Ground-Water Flow

When developing a site-specific ground-water sampling program, it is critical to have an accurate, three-dimensional understanding of the ground-water hydrology of the site under investigation. Investigators must understand how ground water will move across any given site and must identify what factors influence that movement, such as pumping centers, areas of artificial recharge, water-level fluctuations in an adjacent surface–water system, or tides. Samplers must also develop an understanding of how ground water behaves within the sampling point or monitoring well, both between sampling events and during purging and sample collection. Failure to develop this understanding typically results in selection and use of sampling protocols that are not well suited to the hydrogeology or geochemistry of the site. This, in turn, can result in generation of nonrepresentative data.

As illustrated in Figure 1.1, ground water moves through the subsurface in response to differences in hydraulic head, under laminar flow conditions in most hydrogeologic systems. Provided that the difference in hydraulic head is such that horizontal flow dominates in the formation, flow should continue in the same manner through well screens installed in the formation. Robin and Gillham (1987) demonstrated with tracer solutions that formation water moves through the well screen and that this water does not mix with the stagnant water that remains in the casing between sampling rounds. Visual observations of movement of colloidal particles through well screens installed in granular aquifers made by Kearl et al. (1992) demonstrate that horizontal laminar flow in a formation does, in fact, continue through well screens and that water in the screen does not mix with that in the casing. Powell and Puls (1993), using dual tracer tests, also showed that ground water moves through the well screen with little interaction or mixing with the overlying stagnant water in storage in the well casing. Electrical conductivity data presented by Michalski (1989) clearly showed a fresh water zone in the well screen

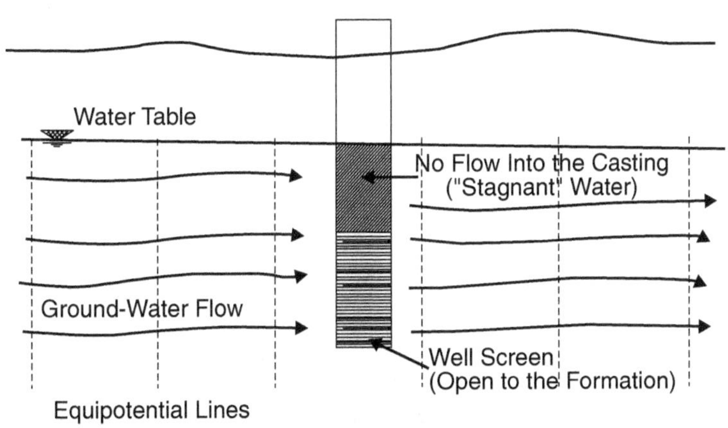

FIGURE 1.1
Movement of ground water in a formation in which horizontal flow dominates. Note that the horizontal flow continues through the screen in a properly designed, constructed, and developed well, but water in the screen does not mix with water in the casing, which is stagnant between sampling events.

separate from the stagnant water in the casing, further supporting the observations made by Robin and Gillham. Robin and Gillham (1987) also theorized that the continual flow of water through the screen allows chemical reactions, such as desorption and adsorption, between well construction materials (well screen and filter pack) and constituents in ground water to approach equilibrium. Work done by Palmer et al. (1987) corroborates this and demonstrates the need to allow time for this equilibration to occur prior to the initial sampling event for a well, which is also suggested by Walker (1983). These studies suggest that for wells in which horizontal flow dominates: (1) water in the screen at any point in time is indeed representative of water in the formation adjacent to the screen; (2) water samples taken directly from the screened interval are representative of ground water in the surrounding formation; and (3) provided that samplers can gain access to the water in the screened interval while minimally disturbing the water column in the well, purging multiple well volumes of water prior to sample collection is unnecessary.

However, where there is a difference in hydraulic head in the formation that results in vertical movement of ground water, and a well is screened across that zone, the well screen effectively acts to short-circuit ground-water flow and directly channels water from the zone with highest hydraulic head to the zone with lowest hydraulic head (Figure 1.2). Several recent studies (i.e., McIlvride and Rector, 1988; Reilly et al., 1989; Church and Granato, 1996; Hutchins and Acree, 2000; Elci et al., 2001, 2003) have documented that, in areas with vertical hydraulic gradients, installation of a monitoring well with a long well screen (i.e., >10 ft long) may set up a localized vertical flow system that renders the well almost useless for sampling because the dilution that occurs in such a well would yield misleading and ambiguous data concerning contaminant concentrations and plume geometry. In some scenarios, installation of a well in this type of setting

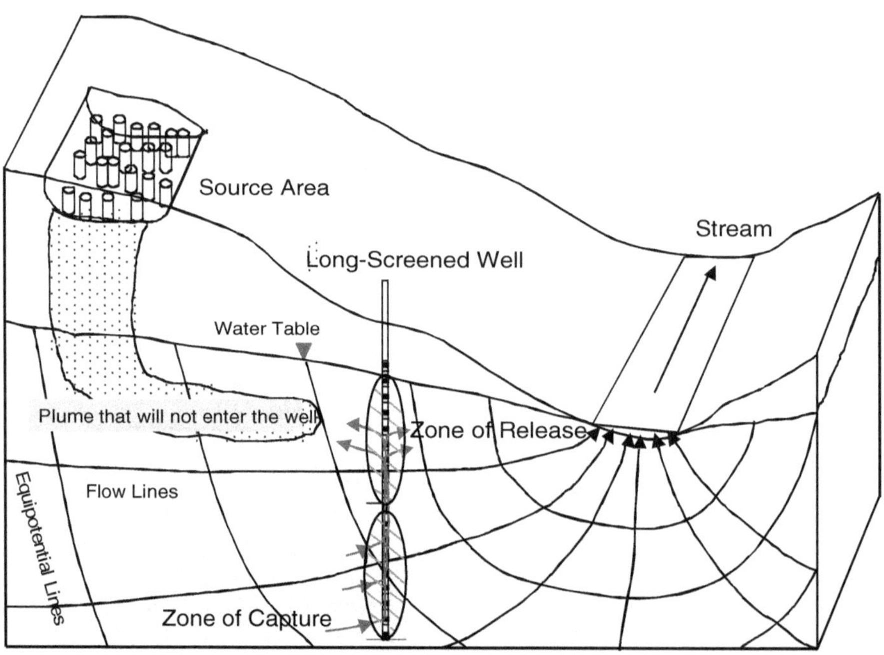

FIGURE 1.2
Movement of ground water in a formation in which there is a vertical difference in hydraulic head. Note that the well screen effectively acts to short-circuit ground-water flow and directly channels water from the zone in the formation with highest hydraulic head to the zone with lowest hydraulic head. (*Source*: McIlvride and Rector, 1988.)

may result in the spread of contamination to parts of a formation that would not otherwise have become contaminated had the well not been installed. Thus, samplers must be keenly aware of the hydraulic conditions that exist in a well so they can make appropriate decisions on whether a well should be sampled, and, if it is sampled, how to interpret the data generated by each sampling event. More detailed information on this topic is available in Nielsen and Schalla (2006) and Einarson (2006).

Hydraulics and Water Chemistry Between Sampling Events

It is well recognized that water in storage in a well casing between sampling events (not including water in the well screen) is not representative of formation water quality (Marsh and Lloyd, 1980; Miller, 1982; Gillham et al., 1985; Barcelona and Helfrich, 1986) and thus should not be collected as part of the sample (Barcelona et al., 1985). Water in storage in a well casing between sampling events is physically isolated by the well casing from ground water in the formation and cannot interact with "fresh" formation water and is considered to be "stagnant." Changes in water chemistry in the cased portion of the well are caused by a variety of factors as summarized in Table 1.2 and illustrated in Figure 1.3. Purging has historically been implemented as part of routine ground-water sample collection procedures to minimize the bias and error associated with incorporating any portion of this column of stagnant water into samples submitted for analysis.

However, while water in the well casing may be chemically nonrepresentative of formation water, the research cited in the preceding section has concluded that the water within the screened interval of a short-screened well is representative of formation water quality, provided that the well has been designed, installed, developed, and maintained properly. In such wells, the water in the formation and that within the well screen are able to exchange freely and the well screen is continually "flushed" (Gillham et al., 1985; Robin and Gillham, 1987). That is, water passes continuously through the well screen and surrounding filter pack because they have a higher hydraulic conductivity than the formation, and the well intake is integrated into the ground-water flow field (Figure 1.1). Even in low hydraulic conductivity formations (clays, silty clays, and clayey silts), ground-water flow is often sufficient to maintain a constant (although slow) exchange of water between the formation and the well screen.

TABLE 1.2

Factors Affecting the Chemistry of Water in Storage within a Well Casing between Sampling Events

The presence of an air–water interface at the top of the water column, which can result the following:
Creation of a dissolved oxygen (DO) concentration gradient (high to low) with depth
Increased aerobic microbial activity because of the presence of DO
Lower pH, due to increased dissolved carbon dioxide (CO_2) at the top of the water column
Loss of volatile constituents from the water column to the headspace in the well casing
Interactions between the well casing and screen materials and ground-water in storage (leaching from, sorption to, or corrosion of the well construction materials)
Contribution of contaminants from sources above the static water level in the well including:
Condensation on the inside surface of the well casing
Water from formations above the zone of interest leaking past joints or cracks in the casing
Introduction of surface contamination as a result of failure of the surface seal, causing leakage into the well
Addition of volatile constituents from vadose zone gases (e.g., landfill gases)

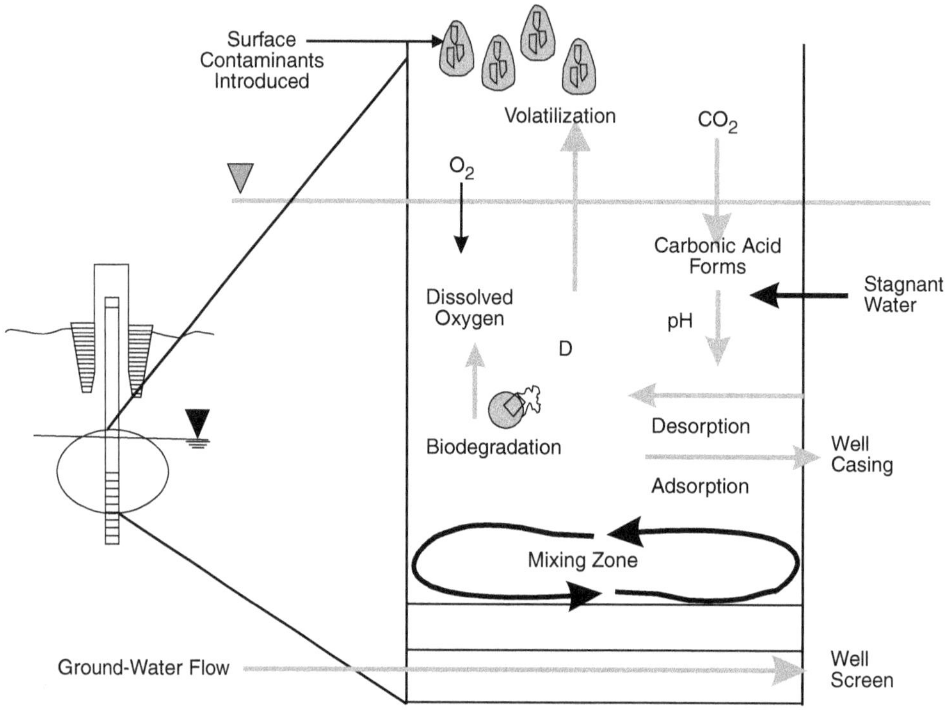

FIGURE 1.3
Changes in water chemistry in the water column within the well casing that occur between sampling events.

Sampling Point Placement, Design, Installation, and Maintenance

Three-Dimensional Placement of the Sampling Point

When collecting ground-water samples from existing wells, it is very important to evaluate the suitability of each individual well with regard to its ability to yield a representative sample from the formation of interest. At the onset of a ground-water sampling program, wells should be evaluated to determine whether they are correctly placed three-dimensionally within the formation to provide the samples required to make decisions at the site. It is critical to understand where the well is placed within the formation and, for contamination investigations, where the well screen is located within the ground-water flow system relative to the known or anticipated source areas of contamination. If the well screen is not in the correct location or if it is too long and not focused on a particular zone of interest, it will be impossible to generate the ground-water data required to satisfy the objectives of the ground-water sampling program, regardless of how carefully the wells are sampled.

If the well is located properly, it is then critical to evaluate the suitability of the design and construction of the well for meeting the objectives of the ground-water sampling program. Design features that must be evaluated include well intake design (i.e., well screen slot size and filter pack grain size), well casing material selection, annular seal construction, surface seal completion, and well security measures. The reader is directed to Nielsen and Schalla (2006) for a detailed discussion of how each of these well design and construction features can affect the chemistry of samples collected from the formation under investigation.

Sampling Point Installation Options

Several sampling point options are available for collecting ground-water samples from a zone of interest in subsurface formation materials. The traditional, and still most common approach, is to install 2 or 4 in. diameter ground-water monitoring wells using a conventional drilling method such as hollow-stem auger drilling (refer to Ruda and Farrar [2006] for more detail). As accelerated site-characterization technologies (refer to Nielsen et al. [2006]) are increasingly used to rapidly develop an understanding of ground-water conditions, direct-push technologies are being more widely used to install sampling points. As discussed in McCall et al. (2006), direct-push technologies can facilitate collection of grab samples of ground water from discrete zones to provide a "snapshot" assessment of ground-water chemistry using sealed-screen samplers such as the HydroPunch (Figure 1.4) and exposed-screen samplers such as the Waterloo Profiler. In addition, direct-push technology has been developed to install long-term ground-water monitoring wells, ranging in diameter from 0.5 to 2 in. Because the means of accessing formations to collect ground-water samples is no longer limited to traditional ground-water monitoring wells, the term "sampling points" is sometimes used to generically refer to any point of access to the saturated zone. The terms "wells" and "sampling points" will both be used throughout this book.

Traditional Drilled Monitoring Wells

An important element of traditional well installation is the drilling method used and the degree of disturbance of the formation that occurs during well installation. As discussed in Ruda and Farrar (2006), a number of drilling methods are available for installation of

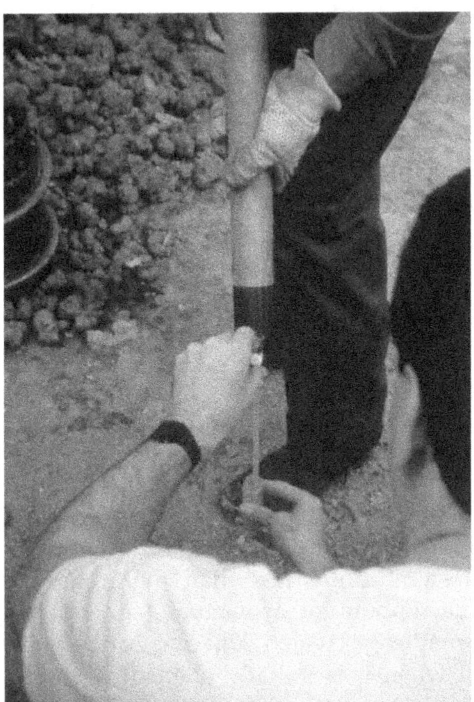

FIGURE 1.4
Direct-push ground-water samplers such as the HydroPunch allow collection of samples representative of formation water chemistry without installing a permanent well.

FIGURE 1.5
Drilling fluid, such as the water-based bentonite fluid used with direct (mud) rotary drilling, can remain in the formation surrounding the well installed in the drilled borehole long after well installation, unless it is properly controlled during drilling and removed during well development. If left in the formation, it can affect ground-water sample chemistry for weeks to months after well installation.

monitoring wells. Selection of a drilling method is dependent on a variety of site-specific constraints such as total borehole depth; character of geologic materials to be penetrated; depth to ground water; requirements for soil sample collection during drilling; potential impact of introducing drilling fluids (e.g., mud and air) into the formation on soil and ground-water samples; ease of mobilization of the rig to and around the site; and cost. Drilling methods can create sampling artifacts through redistribution of formation materials within the borehole, creation of fine particles by disaggregation and crushing of formation materials (granules, grain coatings, cementing agents, and other solids); introduction of foreign materials (e.g., drilling mud [Fetter, 1983; Brobst and Buszka, 1986] [Figure 1.5], air, and water of a chemistry different from formation water) into the formation; and providing a conduit for atmospheric air to contact ground water, which may result in precipitation of metal oxides and hydroxides.

Direct-Push Sampling Tools and Monitoring Well Installations

Direct-push sampling tools may be advanced into the subsurface with direct-push rigs, manually operated vibratory equipment (e.g., jackhammers), CPT equipment, or conventional drilling rigs. The action of advancing any direct-push sampling tool to the desired sampling depth, whether by pushing, hammering, or vibrating the rod string, displaces formation materials, causing some compaction and some disaggregation of granular materials (and associated breakage of grain coatings and cementing agents, such as iron oxyhydroxides and carbonates). The disturbance caused by installation of direct-push sampling tools is significantly less than that caused by conventional drilling methods. However, some amount of artifactual (i.e., resulting from sampling point installation) particulate matter is created adjacent to the sampling tool. This material enters the sampling tool when it is initially opened, resulting in high initial turbidity levels that represent particulate matter that is not mobile under ambient ground-water flow conditions. Many of these particles are highly surface-reactive and, because of their high surface area per unit mass and volume, have very high sorptive capacities and strong binding capabilities for selected groups of analytes. Some of these materials (the largest fraction) will settle out quickly, but much of it will remain in suspension in the water column and could be collected as part of the sample if the sampling tool is not purged first. Developing the sampling point or using low-flow sampling methods to collect samples (although prolonged pumping may be required) generally results in significantly lower turbidity levels and improved sample quality (U.S. EPA, 1996).

Most direct-push technologies are also capable of installing small-diameter wells that can be used for either short-term or long-term monitoring of subsurface conditions. Wells installed using direct-push methods have several advantages over traditional drilled monitoring wells, including:

- Minimal disturbance of formation materials during well installation.
- Well intakes are often shorter and more accurately located with respect to installation within a zone of specific interest because direct-push sampling tools can be used to better define and resolve subsurface conditions.
- Well intakes can be installed using "prepacked" or sleeved well screens (Figure 1.6) that allow for superior control over filter pack placement around the well screen and, in some cases, for use of a significantly finer-grained filter pack material than is possible for conventional wells. This, in turn, results in samples with noticeably lower turbidity than samples collected from conventional wells installed in the same hydrogeologic setting (McCall et al., 1997). This can be a significant advantage in predominantly fine-grained formations, where

FIGURE 1.6
Prepacked well screens commonly used in direct-push well installations allow for use of fine-grained filter pack materials, which can effectively reduce turbidity in samples, even those collected in wells installed in predominantly fine-grained formation materials (i.e., silty clayey sand).

conventional well installation technology is typically unsuccessful in retaining formation fines.

Nominal diameters of well casings and screens for direct-push completions range from 0.5 to 2 in. Although 2 in. completions are more typical of conventionally installed monitoring wells, smaller diameter direct-push wells suffer from several constraints, which include difficulty in well development.

Poor Well Design and Construction

The most common problems with well design and construction that contribute sampling artifacts are: inappropriate selection of well casing and screen materials, which results in compromises in water chemistry (through sorption and desorption, corrosion, and degradation); improper well screen length, which can result in collecting samples that may be significantly diluted; incorrect well-screen slot size and filter-pack grain size selection, which can result in sedimentation of the well; improper filter pack installation, which results in excessive filter pack losses to either the well or the formation during development; and improper annular seal material selection and installation, which can result in grout infiltration into the filter pack and well screen. Poor well construction can permanently bias sampling results and impair the usefulness and integrity of wells as sampling points (Fetter, 1983).

If artifactual turbidity is caused by drilling or improper well design and construction, the turbidity in samples will initially be high and may diminish with each sampling event. The identity of drilling or well construction related artifacts can be confirmed by particle analysis (e.g., via scanning electron microscopy) (Backhus et al., 1993).

Improper Selection of Well Construction Materials

Corrosion of steel well casing and screens (Figure 1.7) and chemical degradation of polyvinyl chloride (PVC) casing and screen may result in contributions of both dissolved and particulate matter to samples. Steel and stainless-steel corrosion products, including Fe, Mn, Cu, Pb, Cd, Ni, Cr, and Mo, have been found in samples collected from wells constructed in both neutral and low pH environments (Parker et al., 1990; Parker, 1991; Hewitt, 1992; Oakley and Korte, 1996). PVC degradation products and PVC cementing agents (Figure 1.8) have been found in samples collected from wells exposed to very high concentrations of organic solvents (Sosebee et al., 1982, 1983; Martin and Lee, 1989; Parker et al., 1990; McCaulou et al., 1995).

Well Screen Length

Wells installed for detection monitoring programs may have screens of variable lengths, with length dependent on the degree of heterogeneity in the hydrogeologic setting. In cases where discrete preferential flow pathways are not apparent, long screens may be employed as a screening tool to maximize the likelihood of detecting a potential release from a facility. This approach serves the intended detection monitoring purpose by biasing the sample collection toward the zones that have higher hydraulic conductivity. In other words, the higher hydraulic conductivity zones provide comparatively more sample volume, thus increasing the likelihood of detecting a release. In cases where discrete high hydraulic conductivity zones or preferential flow pathways are readily identified, well screen lengths are often selected to focus on these zones. If contamination is detected in samples above some action level established for a specific site, the focus of the investigation then shifts to spatial and temporal contaminant plume definition and assessment monitoring. Wells installed for this purpose and for later corrective action

FIGURE 1.7
Corrosion of steel well construction materials (in this case, galvanized steel) can contribute constituents of the casing or screen to ground-water samples. This possibility must be accounted for in interpretation of analytical data from samples.

monitoring (if required) are located, and screen positions and lengths are determined by detailed site characterization (e.g., formation profiling), to locate the specific zones of contamination that require monitoring and, possibly, remediation. These wells generally have short screens designed to focus on specific zones within a formation in which contaminants are preferentially transported. Examples of preferential flow pathways include sand and gravel lenses within a fine-grained matrix or fracture zones or solution channels within otherwise competent rock.

Filter-Pack Grain Size and Well-Screen Slot Size

The purpose of a filter pack is to provide a permeable zone between the well screen and the surrounding formation, with the objective of preventing fine-grained materials from entering, and possibly plugging, the well screen. Filter-pack materials need to be appropriately sized and of an inert chemistry to prevent bias of ground-water sample chemistry (Figure 1.9). To that end, a filter pack should consist of a clean, well-rounded silica sand of a selected grain size and gradation suitable to retain the surrounding formation materials. The filter-pack materials should be installed in the annular space between the well screen and the wall of the borehole using a device such as a tremie pipe to ensure controlled placement of the filter-pack material around the well screen.

Closely related to filter-pack design is the selection of an appropriate slot size for the well screen. A common but misguided approach to well screen design is to use a "one slot size fits all" philosophy for all formations. This often results in samples that are highly turbid, because the combination of slot size and filter-pack grain size used is not appropriate to filter out formation fines. The well-screen slot size should be selected so

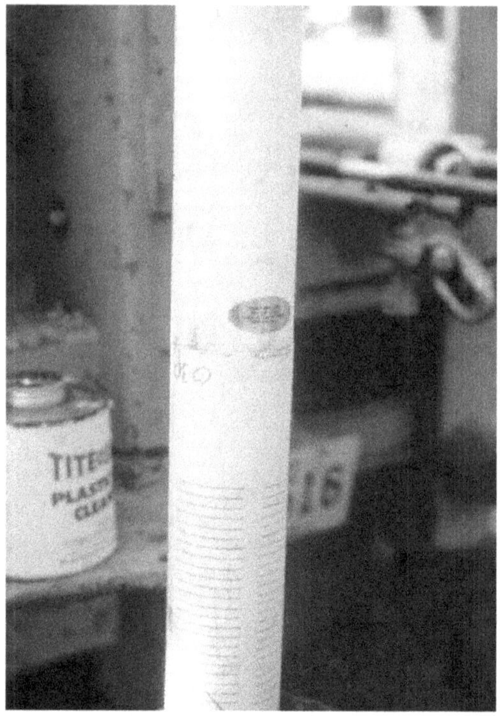

FIGURE 1.8
PVC cementing agents (including tetrahydrofuran, methyl ethyl ketone, cyclohexanone, and dimethylformamide) can leach from casing joints into water standing in the well between sampling events, and have been found in samples even after the wells have been purged. If analytical data from samples show constituents of PVC solvent cement, the possibility of false positives attributed to the cementing agents should be considered.

FIGURE 1.9
Filter-pack materials in conventional monitoring wells may contribute analytes to samples if the materials are not inert to formation water chemistry. Materials should consist of at least 95% silica sand to avoid interactions with ground water and possible false positives attributed to filter-pack materials.

that the open area is maximized to allow rapid sample recovery, effective well development, and proper conduct of formation hydraulic tests and analysis of hydraulic test data. The slot sizes must be selected so the openings will allow the screen to retain the filter pack (or formation material in the case of naturally developed wells) while still permitting efficient well development. Sedimentation of wells is a common problem in wells in which well-screen or filter-pack design is flawed.

Inadequate or Improper Well Development

A common failure of many ground-water monitoring programs is that most wells are not developed properly. In high hydraulic conductivity formations, development is typically not conducted for a long enough period of time. As discussed in Kraemer et al. (2006), wells installed in fine-grained formations dominated by clays and silts should not be developed. In these formations, well development may do more harm than good with respect to facilitating collection of representative ground-water samples (Paul et al., 1988; Nielsen, 1995).

Proper well development achieves several objectives including rectifying drilling damage to the borehole caused during well installation, removing fine materials from the formation adjacent to the borehole, stabilizing the filter pack materials installed adjacent to the well screen, retrieving lost drilling fluid from the formation, and optimizing well efficiency and hydraulic communication between the well screen and the adjacent formation. Proper well development is a critical process required to prevent the inclusion of excess turbidity and particulate artifacts in samples during the life of the well. Well development is a component of the well construction process and should not be confused with well purging, which is a component of the ground-water sampling process.

Well Maintenance

A critical error in many ground-water monitoring programs is the failure to adequately maintain the wells during the operating life of the monitoring network. As a standard procedure during ground-water sampling events, samplers should inspect each well to ensure that its structural integrity has not been compromised since the last sampling event. Proper well maintenance can include repair of damaged surface seals, redevelopment of wells in which siltation has occurred and resulted in reduced well recharge rates, or repair of damaged surface protection or security measures. Adhering to a rigorous well maintenance schedule will ensure that the wells will continue to yield representative samples of sufficient volume for the duration of the sampling program.

Geochemical Changes in Ground-Water Samples

Even if a monitoring well is designed correctly, installed in the most appropriate position within the zone of interest, and adequately developed and maintained, ground-water sample chemistry can be affected by the process of sampling the well. During this process, ground water is brought from an *in situ* environment, where pressure and temperature are stable and relatively uniform, to the surface where atmospheric conditions prevail. The resulting changes in pressure and temperature may result in changes to the sample that are manifested in negative biases (underestimation) for some analytes and positive biases (overestimation) for other analytes. Additional important issues include entrainment of artifactual particulate matter in the sample as a result of purging and sampling practices, and agitation and aeration of the water column in the well during sample collection. These issues, which are closely linked to one another, are discussed briefly. The reader is referred to Gibb et al. (1981), Gillham et al. (1983), Barcelona et al. (1985), and Nielsen and Nielsen (2002) for additional background on these issues.

Pressure Changes

Pressure changes to a sample are virtually unavoidable when a sample is brought to the surface from the ground-water environment. In the subsurface, ground water is, by definition, under a pressure greater than atmospheric pressure. When a sample of ground water is brought to the surface, it is collected at atmospheric pressure. The pressure decrease in the sample caused by exposing it to atmospheric conditions is proportional to the height of the water column above the point from which the sample is retrieved. Removing a sample from a depth of a few tens of meters and subjecting it to atmospheric pressure can represent a change in pressure in the sample of several atmospheres (Gillham et al., 1983). Under decreased pressure conditions, the water loses its ability to retain dissolved gases. For example, if a ground-water sample has a high partial pressure of CO_2 (which is true of most ground water), when the pressure on the sample is lowered, it becomes supersaturated with respect to CO_2. The sample will lose CO_2 gas until it reaches equilibrium with atmospheric conditions. This, in turn, may cause a shift in chemical equilibrium, introducing a secondary bias in pH and those parameters for which a change in pH can induce bias (Gillham et al., 1983). Loss of CO_2 causes an increase in sample pH (by between 0.5 and 1.0 pH units), which may then cause various trace metals, including Fe, Mn, Cd, Pb, As, and Zn, to precipitate (Gibb et al., 1981; Unwin, 1982). If volatile organic compounds (VOCs) are present in the sample, pressure decreases can cause their loss from the sample in one or both of two ways. Volatiles may evolve directly from solution, or they may partition into the headspace formed as other dissolved gases (e.g., CO_2) evolve from solution; more volatiles will evolve as the headspace gets larger. The end result is a sample with potentially significant negative biases for a variety of constituents. The magnitude of possible bias is sample-specific and cannot be predicted quantitatively (Gillham et al., 1983). The pressure decrease inherent in bringing a sample to the surface can be exacerbated by using sampling devices such as suction-lift pumps (e.g., peristaltic pumps [Figure 1.10]), which may impart substantial negative pressure to the sample during sample collection.

There are also situations during sampling events in which samples may be subjected to increases in pressure. For example, some pumping devices (e.g., electric submersible centrifugal pumps) pressurize the sample to drive it to the surface, some flow-through cells with a check valve in the discharge line may impart a back-pressure, and some filtration systems use positive pressure to force water through an appropriately sized filter. Positive pressure may cause gases in the air (e.g., O_2) to dissolve in the sample at a higher rate and concentration when compared with simple exposure to atmospheric air at atmospheric pressure, and result in precipitation of some metal species. However, as Gillham et al. (1983) point out, the chemical alterations induced in water as a result of imposing positive pressure are small when compared with the alterations induced by negative pressure.

Temperature Changes

The temperature of shallow ground water is approximately equal to the mean annual air temperature of the locale and is relatively constant year-round, typically varying by less than 2°C (Driscoll, 1986). This stability in temperature results in stability in shallow ground-water chemistry. During the process of collecting samples, however, there may be several points at which significant and potentially damaging temperature increases may occur to a ground-water sample. For example, temperature increases of as much as 10 to 15°C can occur readily when excess sample tubing at the surface (i.e., on a hose reel) or a

FIGURE 1.10
The negative pressure applied to samples collected with peristaltic pumps, particularly at lifts exceeding 15 ft, can significantly alter sample chemistry, especially for ground water with high levels of dissolved gases, as well as VOCs.

flow-through cell is exposed to direct sunlight or high ambient temperatures. To avoid such temperature increases, lengths of above-ground discharge tubing should be minimized and flow-through cells should be kept in a shaded and cool environment (Figure 1.11). Some sampling devices can also cause temperature increases in a sample. For example, Pohlmann et al. (1994) and Oneacre and Figueras (1996) found that operation of an electric submersible centrifugal pump (Figure 1.12) at low discharge rates caused noticeable (5 to 7°C) increases in sample discharge temperature. Because this type of pump must be cooled by running water over the motor, and this heated water is eventually collected as the sample, this device is poorly suited to collecting samples to be analyzed for temperature-sensitive parameters. The change in water temperature could

FIGURE 1.11
To avoid temperature increases in samples, the length of pump discharge tubing should be minimized and flow-through cells (if used) should be kept in the shade.

FIGURE 1.12
The motors of most electric submersible centrifugal pumps (such as the one pictured here) are cooled by water in the well that moves up over the motor, into the pump intake, and up to the surface to be collected as the sample. Temperature increases as high as 7°C have been noted in samples collected with these pumps. This large increase in sample temperature can alter sample chemistry for temperature-sensitive parameters.

alter sample chemistry in a number of ways. Heating water reduces the solubility of dissolved gases (CO_2 and O_2) in water. The resultant loss of dissolved CO_2 and O_2 can induce a shift in pH and in redox state, which then causes precipitation of carbonates (Ca and Mg) and dissolved metals, most readily Fe. The precipitation of Fe can then cause co-precipitation of other metals such as Ni, Cu, and Cr (Stumm and Morgan, 1996). Heating will also reduce the solubility of VOCs in water, resulting in losses of these constituents of the sample through volatilization (Parker, 1994).

Entrainment of Artifactual Particulate Matter during Purging and Sampling

The significance of disturbance of the water column in the well during purging and sampling caused by the use of sampling devices that agitate and mix the water column (e.g., bailers or inertial-lift pumps) has been the subject of a great deal of research. It is important to understand that a "bulk" ground-water sample should contain not only water obtained from the saturated portion of the formation screened by the well, but also any fine particles, or colloids, that occur under natural conditions and are mobile in the formation under ambient flow conditions. In coarse-grained formations (i.e., coarser than fine sand), fractured bedrock or karst formations, this particulate matter can include colloids that can serve as prime sites for some contaminants to sorb and move with ground-water flow (Dragun, 1988; Mason et al., 1992) and microspheres containing indigenous bacteria (Harvey et al., 1989). This particulate fraction is referred to as the "mobile contaminant load" of a formation and is a legitimate part of a ground-water sample used for assessing contaminant fate and transport and human health risks caused by ingestion or exposure to the water. When a water column is agitated during purging and sampling, however, the samples collected contain not only naturally mobile particulate matter but also larger "artifactual" particles that are not mobile in ground-water systems.

Most investigations are focused strictly on defining the chemistry of the truly dissolved fraction or on determining the presence, concentrations, or fate and transport of specific chemical constituents that are known to be mobile only in the aqueous phase (e.g., highly soluble constituents with very low adsorption coefficients, including low molecular weight organic compounds such as MTBE or some VOCs). However, a number of researchers (O'Melia, 1980, 1990; Eicholz et al., 1982; Robertson et al., 1984; McDowell-Boyer et al., 1986; Cerda, 1987; Gschwend and Reynolds, 1987; Enfield and Bengtsson, 1988; McCarthy and Zachara, 1989; Puls and Barcelona, 1989; Puls, 1990; Ryan and Gschwend, 1990, 1992; Gschwend et al., 1990; Puls and Powell, 1992a, b; Puls et al., 1992; Backhus et al., 1993; Grolimund et al., 1996) have demonstrated that mobile colloid phases (suspended particles in the range of 0.005 to approximately 10 μm) may form in certain subsurface environments and be mobile at ambient ground-water velocities. Many colloidal particles (e.g., secondary clay minerals; hydrous iron, manganese, and aluminum oxides; and dissolved and particulate organic matter), because of their high surface area per unit mass and volume, have enormous sorptive capacities and strong binding capabilities for selected classes of organic and inorganic contaminants (Enfield et al., 1989; Lyman et al., 1992). The colloids present in some aquifers may significantly affect the transport of hydrophobic or strongly sorptive contaminants, including polynuclear or polycyclic aromatic hydrocarbons (PAHs), polychlorinated biphenyls (PCBs), pesticides (e.g., DDT), dioxins and furans, radionuclides (Pu, Am, U, Cs, and Ru), heavy metals and metalloids, and many major ions. Therefore, it is especially important to understand the role of colloids in contaminant transport at sites where low-solubility, highly surface-reactive chemicals are of concern (McCarthy and Zachara, 1989).

Presence and Sources of Artifactual Particulate Matter

For sampling programs in which the primary objective is to determine the "total mobile contaminant load," it is important to ensure that the mobile colloidal fraction is included in the sample. However, while it is important to include naturally mobile colloidal material in samples collected at sites where colloidal transport may be an important contaminant transport mechanism, it is equally important to exclude from all samples particulate or other matter that is clearly artifactual in nature (i.e., due, in some way, to implementing the sampling process). Disturbance caused by practices that significantly agitate and aerate the water column in the well can markedly increase the content of artifactual particulate matter in samples and irreparably change sample chemistry through processes such as oxidation. Sampling artifacts such as these present serious obstacles to proper interpretation of ground-water sampling results.

The most commonly recognized artifact of poor sampling technique is excessive suspended sediment, seen in samples as turbidity. High concentrations of suspended particles in samples collected for chemical analysis are problematic because contaminants potentially present on these particles (e.g., adsorbed to the particle surface) may or may not be mobile in the ground-water system (McCarthy and Zachara, 1989). Because most investigators are interested in determining what is present in the mobile fraction, this compromises the integrity of samples. As Puls et al. (1991) point out, there is a strong inverse correlation between turbidity and representativeness of samples. With many commonly used sampling practices, it is not possible to differentiate between naturally occurring suspended particles and those brought into suspension or created artificially. Therefore, in situations in which it is necessary to assess the total mobile contaminant load, it is important to use sampling practices that do not disturb the water column in the well, to allow collection of representative samples.

Inclusion of artifactual particulate matter in samples can radically alter analytical results, causing spurious increases in concentrations of several classes of analytes. This is particularly true for the major constituents of the aquifer mineral matrix such as Fe, Al, Ca, Mg, Mn, and Si (Powell and Puls, 1997). Maintenance of *in situ* ground-water chemistry conditions during sampling and sample processing at the surface is particularly important in determining mobile colloid concentrations and characteristics. While no effects from artifactual turbidity have been observed for VOCs (Paul and Puls, 1997), large concentration differences between filtered and unfiltered samples have been observed for metals (Puls and Barcelona, 1989; Puls et al., 1992; Pohlmann et al., 1994), PAHs (Backhus et al., 1993), and radionuclides (Buddemeier and Hunt, 1988; Penrose et al., 1990). Failure to address the issue of artifactual particulates can lead to inaccurate estimates of colloid and colloid-associated contaminant concentrations, colloid characteristics, and the distribution of contaminants between the colloidal and dissolved phase (Backhus et al., 1993).

The sources of turbidity caused by implementing the sampling process are many and varied. For example, using a downhole colloidal borescope (similar to a camera), Kearl et al. (1992) observed that a great deal of turbidity measured in the wells they studied was due simply to the initial placement of the sampling device. In this study, colloidal density, quantitatively measured with a turbidimeter and qualitatively measured by visual inspection, was observed to increase significantly with installation of sampling devices and decline exponentially while purging at a low flow rate. The initial disturbance to the well and the high turbidity levels caused by insertion of the sampling device required overnight equilibrium to attain naturally occurring turbidity values and stable flow in the screened interval.

There is substantial evidence to indicate that the activities inherent in bailing (repeated insertion of the bailer in the water column and withdrawal of the bailer to purge the well and collect samples) and pumping at high rates (i.e., in excess of the natural rate of ground-water flow through the well screen) liberate significant quantities of artifactual particles. Using these purging and sampling methods, particulates that would otherwise have remained immobile in the formation under natural ground-water flow conditions are often mobilized into the well. Carefully collected field evidence from a variety of sites in a variety of hydrogeologic settings demonstrates that the use of bailers (Figure 1.13) and high-speed, high-flow-rate pumps (Figure 1.14) disturbs aquifer and filter-pack materials and can disrupt fragile colloidal aggregates, producing samples in which artifactual particulate materials have been entrained (Puls et al., 1992; Backhus et al., 1993; McCarthy and Degueldre, 1993; Powell and Puls, 1993; McCarthy and Wobber, 1996; McCarthy and Shevenell, 1998). Other investigators (Barcelona et al., 1985; Puls and Barcelona, 1989; Puls and Powell, 1992a) have also noted that bailing wells or pumping at high rates during purging or sampling can mobilize particulate matter and significantly increase sample turbidity.

Pohlmann et al. (1994) noted that pumping-induced stresses can cause grain flow within the sand pack of the well and exceed the cohesive forces of aquifer material cementation. Backhus et al. (1993) also cite bailing- and pumping-induced shearing as a source of artifactual turbidity in a series of wells they studied. Shearing, which releases normally immobile solids from the aquifer matrix, is caused by pumping at rates that induce ground-water flow velocities greater than the natural flow velocities; higher pumping rates result in greater shear in the aquifer. Ryan (1988) calculated that in the set of wells he studied, pumping rates greater than 100 ml/min could induce shear capable of mobilizing particles outside of the zone of the aquifer disturbed by well construction. Low pumping rates appeared to provide the best means to minimize collection of artifactual colloids. This was confirmed in tests conducted by Backhus et al. (1993), to determine the effect of higher pumping rates on turbidity levels in several wells at several different sites. In all cases, turbidity increased sharply in response to pumping rate increases after a turbidity plateau had been established at a lower pumping rate. The increase in turbidity reflected greater inclusion of artifactual solids due to increased shear. Their recommendation was that wells not be bailed and that pumping rates used during purging and sampling be minimized (approximately 200 ml/min for the wells they

FIGURE 1.13
Bailers significantly disturb the water column in wells and the formation materials adjacent to the well screen, which results in significant increases in turbidity of samples, much of which is due to the entrainment of artifactual particulate matter. As a result, collection of representative samples for many parameters is very difficult if not impossible using bailers.

FIGURE 1.14
High-speed, high-flow-rate pumps significantly increase the flow velocity in pore spaces in the formation adjacent to the well and create shear that releases substantial amounts of sediment, which is entrained in water collected as samples. (Photo courtesy of QED Environmental Systems.)

sampled). Rather than specify a particular withdrawal rate to be applied to all wells, Ryan et al. (1995) suggested that a minimal drawdown guideline would be the best way to assure that pumping-induced flow rates near the well were not stressing the aquifer and causing mobilization of artificial particulate matter. On the basis of deleterious impacts of bailing and high-flow-rate pumping on ground-water samples, a number of other field studies (Puls and Barcelona, 1989, 1996; Puls et al., 1991, 1992; Puls and Powell, 1992a) also recommend an approach to sampling that includes:

- Using a low flow rate during both purging and sampling
- Placement of the pump intake within the well screen
- Minimal disturbance of the stagnant water column above the screen
- Monitoring water-quality indicator parameters during purging
- Minimization of atmospheric contact with samples
- Collection of unfiltered samples for metals analysis to estimate total contaminant load in the ground-water system

This sampling approach, referred to as low-flow purging and sampling, is described in detail in Chapter 4.

Artifactual colloidal material may also be generated during sampling by exposing ground water containing readily oxidizable materials, such as Fe^{2+}, Mn^{3+}, As^{3+}, Al, and Cd, to atmospheric conditions (Figure 1.15). This increases the redox state of the sample and causes the precipitation of colloid-sized metal oxides and hydroxides (Gillham et al., 1983; Puls et al., 1990; Ryan and Gschwend, 1990; Backhus et al., 1993; McCarthy and Degueldre, 1993). This may occur within the well, as bailing or pumping at rates sufficient to draw water levels down to within the well screen (effectively dewatering the well) exposes anoxic or suboxic ground water to atmospheric air. It may occur at the surface as bailed or pumped water is discharged and exposed to atmospheric conditions. This may also occur in the formation at some distance from a pumped well as dewatering exposes sediments and ground water to atmospheric air and causes mixing of chemically distinct ground waters from different zones. This can impact a much larger segment of the aquifer than the interval of interest (Puls et al., 1991).

The oxidation of iron from a reduced state is particularly important from a sample stability standpoint. As iron is oxidized, precipitating iron oxide and hydroxide, a pH shift (increase) often occurs, which may further impact sample integrity. Iron hydroxide is known to remove other constituents from solution, including Cu, Zn, Co, Cd, As, and Pb (Stoltzenberg and Nichols, 1985, 1986). The kinetics of oxidation-induced precipitation and subsequent sorption processes are such that they can occur within seconds or minutes (Reynolds, 1985; Grundl and Delwiche, 1992; Puls et al., 1992). The end result is that a number of previously dissolved species are removed from solution and may be removed from the sample if it is filtered, resulting in lower concentrations of these analytes than are actually present in ground water.

Inclusion of particulate matter that may have formed or collected in the well between sampling events may also be responsible for increased turbidity in samples. Preventing this material from being included in the sample during a sampling event can be accomplished by minimizing disturbance to the water column during sampling and by positioning the sampling device well above the bottom of the well. Backhus et al. (1993) described a situation where flocculent orange particles visible in water samples collected from a contaminated well were originally settled in the bottom 12 in. of the well and suspended in the top 2 in. of the water column. The material was probably iron oxide

FIGURE 1.15
Exposure of ground-water samples to atmospheric air for even a short period of time can result in oxidation of Fe^{2+} and co-precipitation of other metals and some organic species, which can significantly alter the chemical makeup of samples. Note the iron oxyhydroxide precipitate in the bottom of this VOC vial.

colloids precipitated by oxygenation of the anoxic ground water in the well. In two different wells at the same site, coal tar was found in the bottom of the wells. This was cited as having contributed greatly to observed initial high turbidities and PAH levels in samples collected from these wells by bailers and high-flow-rate pumping. Thus, they recommended cautious insertion of sampling gear into wells and avoidance of contact with the well bottom, to exclude this source of artifactual material.

Agitation and Aeration of Ground-Water Samples during Collection

There are two key times during a ground-water sampling event when agitation and aeration of samples can occur: within wells or direct-push sampling tools as they are purged and sampled using sampling devices that agitate and mix the water column (e.g., bailers or inertial-lift pumps), and at the ground surface, as samples are filtered or sample bottles are filled. Agitation of the sample can upset the balance of dissolved gases, pH, dissolved metals, and semivolatiles and can result in the stripping of VOCs from the sample. Agitation also increases turbidity, which can result in positive bias for the parameters associated with artifactual particulate material discussed previously and interfere with some analytical methods (Puls et al., 1992). Aeration of the sample can cause all of the chemical changes to the sample related to oxidation, as discussed in the previous section.

The most significant sources of bias and error due to sample agitation and aeration are those that occur during the sample retrieval process. For example, using bailers (Figure 1.16) or inertial-lift pumps (Figure 1.17) to purge and sample creates a substantial amount

FIGURE 1.16
The agitation and aeration of the water column caused by bailers is due to the fact that most bailers fit tightly into the well and they must be alternately inserted into and removed from the water column, which creates a surging effect in the well. This results in the collection of samples with high levels of turbidity.

of agitation and mixing of the water column, which results in elevated DO levels in samples (and associated chemical changes) and significant increases in turbidity (predominantly artifactual in nature). The agitation and aeration caused by repeated insertion and removal of a bailer often makes it impossible to measure representative values for DO and turbidity (Pennino, 1987; Pohlman et al., 1994), for VOCs (Muska et al., 1986; Imbrigiotta et al., 1988; Gibs et al., 1994), for trace metals (Puls et al., 1992; Heidlauf and Bartlett, 1993), or for PAHs (Backhus et al., 1993). Other sampling devices, provided they are installed carefully and operated properly, do not cause these problems. However, pumps that are pumped at rates high enough to dewater the sampling point can cause significant agitation and aeration of formation water.

Additional but perhaps less significant sources of agitation and aeration include sample bottle filling practices and sample filtration. A high potential for agitation and aeration occurs during sample bottle filling (particularly for VOC vials) using bailers and inertial-lift pumps, as it is often difficult to control the discharge from these devices (Figure 1.18). Bottom-emptying bailers offer an improvement over top-emptying bailers in this regard. For most pumps, whether continuous or cyclic discharge, it is relatively simple to control the discharge rate from the tubing to minimize agitation and aeration concerns. Bottles should be filled at relatively slow rates — about 250 ml/min or less for VOCs and about 500 ml/min or less for most other parameters (Figure 1.19). These rates provide a good compromise between the potential for agitation caused by higher flow rates and the greater exposure to atmospheric air (resulting in aeration) caused by lower flow rates.

FIGURE 1.17
Inertial-lift pumps also significantly agitate and aerate the water column in wells due to the nature of their operation. Pushing the tubing into the water column and pulling it up repeatedly results in discharge of highly turbid water.

FIGURE 1.18
Discharge from inertial-lift pumps (and top-emptying bailers) is difficult to control, which can result in agitation and aeration of samples during bottle filling.

FIGURE 1.19
Bottle filling should be done at a slow, controlled rate, but not so slow that contact of the sample with atmospheric air is prolonged.

References

Backhus, D.A., J.N. Ryan, D.M. Groher, J.K. MacFarlane, and P.M. Gschwend, Sampling colloids and colloid-associated contaminants in ground water, *Ground Water*, 31(3), pp. 466–479, 1993.

Barcelona, M.J. and J.A. Helfrich, Well construction and purging effects on ground-water samples, *Environmental Science and Technology*, 20(11), pp. 1179–1184, 1986.

Barcelona, M.J., J.P. Gibb, J.A. Helfrich, and E.E. Garske, Practical Guide to Ground-Water Sampling, Contract Report 374, Illinois State Geological Survey, Champaign, IL, 94 pp., 1985.

Barcelona, M.J., H.A. Wehrmann, and M.D. Varljen, Reproducible well-purging procedures and VOC stabilization criteria for ground-water sampling, *Ground Water*, 32(1), pp. 12–22, 1994.

Barcelona, M.J., J.F. Keely, W.A. Pettyjohn, and H.A. Wehrmann, Handbook: Ground Water, EPA/625/687/016, U.S. Environmental Protection Agency, Center for Environmental Research Information, Cincinnati, OH, 212 pp. 1987.

Brobst, R.B. and P.M. Buszka, The effect of three drilling fluids on ground-water sample chemistry, *Ground-Water Monitoring Review*, 6(1), pp. 62–70, 1986.

Buddemeier, R.W. and J.R. Hunt, Transport of colloidal contaminants in ground water: radionuclide migration at the Nevada test site, *Applied Geochemistry*, 3, pp. 535–548, 1988.

Cerda, C.M., Mobilization of kaolinite fines in porous media, *Colloids and Surfaces*, 27, pp. 219–241, 1987.

Cherry, J.A., Ground water monitoring: some current deficiencies and alternative approaches, chap. 13, in *Hazardous Waste Site Investigations: Toward Better Decisions*, R.B. Gammage and B.A. Berven, Eds., Lewis Publishers, Chelsea, MI, pp. 119–133, 1992.

Church, P.E. and G.E. Granato, Bias in ground-water data caused by well-bore flow in long-screen wells, *Ground Water*, 34(2), pp. 262–273, 1996.

Claassen, H.C., Guidelines and Techniques for Obtaining Water Samples That Accurately Represent the Water Chemistry of an Aquifer, Open-File Report 82-1024, U.S. Geological Survey, Lakewood, CO, 49 pp., 1982.

Dragun, J., *The Soil Chemistry of Hazardous Materials*, Hazardous Materials Control Research Institute, Silver Spring, MD, 1988.

Driscoll, F.G., *Ground Water and Wells,* 2nd ed., Johnson Division, St. Paul, MN, 1089 pp., 1986.

Eichholz, G.G., B.G. Wahlig, G.F. Powell, and T.F. Craft, Subsurface migration of radioactive waste materials by particulate transport, *Nuclear Technology,* 58, pp. 511–519, 1982.

Einarson, M., Multilevel Ground-Water Monitoring; Chapter 11 in *The Practical Handbook of Environmental Site Characterization and Ground-Water Monitoring,* Second Edition, D.M. Nielson, Editor, CRC Press/Taylor and Francis Group, Boca Raton, FL, pp. 807–848, 2006.

Elci, A., G.P. Flach, and F. Molz, Detrimental effects of natural vertical head gradients on chemical and water-level measurements in observation wells: identification and control, *Journal of Contaminant Hydrology,* 28, pp. 70–81, 2003.

Elci, A., F. Molz, and W.R. Waldrop, Implications of observed and simulated ambient flow in monitoring wells, *Ground Water,* 39(6), pp. 853–862, 2001.

Enfield, C.G. and G. Bengtsson, Macromolecular transport of hydrophobic contaminants in aqueous environments, *Ground Water,* 26(1), pp. 64–70, 1988.

Enfield, C.G., G. Bengtsson, and R. Lindquist, Influence of macromolecules on chemical transport, *Environmental Science and Technology,* 23(10), pp. 1278–1286, 1989.

Fetter, C.W., Potential sources of contamination in ground-water monitoring, *Ground-Water Monitoring Review,* 3(2), pp. 60–64, 1983.

Gibb, J.P., R.M. Schuller, and R.A. Griffin, Procedures for the Collection of Representative Water Quality Data from Monitoring Wells, Cooperative Ground Water Report 7, Illinois State Water and Geological Surveys, Champaign, IL, 61 pp., 1981.

Gibs, J. and T.E. Imbrigiotta, Well-purging criteria for sampling purgeable organic compounds, *Ground Water,* 28(1), pp. 68–78, 1990.

Gibs, J., T.E. Imbrigiotta, J.H. Ficken, J.F. Pankow, and M.E. Rosen, Effect of sample isolation and handling on the recovery of purgeable organic compounds, *Ground-Water Monitoring and Remediation,* 9(2), pp. 142–152, 1994.

Gillham, R.W., M.J.L. Robin, J.F. Barker, and J.A. Cherry, Ground-Water Monitoring and Sample Bias, American Petroleum Institute Publication, Environmental Affairs Department, Washington, DC, 206 pp., 1983.

Gillham, R.W., M.J.L. Robin, J.F. Barker, and J.A. Cherry, Field Evaluation of Well Purging Procedures, Publication 4405, American Petroleum Institute, Environmental Affairs Department, Washington, DC, 109 pp., 1985.

Grisak, G.E., R.E. Jackson, and J.F. Pickens, Monitoring ground water quality: the technical difficulties, in *Establishment of Water Quality Monitoring Programs,* L.G. Everett and K.D. Schmidt, Eds., American Water Resources Association, 1978.

Grolimund, D., M. Borkovec, K. Barmettler, and H. Sticher, Colloid-facilitated transport of strongly sorbing contaminants in natural porous media: a laboratory column study, *Environmental Science and Technology,* 30(10), pp. 3118–3123, 1996.

Grundl, T. and J. Delwiche, Kinetics of ferric oxyhydroxide precipitation, *Geochimica et Cosmochimica Acta,* 56, 1992.

Gschwend, P.M. and M.D. Reynolds, Monodisperse ferrous phosphate colloids in an anoxic ground water plume, *Journal of Contaminant Hydrology,* 1, pp. 309–327, 1987.

Gschwend, P.M., D.A. Backhus, J.K. MacFarlane, and A.L. Page, Mobilization of colloids in ground water due to infiltration of water at a coal ash disposal site, *Journal of Contaminant Hydrology* 6, pp. 307–320, 1990.

Harvey, R.W., R.L. George, and D.R. LeBlanc, Transport of microspheres and indigenous bacteria through a sandy aquifer: results of natural and forced-gradient tracer experiments, *Environmental Science and Technology,* 23(1), pp. 51–56, 1989.

Heidlauf, D.T. and T.R. Bartlett, Effects of monitoring well purge and sampling techniques on the concentration of metal analytes in unfiltered ground-water samples, Proceedings of the Seventh Outdoor Action Conference and Exposition, National Ground Water Association, Dublin, OH, pp. 437–450, 1993.

Hewitt, A.D., Potential of common well casing materials to influence aqueous metal concentrations, *Ground-Water Monitoring and Remediation,* 12(2), pp. 131–136, 1992.

Hutchins, S.R. and S.D. Acree, Ground-water sampling bias observed in shallow conventional wells, *Ground-Water Monitoring and Remediation,* 20(1), pp. 86–93, 2000.

Imbrigiotta, T.E., J. Gibs, T.V. Fusillo, G.R. Kish, and J.J. Hochreiter, Field evaluation of seven sampling devices for purgeable organic compounds in ground water, *Ground-Water Contamination: Field Methods,* A.G. Collins and A.I. Johnson, Eds., Special Technical Publication 963, American Society for Testing and Materials, Philadelphia, PA, pp. 258–273, 1988.

Kearl, P.M., N.E. Korte, and T.A. Cronk, Suggested modifications to ground-water sampling procedures based on observations from the colloidal borescope, *Ground-Water Monitoring Review,* 12(2), pp. 155–161, 1992.

Kraemer, C.A., J.A. Schultz, and J.W. Ashley, Monitoring Well Post-Installation Considerations; Chapter 12 in The Practical Handbook of Environmental Site Characterization and Ground-Water Monitoring, Second Edition, D.M. Nielsen, Editor, CRC Press/Taylor and Francis Group, Boca Raton, FL, pp. 849–882, 2006.

Lyman, W.J., P.J. Reidy, and B. Levy, Contaminants sorbed onto colloidal particles in water in either the unsaturated or saturated zone, Chap. 9 in *Mobility and Degradation of Organic Contaminants in Subsurface Environments,* C.K. Smoley, Inc, pp. 239–258, 1992.

Makeig, K.S. and D.M. Nielsen, Regulatory Mandates for Ground-Water Monitoring; Chapter 1 in *The Practical Handbook of Environmental Site Characterization and Ground-Water Monitoring,* Second Edition, D.M. Nielsen, Editor, CRC Press/Taylor and Francis Group, Boca Raton, FL, pp. 1–34, 2006.

Marsh, J.M. and J.W. Lloyd, Details of hydrochemical variations in flowing wells, *Ground Water,* 18(4), pp. 366–373, 1980.

Martin, W.H. and C.C. Lee, Persistent pH and tetrahydrofuran anomalies attributable to well construction, Proceedings of the Third National Outdoor Action Conference on Aquifer Restoration, Ground Water Monitoring and Geophysical Methods, National Water Well Association, Dublin, OH, pp. 201–213, 1989.

Mason, S.A., J. Barkach, and J. Dragun, Discussion of literature review and model (COMET) for colloidal/metals transport in porous media, *Ground Water,* 30(1), pp. 104–106, 1992.

McCall, W., D.M. Nielsen, S.P. Farrington, and T.M. Christy, Use of Direct-Push Technologies in Environmental Site Characterization and Ground-Water Monitoring; Chapter 6 in *The Practical Handbook of Environmental Site Characterization and Ground-Water Monitoring,* Second Edition, D.M. Nielsen, Editor, CRC Press/Taylor and Francis Group, Boca Raton, FL, pp. 345–472, 2006.

McCarthy, J. and L. Shevenell, Obtaining representative ground-water samples in a fractured and karstic formation, *Ground Water,* 36(2), pp. 251–260, 1998.

McCarthy, J.F. and C. Degueldre, Sampling and characterization of colloids and particles in ground water for studying their role in contaminant transport, *Environmental Particles,* J. Buffle and H.P. van Leeuwen, Eds., Lewis Publishers, Chelsea, MI, pp. 247–315, 1993.

McCarthy, J.F. and F.J. Wobber, Transport of Contaminants in the Subsurface: The Role of Organic and Inorganic Colloidal Particles, International Series of Interactive Seminars, Summary Report, ORNL/M-349, Oak Ridge National Laboratory, Oak Ridge, TN, 1996.

McCarthy, J.F. and J.M. Zachara, Subsurface transport of contaminants: binding to mobile and immobile phases in ground water aquifers, *Environmental Science and Technology,* 23(5), pp. 496–504, 1989.

McCaulou, D.R., D.G. Jewett, and S.G. Huling, Non-aqueous Phase Liquids Compatibility with Materials Used in Well Construction, Sampling and Remediation, EPA/540/S-95/503, Ground Water Issue, U.S. Environmental Protection Agency, Office of Research and Development, Washington, DC, 14 pp., 1995.

McDowell-Boyer, L.M., J.R. Hunt, and N. Sitar, Particle transport through porous media, *Water Resources Research,* 22(13), pp. 1901–1921, 1986.

McIlvride, W.A. and B.M. Rector, Comparison of short and long-screened wells in alluvial sediments, Proceedings of the Second Outdoor Action Conference, National Ground-Water Association, Dublin, OH, pp. 375–390, 1988.

Michalski, M., Application of temperature and electrical conductivity logging in ground-water monitoring, *Ground-Water Monitoring Review,* 9(3), pp. 112–118, 1989.

Miller, G.D., Uptake and release of lead, chromium, and trace-level volatile organics exposed to synthetic well casings, Proceedings of the Second National Symposium on Aquifer Restoration and Ground-Water Monitoring, National Water Well Association, Dublin, OH, pp. 236–245, 1982.

Muska, C.F., W.P. Colven, V.D. Jones, J.T. Scogin, B.B. Looney, and V. Price, Jr., Field Evaluation of ground-water sampling devices for volatile organic compounds, Proceedings of the Sixth National Symposium and Exposition on Aquifer Restoration and Ground Water Monitoring, National Water Well Association, Worthington, OH, pp. 235–246, 1986.

Nielsen, D.M., The Relationship of Monitoring Well Design, Construction and Development to Turbidity in Wells, and Related Implications for Ground-Water Sampling, Ground Water Sampling — A Workshop Summary, EPA/600/R-94/205, U.S. Environmental Protection Agency, Office of Research and Development, Washington DC, 16 pp., 1995.

Nielsen, D.M. and G.L. Nielsen, Technical Guidance on Low-Flow Purging and Sampling and Minimum-Purge Sampling, NEFS-TG001-02, Nielsen Environmental Field School, Galena, OH, 59 pp., 2002.

Nielsen, D.M., G.L. Nielsen, and L.M. Preslo, Environmental Site Characterization; Chapter 2 in *The Practical Handbook of Environmental Site Characterization and Ground-Water Monitoring, Second Edition*, D.M. Nielsen, Editor, CRC Press/Taylor and Francis Group, Boca Raton, FL, pp. 35–206, 2006.

Nielsen, D.M. and R. Schalla, Design and Installation of Ground-Water Monitoring Wells; Chapter 10 in *The Practical Handbook of Environmental Site Characterization and Ground-Water Monitoring, Second Edition*, D.M. Nielsen, Editor, CRC Press/Taylor and Francis Group, Boca Raton, FL, pp. 639–806, 2006.

O'Melia, C.R., Aquasols: the behavior of small particles in aquatic systems, *Environmental Science and Technology*, 14(9), pp. 1052–1060, 1980.

O'Melia, C.R., Kinetics of colloid chemical processes in aquatic systems, in *Aquatic Chemical Kinetics*, W. Stumm, Ed., Wiley & Sons, New York, NY, pp. 447–474, 1990.

Oakley, D. and N.E. Korte, Nickel and chromium in ground-water samples as influenced by well construction and sampling methods, *Ground-Water Monitoring and Remediation*, 16(1), pp. 93–99, 1996.

Oneacre, J. and D. Figueras, Ground-water variability at sanitary landfills: causes and solutions, Proceedings, ASCE Conference on Uncertainty in the Geologic Environment, Madison, WI, July 31–August 3, 1996, pp. 965–987, 1996.

Palmer, C.D., J.F. Keely, and W. Fish, Potential for solute retardation on monitoring well sand packs and its effect on purging requirements for ground-water sampling, *Ground-Water Monitoring Review*, 7(2), pp. 40–47, 1987.

Parker, L., Suggested guidelines for the use of PTFE, PVC and stainless steel in samplers and well casings, *Current Practices in Ground Water and Vadose Zone Investigations*, D.M. Nielsen and M.N. Sara, Eds., Special Technical Publication 1118, American Society for Testing and Materials, Philadelphia, PA, pp. 217–229, 1991.

Parker, L., The effects of ground-water sampling devices on water quality: a literature review, *Ground-Water Monitoring and Remediation*, 14(2), pp. 130–141, 1994.

Parker, L., A.D. Hewitt, and T.F. Jenkins, Influence of casing materials on trace-level chemicals in well water, *Ground-Water Monitoring and Remediation*, 10(2), pp. 146–156, 1990.

Paul, C.J. and R.W. Puls, Impact of turbidity on TCE and degradation products in ground water, *Ground-Water Monitoring and Remediation*, 17(1), pp. 128–133, 1997.

Paul, D.G., C.D. Palmer, and D.S. Cherkauer, The effect of construction, installation and development on the turbidity of water in monitoring wells in fine-grained glacial till, *Ground-Water Monitoring Review*, 8(1), pp. 73–82, 1988.

Pennino, J., Dissolved Oxygen Changes in Well Water During Purging and Sampling, Unpublished Report, Ohio Environmental Protection Agency, Dayton, OH, 1987.

Penrose, W.R., W.L. Polzer, E.H. Essington, D.M. Nelson, and K.A. Orlandin, Mobility of plutonium and americium through a shallow aquifer in a semi-arid region, *Environmental Science and Technology*, 24, pp. 228–234, 1990.

Pohlmann, K.F., G.A. Icopini, R.D. McArthur, and C.G. Rosal, Evaluation of Sampling and Field-Filtration Methods for Analysis of Trace Metals in Ground Water, EPA/600/R-94/119 U.S. Environmental Protection Agency, Office of Research and Development, Las Vegas, NV, 79 pp., 1994.

Powell, R.M. and R.W. Puls, Hitting the bull's-eye in ground-water sampling, *Pollution Engineering*, pp. 51–54, 1997.

Powell, R.M. and R.W. Puls, Passive sampling of ground-water monitoring wells without purging: multilevel well chemistry and tracer disappearance, *Journal of Contaminant Hydrology*, 12, pp. 51–77, 1993.

Puls, R.W., Colloidal considerations in ground-water sampling and contaminant transport predictions, *Nuclear Safety*, 31, pp. 58–65, 1990.

Puls, R.W. and R.M. Powell, Acquisition of representative ground-water quality samples for metals, *Ground-Water Monitoring Review*, 12(3), pp. 167–176, 1992a.

Puls, R.W. and R.M. Powell, Transport of inorganic colloids through natural aquifer material: implications for contaminant transport, *Environmental Science and Technology*, 26(3), pp. 614–621, 1992b.

Puls, R.W., J.H. Eychaner, and R.M. Powell, Facilitated Transport of Inorganic Contaminants in Ground Water: Part I. Sampling Considerations, EPA/600/M-90/023, U.S. Environmental Protection Agency, Robert S. Kerr Laboratory, Ada, Oklahoma, 12 pp., 1990.

Puls, R.W. and J.F. McCarthy, Well Purging and Sampling (Workshop Group Summary), Ground Water Sampling — A Workshop Summary, EPA/600/R-94/205, U.S. Environmental Protection Agency, Office of Research and Development, Washington, DC, pp. 82–87, 1995.

Puls, R.W. and M.J. Barcelona, Ground-Water Sampling for Metals Analysis, EPA/540/4-89/001, Superfund Ground Water Issue, U.S. Environmental Protection Agency, Office of Solid Waste and Emergency Response, Washington, DC, 6 pp., 1989.

Puls, R.W. and M.J. Barcelona, Low-Flow (Minimal-Drawdown) Ground-Water Sampling Procedures, EPA/540/5-95/504, Ground Water Issue, U.S. Environmental Protection Agency, Office of Solid Waste and Emergency Response, Washington, DC, 12 pp., 1996.

Puls, R.W., D.A. Clark, B. Bledsoe, R.M. Powell, and C.J. Paul, Metals in Ground Water: Sampling Artifacts and Reproducibility, *Hazardous Waste and Hazardous Materials*, 9(9), pp. 149–162, 1992.

Puls, R.W., R.M. Powell, D.A. Clark, and C.J. Paul, Facilitated Transport of Inorganic Contaminants in Ground Water: Part II, Colloidal Transport, EPA/600/M-91/040, U.S. Environmental Protection Agency, Robert S. Kerr Environmental Research Laboratory, Ada, OK, 1991.

Reilly, T.E., O.L. Franke, and G.D. Bennett, Bias in ground-water samples caused by well-bore flow, *Journal of Hydrologic Engineering*, 115, pp. 270–276, 1989.

Reynolds, M.D., Colloids in Ground Water, Masters thesis, Department of Civil Engineering, Massachusetts Institute of Technology, Cambridge, MA, 1985.

Robertson, W.D., J.F. Barker, Y. LeBeau, and S. Marcoux, Contamination of an unconfined sand aquifer by waste pulp liquor: a case study, *Ground Water*, 22(4), pp. 192–197, 1984.

Robin, M.J.L. and R.W. Gillham, Field evaluation of well purging procedures, *Ground-Water Monitoring Review*, 7(4), pp. 85–93, 1987.

Ronen, D., M. Magaritz, H. Gvirtzman, and W. Garner, Microscale chemical heterogeneity in ground water, *Journal of Hydrology*, 92(1/2), pp. 173–178, 1987.

Ruda, T. and J. Farrar, Environmental Drilling for Soil Sampling, Rock Caring, Boreole Logging, and Monitoring Well Installation; Chapter 5 in The Practical Handbook of Environmental Site Characterization and Ground-Water Monitoring, Second Edition, D.M. Nielsen, Editor, CRC Press/Taylor and Francis Group, Boca Raton, FL, pp. 297–344, 2006.

Ryan, J.N. and P.M. Gschwend, Effect of iron diagenesis on clay colloid transport in an unconfined sand aquifer, *Geochimica et Cosmochimica Acta*, 56, pp. 1507–1521, 1992.

Ryan, J.N. and P.M. Gschwend, Colloidal mobilization in two atlantic coastal plain aquifers: field studies, *Water Resources Research*, 26, pp. 307–322, 1990.

Ryan, J.N., Ground Water Colloids in Two Atlantic Coastal Plain Aquifers: Colloid Formation and Stability, Masters thesis, Department of Civil Engineering, Massachusetts Institute of Technology, Cambridge, MA, 1988.

Ryan, J.N., S. Mangion, and R. Willey, Turbidity and Colloid Transport (Working Group Summary), EPA/600/R-94/205, Ground Water Sampling — A Workshop Summary, U.S. EPA Office of Research and Development, Washington, DC, pp. 88–92, 1995.

Schuller, R.M., J.P. Gibb, and R.A. Griffin, Recommended sampling procedures for monitoring wells, *Ground-Water Monitoring Review*, 1(2), pp. 42–46, 1981.

Sosebee, J.B., P.C. Geiszler, D.L. Winegardner, and C.R. Fisher, Contamination of Ground-Water Samples with PVC Adhesives and PVC Primer from Monitoring Wells, *Technical Paper*, Environmental Science and Engineering, Englewood, CO, 24 pp., 1983.

Sosebee, J.B., P.C. Geiszler, D.L. Winegardner, and C.R. Fisher, Contamination of Ground-Water Samples with PVC Adhesives and PVC Primer from Monitoring Wells, Proceedings of the ASTM Symposium on Hazardous and Solid Waste, American Society for Testing and Materials, West Conshohocken, PA, 23 pp., 1982.

Stolzenburg, T.R. and D.G. Nichols, Preliminary Results on Chemical Changes in Ground-Water Samples Due to Sampling Devices, EPRI EA-4118, Project 2485–7, Electric Power Research Institute, Palo Alto, CA, 53 pp., 1985.

Stolzenburg, T.R. and D.G. Nichols, Effects of Filtration Method and Sampling on Inorganic Chemistry of Sampled Well Water, Proceedings of the Sixth National Symposium and Exposition on Aquifer Restoration and Ground Water Monitoring, National Water Well Association, Dublin, OH, pp. 216–234, 1986.

Stumm, W. and J.J. Morgan, *Aquatic Chemistry: Chemical Equilibria and Rates in Natural Waters*, John Wiley and Sons, Inc, New York, NY, 1996.

Thornton, D., S. Ita, and K. Larsen, Broader Use of Innovative Ground Water Access Technologies, Proceedings of the HazWasteWorld Superfund XVIII Conference, Vol. 2, E.J. Krause Co, Washington, DC, pp. 639–646, 1997.

U.S. EPA, Handbook for Sampling and Sample Preservation of Water and Wastewater, U.S. Environmental Protection Agency, Environmental Monitoring and Support Laboratory, Cincinnati OH, pp. 386–397, 1982.

U.S. EPA, Handbook, Ground Water, Volume II: Methodology, EPA/625/6-90/016b, U.S. Environmental Protection Agency, Office of Research and Development, Center for Environmental Research Information, Washington, DC, 141 pp., 1991.

U.S. EPA, Low Stress (Low Flow) Purging and Sampling Procedures for the Collection of Ground-Water Samples from Monitoring Wells, SOP#GW0001, U.S. Environmental Protection Agency, Region 1, Boston, MA, 13 pp., 1996.

Unwin, J.P., A Guide to Ground-Water Sampling, Technical Bulletin 362, National Council of the Paper Industry for Air and Stream Improvement, New York, NY, 1982.

Walker, S.E., Background Ground-Water Quality Monitoring: Well Installation Trauma, Proceedings of the Third National Symposium on Aquifer Restoration and Ground Water Monitoring, National Water Well Association, Worthington, OH, pp. 235–246, 1983.

2

The Ground-Water Sampling and Analysis Plan (SAP): A Road Map to Field Sampling Procedures

Gillian L. Nielsen

CONTENTS

Objectives of the SAP

As discussed in Chapter 1, there is a great deal of complex science behind ground-water sampling. Much of this science drives the selection of sampling methods and equipment for each site. To ensure that sampling teams are aware of requirements for sample collection, a site-specific SAP must be written by an experienced practitioner and followed by the sampling team. The objectives of a typical SAP are summarized in Table 2.1. Implementation of a comprehensive and well thought out SAP should ensure that ground-water sample collection procedures are consistent from one sampling event to the next, thus reducing the potential for sampling team related error and bias. This, in turn, should ensure that the data generated, both in the field and as a result of laboratory analysis of samples, are comparable and without peaks and valleys referred to as "data bounce."

TABLE 2.1

Objectives of a Sampling and Analysis Plan

Provide a written statement of objectives of the sampling program
Provide a schedule for sample collection (number of wells, when and how frequently they should be sampled)
Provide detailed procedures for all aspects of the ground-water sample collection process to be implemented by the sampling team to ensure sampling accuracy and precision including:
– Well integrity inspection
– Wellhead screening for volatile or combustible vapors
– Water-level and product-thickness measurement
– Field QA/QC measures
– Purging and sampling device selection and operation
– Well purging procedures and management of purge water
– Parameters required for ground-water sample analysis
– Field equipment calibration procedures
– Field parameter measurement procedures
– Field equipment decontamination procedures
– Sample collection procedures, including sample container selection, sample pretreatment requirements, order of sample collection, and sample container filling
– Sample handling and preparation for shipment
– Documentation of sampling event activities
Provide written documentation of field procedures for outside evaluation
Provide a vehicle for project management and budgeting

Preparation of the SAP

The SAP should be written by an experienced practitioner who has relevant field experience and who can identify potential sources of error and bias in each component of the ground-water sample collection process. In addition, working experience in a laboratory or working closely with laboratory personnel is an advantage.

Prior to submitting the SAP to regulatory agency personnel for their approval, it is recommended that field sampling team personnel have the opportunity to review and provide honest input on procedures contained within the SAP. Communication between all levels of personnel involved in the sampling program will ensure that the final SAP is not only technically correct for the site-specific program, but that field protocols are practical and realistic and based on good science. All SAP writers are encouraged to remember this saying: "nothing is impossible for the person who does not have to do it." A procedure may sound good on paper, but if the sampling team finds the procedure to be too cumbersome or too confusing, shortcuts will be made in the field, which can introduce substantial sampling error, imprecision, and bias. Project personnel need to remember that for most ground-water SAPs, once the document has received regulatory approval for implementation, procedural changes cannot be made randomly in the field. Deviation from the documented protocol will call into question all results from field measurements and subsequent data generated by laboratory analysis of collected samples.

Selection of Field Protocols to Be Incorporated into the SAP

As illustrated in Table 2.1, the SAP details specific standard operating procedures for a number of work tasks to be implemented during the sampling event. While preparing the SAP, a great deal of thought needs to go into the selection of protocols for each of the work tasks most appropriate for any individual site. This includes methods and equipment for wellhead screening, water-level and product-thickness measurement, field quality

FIGURE 2.1
Well headspace screening must be done with the appropriate instrument. This PID is used to detect volatile organic compounds (VOCs) associated with petroleum hydrocarbons and organic solvents.

assurance (QA) and quality control (QC), sampling point purging and sample collection, field parameter measurement, and field equipment decontamination.

Well Headspace Screening

The first task implemented by the sampling team at each sampling point is opening the well and screening the headspace above the water column in the well for the presence of volatile or combustible gases and vapors. The SAP must detail which instrumentation is required for screening the well headspace (e.g., photoionization detectors [PIDs] [Figure 2.1], flame ionization detectors [FIDs], or combustible gas indicators) and how the data are to be used (e.g., health and safety personal protective equipment [PPE] selection). It must also indicate how the equipment is to be calibrated, operated, and maintained throughout the sampling event. Equipment for headspace screening is selected based on the types of contaminants known to be present at the site. Table 2.2 lists the types of equipment that may be used for well headspace screening and their applications. Nielsen et al. (2006) describes the use of PIDs and FIDs in more detail for headspace screening, and Maslansky and Maslansky (2006) describes how to use data from PIDs and FIDs to select the most appropriate PPE.

Factors that must be considered in the selection of headspace screening equipment include: type of data generated (qualitative versus quantitative), ability to detect the parameters of concern, ease of calibration; sources of interference, and ease of use. Readers are directed to Maslansky and Maslansky (1993, 1997, 2006) for detailed discussions on the applications and limitations of each of the types of instrumentation presented in Table 2.2 and for guidance on how to incorporate and interpret data from wellhead screening in a health and safety plan.

Water-Level and Product-Thickness Measurement

Following well headspace screening, the next task is to take water-level or product-thickness measurements (Figure 2.2). In this portion of the SAP, specific procedures for

TABLE 2.2

Examples of Instrumentation Appropriate for Well Headspace Screening

Instrumentation	Class of Contaminant of Concern	Example Application
PID	Volatile organic vapors and gases and some inorganics	Dry cleaning facility for PCE and TCE
FID	Volatile to semi-volatile organic vapors	Underground storage tanks with jet fuel
Combustible gas indicator (may be combined with an oxygen meter)	Flammable vapors	Landfills where methane may accumulate in wells
Toxic gas sensors	Vapors containing known toxics	Industrial facility where H_2S is generated

where, when, and how to take water-level and, if applicable, separate-phase product-measurements (LNAPL or DNAPL) must be described in detail. These measurements must be taken in all sampling points prior to any purging and sampling activities, to ensure that the data are collected under as close to the same environmental and atmospheric conditions as possible. This is of particular concern when taking water-level measurements, which are potentially affected by a number of environmental variables including changes in ambient air pressure (especially problematic for shallow, unconfined formations), tides, changes in levels of nearby rivers, precipitation events, and operation of nearby pumping wells.

A number of different methods are available for water-level measurement, including ploppers (or poppers), chalked tapes, electronic water-level gauges, pressure transducers, bubblers, sonic devices, floats, and acoustic probes. The method appropriate for any given site depends on the measurement accuracy required, the depth to static water level

FIGURE 2.2

Water-level measurements should be taken in all wells in a sampling program prior to purging and sampling of the wells (to collect data for piezometric surface mapping) and again just before purging (to determine the well volume for well-volume purging or the initial static water level for low-flow purging and sampling). This water-level gauge will be used to track drawdown in a well during low-flow purging and sampling.

in the wells included in the sampling program, well diameter, whether single measurements or continuous data are required, and the possible sources of interference present in the well. ASTM Standard D 4750 (ASTM, 2006a) discusses some of the practices available for measurement of water levels in monitoring wells including ploppers, carpenter's chalk on a steel tape, and a variety of electronic water-level gauges. Chapter 7 discusses these and other water-level measurement methods in detail.

Water-level data are used for a variety of applications including determining the volume of water in the sampling point, determining the direction of ground-water flow and the hydraulic gradient (horizontal and vertical), and calculating the rate of ground-water flow. Readers are directed to Chapter 7 for a detailed discussion on water-level data interpretation and presentation. ASTM Standard D 6000 (ASTM, 2006b) also provides a discussion on methods for presentation of water-level information obtained from ground-water investigations.

At sites with petroleum hydrocarbon contamination, some monitoring wells and direct-push tools may contain a separate-phase layer of product (LNAPL), which must be measured in addition to the water level. The thickness of this layer is referred to as the "apparent product thickness," which reflects the amount of hydrocarbon that has accumulated in the well — not the actual thickness of product in the formation.

To measure the thickness of LNAPL, one of three methods can be used: (1) hydrocarbon-sensitive and water-sensitive pastes applied to a measuring tape; (2) an electronic oil–water interface probe; and (3) a clear, acrylic hydrocarbon bailer (API, 1989). The first method requires the application of a hydrocarbon-sensitive paste to one side of a tape and a water-sensitive paste to the other side. The tape is lowered into the well and as the LNAPL and water contact the pastes, color changes occur. The depths of each color change are measured, with the difference between the changes indicating LNAPL thickness. The depth to ground water, which must be adjusted because of the presence of LNAPL, is indicated by the color change on the side of the tape coated with the water-sensitive paste. Compensation factors and the significance of the LNAPL in the well are discussed in API (1989) and Testa and Winegardner (1991).

Electronic oil–water interface probes are capable of determining the depth to ground water using a conductive probe similar to that used in an electronic water-level gauge. In addition, they are equipped with one of three hydrocarbon-detecting sensors: (1) infrared; (2) optical; or (3) a float. The meter emits different audible tones or different indications via lights (i.e., continuous versus flashing) when the probe contacts the LNAPL and then the LNAPL–water interface. This method is quicker than using pastes but is much more expensive. It is also less accurate than the measuring tape with pastes when measuring thin layers of hydrocarbon. Quoted accuracies and detection capabilities of most oil–water interface probes are usually between 0.12 and 0.06 in.

Clear hydrocarbon bailers are useful for collecting a grab sample of the LNAPL layer, which can be visually examined and measured within the bailer. To measure LNAPL thickness, the bailer is lowered to the point at which the first fluid in the well is encountered, then lowered an amount slightly less than the length of the bailer. To ensure a measurement of the full LNAPL thickness, the bailer must be long enough to ensure that its top will be above the air–LNAPL interface, when the bottom check valve is below the LNAPL–water interface. Of the three methods, this is the least accurate method for LNAPL measurement because the thickness of product in the bailer is never an exact indication of what is in the well. It is always less than the product thickness in the well by a factor that depends on the diameter of the bailer (compared with the diameter of the well) and the design of the check valve.

Field Quality Assurance and Quality Control

A ground-water sampling event generates information and measurements used for making important site-specific decisions such as determining whether a facility design is effective at preventing impacts to the environment, determining whether ground-water contamination is present at a site, determining whether contamination poses an exposure risk to off-site receptors, determining whether natural attenuation is an appropriate remedial alternative for a site, or determining the most effective remedial design for a site. Given the significance of these decisions and their potential associated costs, it is critical that all analytical and field measurement data are not only technically sound but also scientifically and legally defensible. An important component of the SAP is a detailed field QA/QC program which, when implemented, provides sampling teams with the confidence that the results of the sampling event will be technically correct and defensible.

A QA program may be defined as those operations and procedures that are undertaken in the field to provide measurement data of a stated quality with a stated probability of being correct (Taylor, 1985). The QA program documents administrative and field procedures that are designed to monitor management of the project as well as field sample collection and measurement activities. Table 2.3 summarizes the key administrative and field elements of a field QA program for ground-water sampling.

When followed, a field QA/QC program will ensure that all data generated through field measurement and analysis are accurate and precise with a minimum amount of error and bias. The QA/QC program ensures that the data produced are defensible if the project is subject to litigation, because it provides clear documentation of field procedures and outlines the system of checks and balances that were used to verify sampling and field measurement accuracy and precision. ASTM Standard D 7069 provides a standard guide for field QA in a ground-water sampling event (ASTM, 2006c).

Samples referred to as field QC samples should be collected during every sampling event (Table 2.4). The purpose of field QC samples varies with the type of sample collected, but these samples provide a formal means of verifying the precision and accuracy of various components of field sample collection procedures. Field QC samples are commonly collected at one or more designated sampling points after all laboratory samples have been collected at that (those) point(s). This ensures that sufficient sample volume is available for collection of samples for laboratory analysis.

TABLE 2.3

Key Elements of a Field QA Program

Administrative elements
Project description and definition of project objectives
Project fiscal information (travel, support services, expendable supplies, equipment needs)
Schedule of tasks and products (field activities, analysis, data review, reporting)
Project organization and responsibility
Selection of appropriately trained and experienced personnel for field and management roles
Field elements
Implementation of technically sound SOPs
Documentation of protocols for operation, calibration, and maintenance of all field instrumentation
Collection of field QC samples (which, when, how, and how many)
Adherence to required sample pretreatment methods and holding times
Use of chain-of-custody procedures
Record keeping procedures that incorporate good laboratory practices (GLPs)
Methods for checking accuracy of field parameter measurements
Description of corrective actions to be implemented if an error is detected at any point in the field

TABLE 2.4

Common Field QC Samples

Type of QC Sample	Recommended Frequency of Collection
Trip blank	One per shipment[a] per day per laboratory
Temperature blank	One per shipping container
Field blank	One per waste management unit
Equipment blank	One for every ten sampling points
Blind duplicate sample	One for every ten sampling points
Spiked sample	One for every ten samples submitted for analysis
Field split sample	Variable, but at least one upgradient location and two downgradient locations should be sampled

[a]The term "shipment" must be defined for each project. For example, if ten coolers are being sent to a laboratory, under one definition, all ten coolers might be defined as a single shipment. In another regulatory program, each cooler could be considered to be an individual shipment.

In many cases, regulatory programs dictate which QC samples are required to be included in the sampling protocol. Unfortunately, in many instances, only minimum attention is paid to incorporating these samples into a sampling program. When an SAP is prepared for any site, it is important to incorporate not only those QC elements required to satisfy regulatory requirements, but also those system checks that will facilitate data validation. When determining the level of field QA/QC independently of regulatory guidelines, the political sensitivity of the site being monitored needs to be evaluated. A higher level of QA/QC is generally warranted if the site is operating under a consent order, if there is a strong public opinion about detrimental site impacts, or if the site is about to be sold. QA/QC levels should also be higher than normal at sites where the concentration of contaminants in samples is very close to action levels or analytical method detection limits. Under these circumstances, because critical decisions are made based on comparison of sampling results to action levels, it is important to have the utmost confidence in the accuracy and precision of the samples and the field data collected. In long-term monitoring programs or in programs where there are sufficient numbers of sampling points to warrant multiple sampling teams, field QA/QC efforts should be elevated to ensure consistency between sampling teams. Without good field QA/QC, it can be difficult to directly compare results generated by two different sampling teams at a single site, even if they are implementing the same sampling protocol.

Table 2.4 presents a list of the various QC samples that may be included in a field QC program, with the typical frequency of collection for each type of sample. If a ground-water sampling program is subject to scrutiny by outside groups, such as attorneys or regulatory agency personnel, the ratio of QC samples to sampling locations will decrease (i.e., instead of a 1:10 ratio, some QC samples may be required at a 1:5 ratio). There is a tendency to try to increase the ratio of sampling locations to QC samples in an effort to save costs. As an example, the authors recently reviewed an SAP that incorporated equipment blanks to verify the effectiveness of field equipment decontamination procedures. The plan indicated that an equipment blank would be collected after every 30 sampling locations (rather than every ten locations) to save money. When laboratory results were reported, a contaminant was detected not only in each of the 30 samples collected but also in the equipment blank. As a consequence, all 30 of the sampling points had to be resampled (rather than ten) to determine whether the contaminant concentrations were real or whether the data indicated that there was a problem with field equipment cleaning procedures.

In addition to selecting which and how many QC samples will be incorporated into the sampling program, it is important to select which parameters should be analyzed on each of the QC samples so that the information derived from their analysis will be meaningful. Ideally, each QC sample will be analyzed for the same parameters as the ground-water samples. This will provide the most meaningful information and will permit identification of any potential problems for any of the analytes of interest. This, however, is also the most expensive strategy and, consequently, the least likely to be implemented. More likely, regulatory guidance will be referenced to put together a short "required" list of analytes. In many cases, regulatory guidance requires that field QC samples be analyzed for volatile constituents because that group of analytes is most sensitive to handling. However, not all sites are concerned with volatiles, so important resources may be expended on potentially inconsequential analyses.

Purposes of Field Quality Control Samples

It is important that sampling team members understand the purpose of each type of QC sample to understand the importance of collecting the samples correctly. The purpose of each of the QC samples presented in Table 2.4 will be described herein. Details on how to collect each QC sample are found in Chapter 6.

A trip blank is a blank designed to determine whether anything associated with the preparation of the sample containers by the laboratory, shipment of the empty containers to the field, traveling in the field with the sampling team during the sampling event, and return shipment to the laboratory could have any impact on sample integrity. This blank is prepared by the laboratory and requires no preparation or handling by the field sampling team.

The purpose of a temperature blank is to provide a formal mechanism for the laboratory to confirm actual sample temperatures upon arrival at the laboratory. When samples arrive at the laboratory, temperatures should be checked in sample containers of differing volume and placement within the shipper. This can be done using: (1) a certified thermometer inserted into one or more sample containers; (2) a calibrated infrared gun to determine the surface temperature of individual containers; or (3) a specially prepared 40 ml sentry vial (Figure 2.3) that contains a permanently affixed certified thermometer,

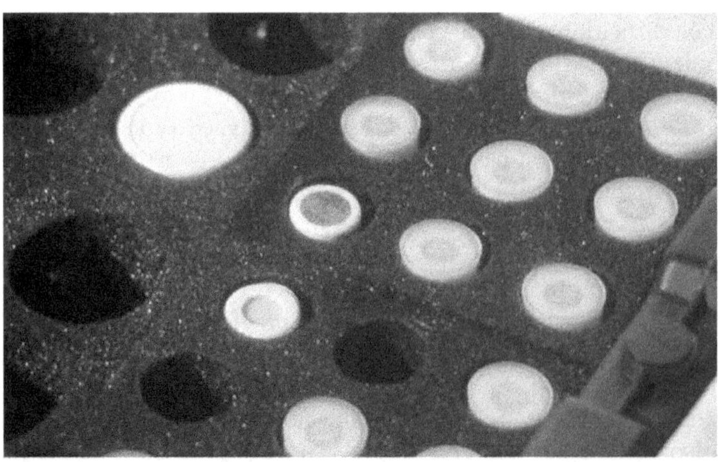

FIGURE 2.3
Temperature blanks (in this case, the vial with the dark dot on the top) are used to provide the laboratory with a convenient means of checking sample temperatures upon arrival of a shipment of samples at the lab.

usually provided by the laboratory. The latter method, called a temperature blank, is preferred by some laboratories because it does not require that a foreign object (a thermometer) be inserted into the sample container (which could result in sample contamination). The infrared method is preferred by other laboratories because it is fast, does not require that a thermometer be inserted into the sample container, and permits evaluation of a number of sample containers of differing volumes throughout the sample shipper. Regardless of which method is used by the laboratory, when sample temperatures are checked, the laboratory will look for an arrival temperature of $4 \pm 2°C$. If the arrival temperature is outside of this range, the sampling team will be contacted to discuss an appropriate course of action: resample the well or analyze the received sample, with a disclaimer stating that the analyzed sample was not appropriately preserved with respect to temperature.

At some field sites, there is a concern that ambient air contamination levels may be sufficiently high to influence concentrations of contaminants detected in ground-water samples. In some instances, trace quantities of VOCs may be gained by a ground-water sample through contact with atmospheric air, causing a positive bias in their determination (Gillham et al., 1983). A good example of a site where this is a common problem is the corner service station. At these sites, ground-water monitoring wells are commonly installed in the vicinity of pump islands. When samplers purge and sample these wells, the station typically remains open for business, meaning that as wells on one side of the pump island are sampled, station customers may dispense fuel into their vehicle on the other side of the island. Under these circumstances, ambient air is contaminated by fumes from the dispenser and exhaust fumes from adjacent vehicles, which contain the same volatile constituents for which ground-water samples will be analyzed. To quantify the extent to which ambient conditions may influence sample chemistry, a field blank is collected.

A fourth important field QC sample is the equipment blank. Whenever a piece of equipment is used in more than one sampling location, it should be cleaned between locations (Figure 2.4). It will be necessary to collect an equipment blank to verify the effectiveness of field equipment cleaning procedures (see Chapter 8 for more detailed discussions on field equipment cleaning procedures). Two forms of equipment blank can be collected. The first, sometimes referred to as a rinseate blank, is designed to determine whether, following equipment cleaning, soluble contaminants remain on the surfaces of the equipment. In some cases, contaminants may not readily solubilize in control water, but may remain on the surface of cleaned equipment as a residue. To determine whether a residue remains on a surface following cleaning, a wipe or swipe sample may be collected.

To test the precision of sampling teams, a blind duplicate sample is collected as part of the field QC program. The objective of the blind duplicate is to collect two samples of ground water that are as close to chemically identical as possible. There are two procedures for collecting blind duplicates — one procedure for nonvolatile parameters and a second procedure for parameters requiring the use of 40 ml VOC vials. Detailed procedures on how to collect each type of duplicate sample are provided in Chapter 6.

In many regulatory agency guidance documents, a field QC sample called a spiked sample is presented as a means to determine whether there is any matrix activity (most often in the form of microbial activity) that might alter sample chemistry from the time a sample is collected to the time it is extracted or analyzed by the laboratory. Samplers must remember that once filled, sample containers continue to be living, breathing environments that may undergo chemical and physical changes, especially where no chemical preservatives are used. Spiked samples can be an effective means of determining whether field chemical and physical preservation methods are appropriate. Field spiked samples should not be confused with spiked samples used by the laboratory as part of its internal QC program.

FIGURE 2.4
An equipment blank is collected by passing a rinse of final control water over the surface of a piece of equipment after it has been used and then cleaned.

The final QC sample that is commonly incorporated into a ground-water sampling program is a field split sample. Field split samples are collected for the purpose of verifying the performance of one laboratory against a laboratory of known performance levels. Typically, field split samples are collected when a regulatory agency wishes to evaluate the performance of a new or unknown laboratory against the regulatory agency's approved laboratory to ensure the accuracy of sample analysis.

Purging and Sampling Device Selection and Operation

Many devices are available for purging and sampling ground water. These devices and their operational characteristics are described in detail in Chapter 3. The SAP must include a description of the devices selected for use on a task-specific basis. It should also include information on how to operate the selected devices and how to maintain the equipment to ensure proper operation in the field. While not always required by regulatory agencies when reviewing SAPs, it is highly recommended that the SAP includes, as an appendix, the operations manuals for each device specified for purging and sampling. This provides sampling team members with critical equipment-specific information such as operating procedures, calibration procedures, troubleshooting tips, spare parts lists, and contact information for equipment repair.

Sampling Point Purging Methods

Most traditional approaches to ground-water sampling are based upon the assumption that all water that resides in sampling points between sampling events is stagnant (i.e., does not interact with formation water) and does not represent the chemistry of water in the formation. Thus, the historical means of meeting the objectives of sampling programs has been to remove all of the water from the well and to induce fresh formation water to enter the well so it can be collected as a sample. This process is referred to as well purging. Several purging strategies, described in detail in Chapter 4, are in common use.

The SAP must specify which purging strategy or strategies will be used at a site for wells with high yield, as well as for wells with low yield. The SOPs for purging must provide step-by-step procedures that include specifying: water-level measurement requirements and methods; well-depth measurement requirements and methods; the equation used to calculate a well volume (if required); the device used for purging and guidance on placement of the device within the water column, operation of the device, and cleaning of the device if used portably; when and how to measure any required field parameters during purging; and how stabilization of field parameter measurements is defined, if required as part of the purging protocol.

Field Measurement of Water-Quality Indicator Parameters and Turbidity

As discussed in Chapter 4, monitoring of selected chemical and physical indicator parameters in ground water in the field is an integral component of some strategies used to purge high-yield wells. Chemical parameters most commonly measured include pH, specific conductance (or conductivity) and, more recently, DO and ORP (redox potential, also measured as Eh). In addition, some sampling programs include the measurement of temperature, and some include turbidity, a physical parameter that is an indicator of the disturbance caused to the water column by the purging and sampling method used.

The SAP must detail exactly which indicator parameters are to be measured in the field, and where, when (in the context of well purging and sample collection), and how (related to the type of equipment used) those parameters are to be measured. For applications involving measurement to stabilization, the SAP must also provide an accurate definition of "stabilization" for each parameter, that is specific to the instrumentation used. It is the experience of the authors that this is a common deficiency of many SAPs, which results in a great deal of error in field parameter measurement. Table 4.5 (see Chapter 4) summarizes the common errors in field parameter measurement that lead to inaccuracy and imprecision of data generated in the field.

Sample Pretreatment Options

Another group of parameter-specific field protocols that must be evaluated and included in the SAP are methods for sample pretreatment, including sample filtration and physical and chemical preservation. Sample pretreatment must be performed at the wellhead at the time of sample collection to ensure that physical and chemical changes do not occur in the samples during the time that the sample is collected and after the sample container has been filled and capped. ASTM International has published Standard Guides that address both types of sample pretreatment. ASTM Standard D 6564 (ASTM, 2006d) provides a detailed guide for field filtration of ground-water samples, and ASTM Standard D 6517 (ASTM, 2006e) discusses physical and chemical preservation methods for ground-water samples. Sample pretreatment is discussed in detail in Chapter 5.

Filtration

Ground-water sample filtration is a sample pretreatment process implemented in the field for some constituents, when it is necessary to determine whether a constituent is truly "dissolved" in ground water. Filtration involves passing a raw or bulk ground-water sample directly through a filter medium of a prescribed filter pore size either under negative pressure (vacuum) or under positive pressure. Particulates finer than the filter pore size pass through the filter along with the water to form the filtrate, which is submitted to the laboratory for analysis. Particulate matter larger than the filter pore size is retained by the filter medium. In the case of most ground-water monitoring programs, this material is rarely analyzed, although it may be possible to analyze the retained fraction for trace metals or for some strongly hydrophobic analytes such as PCBs or PAHs.

After a decision is made to field filter ground-water samples to meet DQOs for an investigation, decisions must be made regarding selection of the most appropriate field filtration method. The ground-water sample filtration process consists of several phases: (1) selection of a filtration method; (2) selection of filter media (materials of construction, surface area, and pore size); (3) filter preconditioning; and (4) implementation of field filtration procedures. Information on each part of the process must be presented in detail in the SAP to provide step-by-step guidance for sampling teams to implement in the field.

The filtration method used for any given sampling program should be documented in the site-specific SAP and should be consistent throughout the life of the sampling program to permit comparison of data generated. If an improved method of filtration is determined to be appropriate for a sampling program, the SAP should be revised in lieu of continuing use of the existing filtration method. In this event, the effect on comparability of data needs to be examined and quantified to allow proper data analysis and interpretation. Statistical methods may need to be used to determine the significance of any changes in data resulting from a change in filtration method.

Physical and Chemical Preservation of Ground-Water Samples

The second form of pretreatment of ground-water samples is physical and chemical preservation. As described in ASTM Standard D 6517 (ASTM, 2006e), ground-water samples are subject to unavoidable chemical, physical, and biological changes relative to *in situ* conditions when samples are brought to ground surface during sample collection. These changes result from exposure to ambient conditions, such as pressure, temperature, ultraviolet radiation, atmospheric oxygen, and atmospheric contaminants, in addition to any changes that may be imparted by the sampling device as discussed earlier in Chapter 3.

Field Equipment Cleaning Procedures

The SAP must provide a written protocol appropriate for cleaning field equipment used throughout the ground-water sampling program to prevent cross-contamination of sampling locations and collection of unrepresentative samples. Even in situations where dedicated equipment is being used for purging and sample collection, there will be pieces of equipment taken from sampling point to sampling point, which will require cleaning between locations (Figure 2.5 and Figure 2.6). Examples of equipment that falls into this category are electronic water-level gauges, oil–water interface probes, flow-through cells, and instrumentation used to measure water-quality indicator parameters (e.g., pH meters and multiparameter sondes). Equipment cleaning is also necessary to ensure that equipment will continue to operate properly in environments with high levels of suspended sediment and aggressive chemical constituents and to ensure that sampling team members are not accidentally exposed to contaminants that may be present on the surfaces of equipment following its use.

To assist in the development of streamlined approaches to equipment cleaning, ASTM International has produced standards that address procedures for cleaning field

FIGURE 2.5
All pieces of field equipment that are used in wells should be thoroughly cleaned before use in the well. Cleaning should also be done as part of periodic maintenance to keep equipment functioning properly.

equipment used at nonradioactive sites (ASTM Standard D 5088 [ASTM, 2006f]) and low-level radioactive waste sites (ASTM Standard D 5608 [ASTM, 2006g]). In these standards, field equipment cleaning protocols are described for equipment that contacts samples submitted to the laboratory (such as a pump) and equipment that facilitates sample collection but that does not actually contact the sample (such as a reel used to hold pump tubing). The reader is directed to Chapter 8 of this text for a more detailed discussion of field equipment cleaning protocols.

Documenting a Sampling Event

The procedures and equipment used and the data generated during a ground-water sampling event must be documented in the field at the time of data generation and sample collection. The SAP must detail procedures required for recording field observations and measurements made throughout the sampling event. These records are used to document field conditions that may be important to refer to when interpreting laboratory data and to document exactly what was done in the field during the sampling event, including how the well headspace was screened, how water levels were taken, how sampling points were purged, how samples were collected, what samples were collected at each sampling point, results of all field measurements and field parameter analyses, and how samples were pretreated and prepared for shipment. Specific guidance on what to include in field notes for a ground-water sampling event is provided in ASTM Standard D 6089 (ASTM, 2006h). The suggested content of ground-water sampling event documentation is summarized in Table 2.5.

Three primary mechanisms are available for recording information collected during a sampling event: (1) written records; (2) electronic records; and (3) audio-visual records. When writing the section of the SAP devoted to documentation, project managers are encouraged to seek counsel from in-house or client legal staff for guidance on what are considered to be acceptable practices for field documentation. This is particularly important in cases where a site is undergoing litigation or could be the subject of a corporate sale or merger. It is important to remember that the political climate of a site can

FIGURE 2.6
A typical setup for cleaning downhole equipment (such as pumps), with a bucket for an initial rinse with control water (foreground), followed by a bucket for washing with a phosphate-free liquid detergent (e.g., Liquinox) (center) and, finally, a bucket for the final rinse with distilled or deionized water (background). For some stringent cleaning programs, additional steps may be required.

change during the course of a ground-water sampling program, so it is often wise to have a higher level of QA/QC associated with field record keeping than might be thought necessary at the time of sample collection.

Written Records

Written records are the most common form of documentation. Written records generated in the field include site-specific field notebooks (Figure 2.7), sample container labels, sample container security tags and seals, chain-of-custody forms, visitor logs, field equipment calibration and maintenance logs, and commercial sample shipment manifests. In general, all field observations and measurements made during a sampling event should be recorded using black indelible ink on a bound, site-specific field notebook and not on loose pages or forms. Field notebooks, commonly made of water-proof paper, provide a secure and relatively tamper-proof mechanism for retaining all field records. Field notebooks should have preprinted page numbers throughout or should have a printed statement in the front of the book indicating how many pages are contained in the book. This is a critical means of ensuring that pages cannot be removed from or added to the field notes without detection — the major limitation of using a collection of loose-leaf forms.

Many companies have spent a great deal of time developing field forms (Figure 2.8) for sampling teams to complete. The objective of these forms is to ensure precision in information collection during ground-water sampling programs. The result is that many samplers have become dependent on forms to dictate what must be done and which measurements must be taken in the field, so there is sometimes resistance to switch to a

TABLE 2.5

Information to Document during a Ground-Water Sampling Event

Facility or site name and well identification
Weather conditions
Names and affiliations of sampling team members and others present during the sampling event
Instrumentation calibration results
Well integrity inspection results
Changes in land use or physical conditions at the site since the last sampling event
Results of well headspace screening and details on how screening was performed
Water-level measurement results and product-thickness measurements (if taken), indicating what equipment was used to take the measurements and procedures followed
Well-depth measurement results (if taken)
Description of the well purging method, equipment used, how the equipment was operated, and time required to purge each well
Equation used to calculate a well volume (if relevant) and results of calculations for each well
Total volume of water removed during purging of each well
Description of how purge water was managed
Results of all field parameter measurements and the definition of stabilization used
Description of how it was determined when each well was ready to sample
Description of the sampling device, if different from the device used to purge the well, and a description of how the device was operated during sample collection
Description of the sample containers filled and the order in which containers were filled
Description of the water collected as samples (appearance, odor, and turbidity)
Description of sample pretreatment methods (filtration and chemical or physical preservation) for specific parameters
Description of any problems encountered in the field during the sampling event
Description of the temporary storage method used for samples during the sampling event (including use of sample security seals and tags)
Description of all QC samples collected — types, and how many, and how they were collected
Description of sample preparation for shipment, shipment method, security tag serial numbers, time of sample delivery to the lab (if hand delivered) or to the commercial carrier, and forms accompanying the samples (e.g., chain-of-custody and analysis request form)
Description of any photographs taken during the sampling event

field notebook. However, from a legal perspective, field forms are not recommended because they are easy to lose and information is easy to alter and, thus, can be challenged in a legal proceeding. Two alternatives to this problem are: (1) have forms printed on water-proof paper with page or form numbers and bound using saddle-stitching or spiral wire binding (Figure 2.9); or (2) record all original notes in a field notebook and then transfer the information to the form (never the other way around) for easy dissemination of field information back at the office.

Because entries in field notebooks are made in ink, it is important that sampling team members resist the temptation to scratch out or blacken out any errors when entering a correction. Instead, good laboratory practices (GLPs) should be followed. Using GLPs, it is appropriate to use a single stroke through the incorrect information. Next to the line containing the error, the samplers should put their initials, the date of the correction, and an appropriate error code to explain when and why the change was made (Garner et al., 1992). This makes it much easier to explain entry changes at a later date, as is necessary when field notes are obtained as part of a court case as evidence or through discovery. Table 2.6 provides some examples of common error codes that can be used in written ground-water sampling records.

When recording field notes, samplers must be certain that the information recorded is accurate, factual (not opinionated), and detailed enough so that others can reconstruct and understand what occurred during the sampling event. It is also critical that notes are

FIGURE 2.7
Field notebooks are the preferred method for recording written records in the field.

neat and legible. A good practice for the sampling team to adopt is to submit the field book to the project manager for review immediately at the conclusion of each sampling event so the project manager can review field notes for completeness, clarity, and errors. By conducting this review immediately, the project manager can question the sampling team while memories are still fresh and before they head to another site for field work.

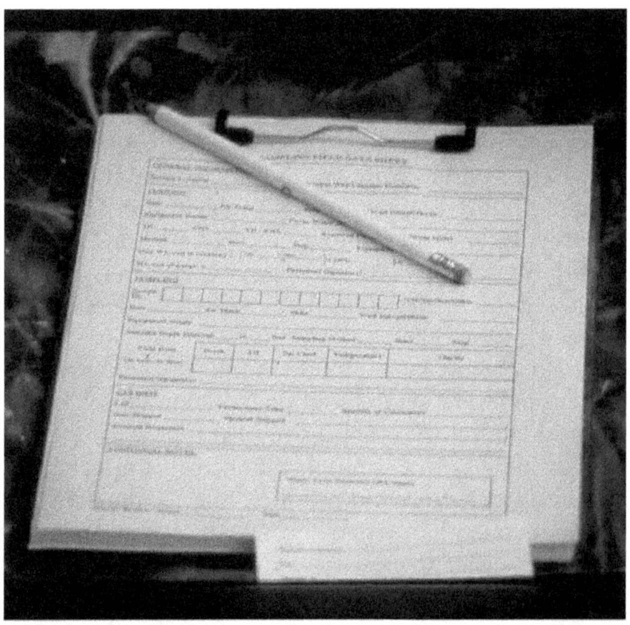

FIGURE 2.8
Field forms are commonly used to record field data, because they prompt the user to collect specific information and they are convenient. However, they are not recommended for recording original data because of the ease with which data can be altered undetectably. Note that, in this case, a pencil is being used to record data — this is strongly discouraged. Field notes should always be recorded in black indelible ink.

FIGURE 2.9
If field forms are used, it is preferred that they be printed on water-proof paper and bound into a book like the one pictured here, with pages numbered to prevent the possibility of alteration of information recorded for a sampling event.

Chain-of-Custody: There are exceptions to the "no loose paper" rule — specifically the chain-of-custody form and shipping manifests. The chain-of-custody form (Figure 2.10) is a loose form typically provided by the laboratory, which is used to document possession of samples in the field during sample collection and to document samples being shipped or delivered to the laboratory. From a legal perspective, the objective of chain-of-custody forms is to provide sufficient evidence of sample integrity to assure legal defensibility of the samples (ASTM, 2006i). Depending upon its design, the chain-of-custody form may also act as a sample analysis request form, or the latter may be a separate form that is sent

TABLE 2.6
Examples of Common Error Codes

RE	Recording error
CE	Calculation error
TE	Transcription error
SE	Spelling error
CL	Changed for clarity
DC	Original sample description changed for further clarity
WO	Write over
NI	Not initialed and dated at time of entry
OB	Observation not recorded at time of initial observation

Note: Error codes should be circled when recorded.
Source: Garner et al., 1992. With permission.

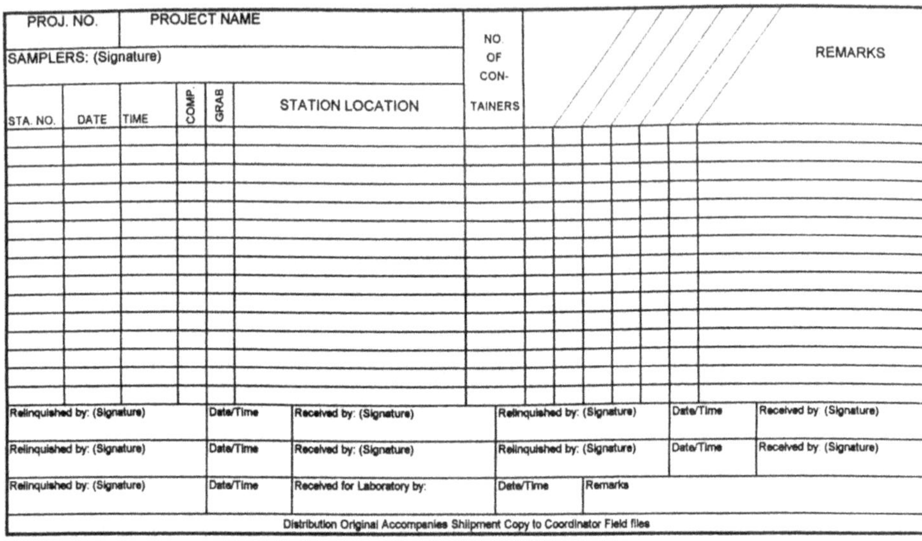

PROJ. NO.	PROJECT NAME		NO. OF CON- TAINERS		REMARKS
SAMPLERS: (Signature)					
STA. NO.	DATE	TIME	COMP.	GRAB	STATION LOCATION

Relinquished by: (Signature)	Date/Time	Received by: (Signature)	Relinquished by: (Signature)	Date/Time	Received by: (Signature)
Relinquished by: (Signature)	Date/Time	Received by: (Signature)	Relinquished by: (Signature)	Date/Time	Received by: (Signature)
Relinquished by: (Signature)	Date/Time	Received for Laboratory by:	Date/Time	Remarks	

Distribution Original Accompanies Shipment Copy to Coordinator Field files

FIGURE 2.10
A typical chain-of-custody form that should accompany all filled sample containers during the sampling event and while samples are in transit to the laboratory.

along with samples to the laboratory. ASTM Standard D 4840 (ASTM, 2006i) provides an example of a chain-of-custody form that has been widely adopted by many laboratories. A common mistake made by sampling teams using chain-of-custody forms is that they wait until the end of the day to complete the form rather than complete it as samples are collected. This defeats one of the purposes of having the chain-of-custody form (to document possession of samples during the sampling event) and can result in errors being made when completing the forms at the end of the day. The original chain-of-custody form must accompany the samples to the laboratory and a copy should be retained by the sampling team.

Shipping Manifests: Shipping manifests are used when commercial couriers or carriers are used to transport samples from the field to the laboratory. In most cases, the tracking system used by commercial carriers is considered to be sufficient to document possession of a shipment of samples provided samplers retain copies of manifests. It is the responsibility of the sampling team to ensure that the correct type of manifest is used for the samples being shipped and that the manifest is completed correctly. Failure to do this can result in refusal of the carrier to deliver the samples, loss of the samples, or legal action. This is especially critical when samples contain sufficient concentrations of contaminants to require that they be classified as regulated substances for shipping (e.g., a ground-water sample that contains non-aqueous phase gasoline could be considered to be a flammable liquid for shipping purposes).

Electronic Records

Many instruments used in the field for field parameter measurement contain a data-logging component to electronically record all measurements made. While this is certainly convenient for samplers and theoretically should be a more accurate way to record numerical data, there are some concerns from a legal perspective. The greatest concern is that it is possible to change, or lose entirely, an electronic file without keeping a permanent record of the original file. For this reason, in many situations, it is required that original data be recorded in the field notebook and that the data-logging system be used as an electronic backup recording system (not the other way around). For most ground-water sampling applications, this is not problematic due to the timeframe involved between measurements. In situations where pumping tests are being conducted, however, water-level measurements may need to be recorded in a timeframe too fast to be done manually, so exceptions to this rule may be made.

Audio-Visual Records

Audio-visual record keeping is a third mechanism for documenting field activities during ground-water sampling. This may include audio recording of field activities, but more commonly involves taking photographs or video clips of site activities. Written permission must be obtained from authorized facility personnel by sampling teams prior to taking any audio-visual records. This is especially a concern at active facilities where audio-visual records may pose a security risk (e.g., at U.S. Department of Defense or U.S. Department of Energy sites) or breach confidentiality agreements (e.g., at active manufacturing facilities). Options for photographic recording of field activities include 35 mm cameras, digital cameras, cameras with self-developing film, and video cameras (tape and digital formats). In politically sensitive cases, counsel should be consulted during the preparation of the SAP to select the most appropriate type of camera, recording medium (film type or digital disk), camera lens (e.g., fixed, macro or zoom, and wide angle), and method of film development or disk imaging. In addition, some cameras have the capability to date- and time-stamp images, which might be of value in some applications provided the feature is set up correctly.

References

API, A Guide to the Assessment and Remediation of Underground Petroleum Releases, API Publication 1628, American Petroleum Institute, Washington, DC, 82 pp., 1989.

ASTM, Standard Test Method for Determining Subsurface Liquid Levels in a Borehole or Monitoring Well, ASTM Standard D 4750, ASTM International, West Conshohocken, PA, 5 pp., 2006a.

ASTM, Standard Guide for Presentation of Water-Level Information from Ground-Water Sites, ASTM Standard D 6000, ASTM International, West Conshohocken, PA, 16 pp., 2006b.

ASTM, Standard Guide for Field Quality Assurance in a Ground-Water Sampling Event, ASTM Standard D 7069, ASTM International, West Conshohocken, PA, 4 pp., 2006c.

ASTM, Standard Guide for Field Filtration of Ground-Water Samples, ASTM Standard D 6564, ASTM International, West Conshohocken, PA, 5 pp., 2006d.

ASTM, Standard Guide for Field Preservation of Ground-Water Samples, ASTM Standard D 6517, ASTM International, West Conshohocken, PA, 6 pp., 2006e.

ASTM, Standard Practice for Decontamination of Field Equipment Used at Nonradioactive Waste Sites, ASTM Standard D 5088, ASTM International, West Conshohocken, PA, 8 pp., 2006f.

ASTM, Standard Practice for Decontamination of Field Equipment Used at Low-Level Radioactive Waste Sites, ASTM Standard D 5608, ASTM International, West Conshohocken, PA, 8 pp., 2006g.

ASTM, Standard Guide for Documenting a Ground-Water Sampling Event, ASTM Standard D 6089, ASTM International, West Conshohocken, PA, 3 pp., 2006h.

ASTM, Standard Guide for Sample Chain-of-Custody Procedures, ASTM Standard D 4840, ASTM International, West Conshohocken, PA, 8 pp., 2006i.

Garner, W.Y., M.S. Barge, and J.P. Ussary, Eds., *Good Laboratory Practice Standard: Applications for Field and Laboratory Studies*, ACS Professional Reference Book, American Chemical Society, Washington, DC, 1992.

Gillham, R.W., M.J.L. Robin, J.F. Barker, and J.A. Cherry, Ground-Water Monitoring and Sample Bias, American Petroleum Institute Publication, Environmental Affairs Department, Washington, DC, 206 pp., 1983.

Maslansky, C.J. and S.P. Maslansky, *Air Monitoring Instrumentation: A Manual for Emergency, Investigatory and Remedial Responders*, Van Nostrand Reinhold, New York, NY, 304 pp., 1993.

Maslansky, S.P. and C.J. Maslansky, *Health and Safety at Hazardous Waste Sites: An Investigator's and Remediator's Guide to HAZWOPER*, Van Nostrand Reinhold, New York, NY, 612 pp., 1997.

Maslansky, S.P. and C.J. Maslansky, Health and Safety Considerations in Ground-Water Monitoing Investigations; Chapter 19 in *The Practical Handbook of Environmental Site Characterization and Ground-Water Monitoring*, Second Edition, D.M. Nielsen, Editor, CRC Press/Taylor and Francis Group, Boca Raton, FL, pp. 1219–1262, 2006.

Nielsen, D.M., G.L. Nielsen, and L.M. Preslo, Environmental Site Characterization; Chapter 2 in *The Practical Handbook of Environmental Site Characterization and Ground-Water Monitoring*, Second Edition, D.M. Nielsen, Editor, CRC Press/Taylor and Francis Group, Boca Raton, FL, pp. 35–206, 2006.

Taylor, J.K., What is Quality Assurance? *Quality Assurance for Environmental Measurements,* Special Technical Publication 867, J.K. Taylor and T.W. Stanley, Eds., American Society of Testing and Materials, Philadelphia, PA, 1985.

Testa, S.M. and D.L. Winegardner, *Restoration of Petroleum-Contaminated Aquifers*, Lewis Publishers, Inc, Chelsea, MI, 269 pp., 1991.

3

Purging and Sampling Device Selection and Operation

David M. Nielsen

CONTENTS

Introduction

Many devices are available for purging and sampling ground water. The SAP must include a description of the devices selected for use on a task-specific basis. It should also include information on how to operate the selected devices and how to maintain the equipment to ensure proper operation in the field. While not always required by regulatory agencies when reviewing SAPs, it is highly recommended that the SAP includes, as an appendix, the operations manuals for each device specified for purging and sampling. This provides sampling team members with critical equipment-specific information such as operating procedures, calibration procedures, troubleshooting tips, spare parts lists, and contact information for equipment repair.

Selection Criteria

One of the most important yet least considered factors that has an effect on the physical and chemical integrity of ground-water samples is the device used for purging and sample collection. Despite being one of the most critical elements of the sampling program, sampling device selection often receives little attention when a sampling plan is prepared. A properly selected sampling device will provide samplers with the most representative sample possible at a reasonable cost; an inappropriate sampling device could have significant and long-lasting effects on sampling results, which may not be immediately apparent.

A number of criteria must be evaluated on a site-specific basis when selecting devices for purging and sample collection. Based on these criteria (summarized in Table 3.1), each device has a unique set of advantages and limitations that define its suitability for use. Sampling device selection criteria are discussed in detail in Nielsen and Yeates (1985), Parker (1994), and ASTM Standard D 6634 (ASTM, 2006a) and summarized subsequently.

Accuracy and Precision of the Device

As stated earlier, a key objective of ground-water sampling programs is to collect representative samples of water in the formation to permit characterization or monitoring of formation chemistry. To do this, it is essential that sampling devices have the physical capability of moving ground water from *in situ* conditions to ground level in a manner that ensures minimal impact on the physical and chemical properties of the sample (i.e., accuracy). This is essential if analytical data derived from the analysis of samples are

TABLE 3.1

Selection Criteria for Ground-Water Purging and Sampling Devices

Accuracy and precision of samples provided by the device
Materials of construction of the device and accessory equipment
Outside diameter of the device
Lift capability of the device
Flow-rate control and range
Ease of operation and in-field servicing
Portable versus dedicated application
Ease of field decontamination
Reliability and durability
Purchase price and operating costs

to be meaningful. In addition, to ensure that data from different sampling events can be compared, as is essential for long-term monitoring programs, sample collection must be performed in a manner that can be repeated each sampling event, regardless of whether the sampling team members change (i.e., precision). To determine whether this can be accomplished by any individual device, it is important to remember that *in situ* ground water is at a stable temperature and at a pressure that is higher than atmospheric conditions at ground surface.

Materials of Construction

The choice of materials used in the construction of purging and sampling devices should be based on knowledge of the geochemical environment and how the materials may interact with the water collected as a sample via physical, chemical, or biological processes. Materials used in the manufacture of purging and sampling devices and associated tubing, hoses, pipes, and support lines (e.g., rope, cable, or chain) may be a source of bias or error. Materials should not sorb analytes from samples, desorb previously sorbed analytes into samples, leach matrix components of the material that could affect analyte concentrations or cause artifacts, or be physically or chemically degraded by water chemistry.

Materials commonly used in the manufacture of sampling devices include rigid PVC (Type I PVC), stainless steel, polytetrafluorethylene (PTFE),[1] polyethylene (PE), polypropylene (PP), flexible PVC (Type II PVC), fluoroelastomers,[2] polyvinylidene fluoride (PVDF), Buna-N, ethylene–propylene diene monomer (EPDM), and silicone rubber (Figure 3.1). A number of studies indicate the relative sorption and desorption rates of these materials, their potential for alteration of sample chemistry, and their desirability for use in sampling devices (Barcelona et al., 1983; Barcelona et al., 1985; Reynolds and Gillham, 1985; Barcelona and Helfrich, 1986; Holm et al., 1988; Gillham and O'Hannesin, 1990; Parker, 1991).

Other materials-related sources of bias or error include surface traces of organic extrusion aids or mold release compounds used in the extrusion or molding of polymeric materials. In addition, some formulations of polymeric materials may contain fillers or processing additives that can leach from the material and alter sample quality. Traces of cutting oils, solvents, or surface coatings may be present on metallic materials. These should be removed by thorough cleaning and, once removed, should not affect sample chemistry. Metallic materials are subject to corrosion, the residues of which could potentially affect sample quality. Electropolishing or other surface passivation processes can improve corrosion resistance. In all cases, the types of materials that may contact the sample should be selected on the basis of the potential for chemical or physical interactions to occur during both short contact times, as in sample collection, and long contact times, typically associated with dedicated equipment submerged in ground water for the duration of the sampling program. The potential interactions that may occur include the following:

- Corrosion of steel parts in water with a low pH, high O_2 content, high CO_2 content, high total dissolved solids content, high H_2S content, or high Cl^- content, or where iron bacteria are present (Figure 3.2) (generally only a concern in long-term contact situations)

---

[1]PTFE is also commonly known by the trade name Teflon®, which includes other fluoropolymer formulations. Teflon is a registered trademark of E. I. DuPont De Nemours and Company.

[2]Fluoroelastomers (FPM and FKM) are commonly known by the trade name Viton®, a registered trademark of DuPont.

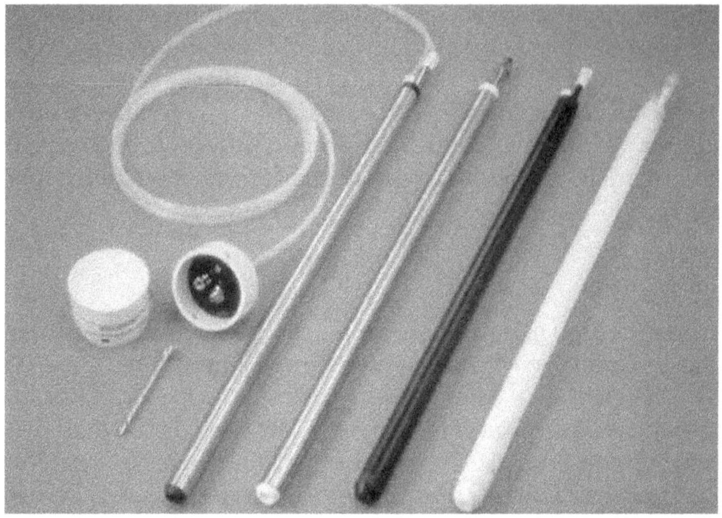

FIGURE 3.1
Some sampling devices can be made of a variety of materials; for others, the materials are fixed. In this case, the dedicated bladder pumps pictured all have PTFE bladders and (left to right) are made of stainless steel and PVC, stainless steel and PTFE, all PVC, and all PTFE. Materials for any particular dedicated application of this type of pump should be selected based on the geochemical environment to which it will be exposed. (Photo courtesy of QED/Environmental Systems.)

- Degradation of plastic materials in the presence of high concentrations of some chlorinated solvents
- Adsorption and later desorption of metals or organic compounds onto or from solid surfaces (including rigid and flexible parts)
- Leaching of compounds from plastic materials

The reader is referred to NWWA/PPI (1981), Barcelona et al. (1983, 1985), Reynolds and Gillham (1985), Barcelona and Helfrich (1986), Holm et al. (1988), Gillham and O'Hannesin (1990), Parker (1991), and Nielsen and Schalla (2006) for additional detailed information on materials compatibility with solutes, relative adsorption and desorption rates, potential for alteration of sample chemistry, and material rankings in terms of desirability for use in different sampling applications.

To further minimize the potential for sample alteration caused by sample contact with materials, the surfaces of all materials should be thoroughly cleaned prior to use or installation in a sampling point using an appropriate protocol to avoid possible cross-contamination concerns (refer to Chapter 8 for additional information on equipment decontamination). Several different protocols may be applied at any given site, depending on the oversight agency, the sampling objectives, and other factors. Some of the more common protocols include those outlined in ASTM Standard D 5088 (ASTM, 2006b) or ASTM Standard D 5608 (ASTM, 2006c), U.S. EPA regional office protocols (e.g., Region 4 EISOPQAM [U.S. EPA, 1997]), and state and local regulatory agency protocols.

The materials used in sample discharge tubing (where sampling pumps are used) provide the greatest potential source of bias or error with respect to possible sample alteration, because contact times with the sampled water and surface contact areas are highest with these materials. The most significant interactions are the potential for sorption of metals and/or organic compounds to the tubing, and possible desorption at a

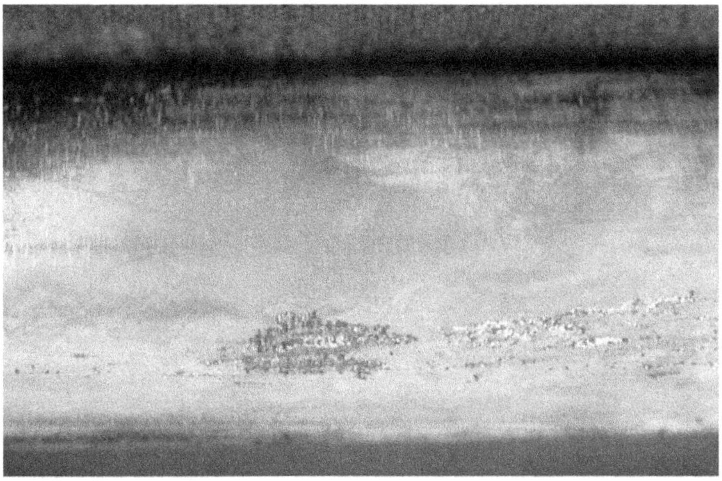

FIGURE 3.2
Where steel materials are selected (even stainless steel), the geochemical environment should be evaluated not only for corrosion potential but also for the presence of iron bacteria. These bacteria secrete a material that causes significant corrosion on steel materials from which some dedicated pumps are made. (Photo courtesy of QED Environmental Systems.)

later time, and leaching of materials out of the tubing into the sampled water. Ho (1983), Devlin (1987), and Barker and Dickhout (1988) demonstrated that the silicone rubber tubing (Figure 3.3) recommended by manufacturers for use in the pump head of most peristaltic pumps (because of its extreme flexibility) was a significant source of bias for many organic compounds, including VOCs, caused by sorption and desorption processes. They recommended keeping the tubing length in the pump head to a minimum and using more inert tubing materials for the remainder of the tubing run to and from the well. Barcelona et al. (1983, 1985) found that flexible (Type II) PVC tubing could leach plasticizers (primarily phthalate esters) into sampled water and sorb contaminants from and later desorb contaminants into samples. Holm et al. (1988) demonstrated that gas diffusion through flexible polymeric tubing could introduce measurable concentrations of oxygen into initially anoxic water and that the amount of gas transferred is proportional to the tubing length and inversely proportional to the pumping rate. Any of these interactions could introduce significant bias into sampling results, so the sample discharge tubing material should be very carefully chosen to minimize or eliminate this source of bias. PTFE, PE, and PP tubing appear to offer superior performance over other materials.

Outside Diameter of the Device

Historically, the majority of ground-water sampling programs have made use of 2 or 4 in. diameter monitoring wells to facilitate ground-water sample collection. For that reason, the majority of commercially available devices used for purging and sampling wells are designed to fit inside a 2 in. diameter sampling point. However, with increasing use of direct-push sampling tools and smaller diameter monitoring wells to characterize ground-water quality, the need has arisen for smaller diameter devices that will fit inside sampling points with diameters as small as 0.5 in. Not only it is important that these devices be small enough, but they must also deliver samples with the same degree of accuracy and precision afforded by larger diameter devices.

FIGURE 3.3
The silicone rubber tubing typically specified for use with peristaltic pumps can negatively bias samples collected for organic compound analysis.

With either traditional monitoring wells or direct-push sampling points, it is also important to consider that casing may not be plumb or may have constrictions limiting the inside diameter (i.e., at rod or casing joints). With direct-push tools, subsurface obstructions, such as boulders, may cause direct-push rod to deflect, resulting in rod that is no longer plumb. In traditional wells, casing may shift over the course of its lifespan to the point where casing joints may fail or where casing is no longer plumb. This is of particular concern at facilities such as landfill sites, where, over time, and under warm ambient subsurface temperatures, well casing can bend or break as wastes settle.

Lift Capability

The term "lift capability" of a device is often misunderstood to mean the maximum depth at which a pump can be installed within a well and still operate. More correctly, the lift capability of a device is linked to the ability of a device to move water from the depth of the static water level within a sampling point (which may be significantly shallower than the intake of the pump) to ground surface. This concept is illustrated in Figure 3.4. The greater the depth to water, the more pumping head the device must overcome to deliver water to the surface. Lift is not related to the depth of the sampling device. The lift capability of the device is a critical selection criterion when deciding whether or not the device is suitable for individual applications, especially when working with lifts greater than 200 ft. The selection of available purging and sampling devices is more limited with increased depth to water.

Flow Rate Control and Range

Consideration should be given to appropriate water removal rates when selecting purging and sampling devices. Ideally, the same device will be used for both purging and sampling, therefore, a device should be capable of operating at flow rates suitable for both applications. For sample collection, it is critical to evaluate whether a device can operate at a flow rate suitable for the parameters analyzed in the samples collected. For example,

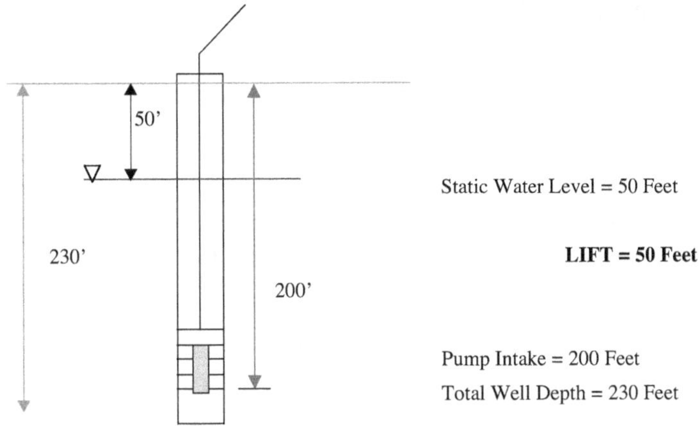

FIGURE 3.4
Lift capability is related to the ability of a pumping device to lift water from the static water level, not the depth at which the pump is set in the well.

samples collected for analysis of some sensitive parameters (i.e., VOCs and trace metals) should be taken at low flow rates (generally less than 250 ml/min). Sampling rates should be high enough to fill sample containers efficiently and with minimal exposure to atmospheric conditions, but low enough to minimize sample alteration via agitation or aeration. Additionally, the use of low-flow purging and sampling techniques may require adjusting the pumping rate to account for the hydraulic performance of the sampling point. This requires that the flow rate be highly controllable between rates of less than 100 ml/min and more than 500 ml/min (Figure 3.5). Throttling down the device using a valve in the discharge line reduces the flow rate, but creates a pressure drop across the valve, and does not necessarily reduce the speed of the device in the well. Therefore, this method of flow rate control is not recommended. A better method of reducing flow rate is to divert a portion of the discharge stream via the use of a "T" or "Y" fitting.

FIGURE 3.5
Flow rate control is important for devices used for both purging and sampling. Control should be exerted directly on the pump driving mechanism, but not through the use of in-line valves.

Ease of Operation and In-Field Servicing

Ease of operation and servicing are important but frequently overlooked practical considerations in the selection of purging and sampling devices. A common source of poor precision in sampling results is a sampling device operating problem (Barcelona et al., 1984). This could be due to any one of several factors including: (1) the device and accessory equipment are too complicated to operate efficiently under field conditions; (2) the operator is not familiar enough with the device to operate it properly; or (3) the operating manual supplied with the device does not clearly outline the procedures for proper use. Thus, it is not only important to select a device that is simple to operate but also to provide proper training for the operators of the device. Because mechanical devices are subject to malfunction or failure, it is desirable to select a device that can be serviced in the field to prevent delays during sampling. An alternative is to have a replacement device available. Some of the devices described subsequently may be too complex for field repairs or may be factory sealed, requiring servicing by the manufacturer or a qualified service facility.

Operational characteristics, such as solids handling capability, ability to run dry, cooling requirements, and intermittent discharge, must be considered in the application of some purging and sampling devices. Some devices may experience increased wear or damage as solids pass through the device, causing reduced output or failure. Solids may also clog check valves or passages, which can reduce discharge rate or, in the case of grab samplers, cause the sample to leak out.

In wells with little water or low yield, it is important to know whether a device must be fully submerged to operate and what the ramifications of operating the device without water may be. Some devices, such as bailers, are not dependent on submergence to operate, while other devices, such as many (but not all) pumps, do require that they be fully submerged in the water column to operate. A pump running dry can occur when the water level in the sampling point is drawn down below the pump intake. In some pump designs, typically those with rotating or reciprocating mechanisms, this can cause permanent damage to or failure of the device, which is often associated with overheating of the device.

Another operational feature of devices to consider is how water is discharged from the device. In some devices, such as bladder pumps, ground water is brought to the surface in an intermittent or cyclic discharge. Intermittent discharge creates some special considerations for sampling teams to ensure the accuracy of sample collection, particularly for volatile constituents and for accurate measurement of indicator parameters with in-line monitoring devices and flow-through cells. For example, sampling teams must learn how to optimize the discharge from devices with intermittent flow to ensure that sufficient sample volume is available during the pump discharge cycle to fill a 40 ml VOC sample vial without headspace. When taking measurements for indicator parameters in a flow-through cell, care must be taken to ensure that measurements are made during pump discharge cycles while water is running over the sensors. This is especially critical when measuring parameters such as DO or oxidation–reduction potential (ORP). When filtering, care should be taken to prevent air from entering the filter during pump refill cycles.

Portable versus Dedicated Applications

As defined in ASTM Standard D 6634 (ASTM, 2006a), a "dedicated" device is one that is permanently installed in a sampling point and does not come into contact with ambient or atmospheric conditions during operation. The only time dedicated equipment should come into contact with atmospheric conditions is during periodic routine maintenance. A bladder pump (Figure 3.6) is an excellent example of a device ideally suited to a dedicated installation. A "designated" device is a device that is assigned for use at

FIGURE 3.6
Bladder pumps such as the one installed in the well pictured here, are ideally suited to dedicated applications, meaning that the pump is not removed from the well after initial installation except for routine maintenance. (Photo courtesy of QED Environmental Systems.)

a single sampling point, but which, by virtue of its design, must come into contact with ambient conditions to operate. Grab sampling devices such as bailers can therefore be a designated device but not a dedicated device. The third category of device is a "portable" device. Portable purging and sampling devices are used in multiple sampling points at one or more sites. The device is in contact with ambient conditions between sampling events or between wells, and is typically stored at a location remote from the field site. The electric submersible gear-drive pump (Figure 3.7) is a popular pump that is used portably.

The decision regarding whether to use portable or dedicated purging and sampling equipment is made on a site-by-site basis with a number of considerations in mind, including the number of sampling events anticipated at the site, the level of QA/QC required in the field, ease of equipment decontamination (contaminant concentrations, equipment design features, and ability to thoroughly clean the equipment), and accessibility of sampling points. At sites where a ground-water sampling program requires that six or more sampling events be implemented, it is generally more cost effective to install a dedicated system of sampling pumps. While, in general, it is more expensive initially to purchase and install dedicated equipment for each sampling point, cost savings will be realized through significantly lower operating costs associated with equipment setup, decontamination, and collection of fewer QC samples (i.e., equipment blanks). In addition, with dedicated equipment, there is a low potential for the sampling team to be exposed to ground-water contaminants, equipment is not exposed to atmospheric conditions, cross-contamination of sampling points is virtually eliminated, and there is no agitation of the water column associated with equipment deployment into the sampling point. The net result is that more accurate and precise samples can be collected from a dedicated sampling system.

FIGURE 3.7
Portable devices are moved from well to well and cleaned in between uses, so the ideal portable device (and accessory equipment) is lightweight and easy to clean. This electric submersible gear-drive pump is a popular portable purging and sampling device.

There will be sites, however, where either due to the limited number of sampling events or the temporary nature of sampling points, portable equipment will be the preferred option. In these situations, a common device is used to purge and sample each sampling point. As a consequence, when selecting equipment for portable use, it is critical to focus on the ease with which the device and all accessory equipment (i.e., tubing or tubing bundles, hose reels, battery packs, generators, compressed air source, controlling devices, decontamination equipment and supplies, purge water containers, etc.) can be moved between sampling points — especially in areas of rough terrain. While some devices can be hand-carried to remote sites, some manufacturers have mounted their equipment on backpack frames, small wheeled carts, or specialized vehicles in an effort to improve portability (Figure 3.8). Other equipment is too bulky and heavy to be transported in the field without being vehicle mounted. Another important consideration is the ease with which the portable equipment can be cleaned between sampling points. Manufacturers often design equipment especially for portable applications so it can be easily disassembled in the field for thorough cleaning.

There are several disadvantages to using portable equipment. The equipment will be exposed to surface contaminants and contaminants from other sampling points which, if not removed during equipment cleaning, can introduce error into samples being collected. In addition, sampling team members are at greater risk of exposure to contaminants in ground water due to their handling of the sampling device and accessory equipment. Of significant concern in many projects is the increase in variability (imprecision) of operation of portable equipment between sampling team members even though they all may be following the same sampling protocol. Issues such as inconsistency in the depth of placement of the pump intake and care with which the device is lowered through the water column can be significant sources of error. As an example, when samplers elect to use portable pumping equipment for low-flow purging and sampling, they must be aware that regardless of how carefully the device is lowered

FIGURE 3.8
Portable devices are often mounted on wheeled carts (a), backpack frames (b), or other convenient means of transport.

into the sampling point, there will be some disturbance of the water column. This may result in resuspension of sediments that have settled to the bottom of the well screen or sump, which will be measured as higher sample turbidity. To help compensate for this source of error, time must be built into the sampling protocol to permit some equilibration time immediately following pump placement. Some sampling teams may be extremely careful in situating the pump and waiting a prescribed timeframe, while others may ignore these considerations entirely.

The issue of increased turbidity due to pump installation in a well was documented in work done by Kearl et al. (1992) and Puls and Powell (1992) using a colloidal borescope. Kearl et al. observed a significant increase in particle size and number of particles when a portable pump was installed in a monitoring well prior to purging and sampling. This same effect was observed as increased turbidity measurements in a study conducted by Puls et al. (1992). In both cases, the effects of sampling device insertion decreased with time. In a field study conducted in Kansas City, Missouri, where low-flow purging methods were compared to traditional purging methods, Kearl et al. (1994) determined that the effect of increased particle size and increased number of particles resulting from the insertion of a borescope into a monitoring well, subsided after 30 min. After that time, laminar flow conditions dominated and particle size and number decreased.

Designated equipment is often used in an attempt to have the benefits of both dedicated and portable equipment. In theory, because the device is used in only one sampling point, the potential for cross-contamination of sampling points and samples is minimized. However, introduction of contaminants as a result of improper storage between sampling events and exposure to atmospheric conditions during operation are potential sources of error. In some cases, equipment is stored inside its assigned sampling point (e.g., a designated bailer may be hung inside a well [Figure 3.9]), while in other cases, the equipment is stored remotely (e.g., wrapped in plastic and stored in a field office). As a consequence, field equipment blanks should always be collected to verify cleanliness of designated equipment.

Portable equipment must be cleaned between uses in each sampling point or discarded after use to avoid cross-contamination of sampling points and samples. Thus, the components of portable sampling equipment must be able to withstand the necessary cleaning processes. Some devices, by virtue of their design, are very easy to disassemble for cleaning (Figure 3.10), while others are more difficult (Figure 3.11) and some are impossible to disassemble for cleaning. It may be more practical to clean these devices by circulating cleaning solutions and rinses through the device and any associated tubing, hose, or pipe in accordance with ASTM Standard D 5088 (ASTM, 2006b) or to replace the associated tubing, hose, or pipe. Field decontamination operations can be difficult due to the need for sufficient decontamination supplies, exposure of the equipment to potential contaminants, and handling and disposal of the decontamination waste water and supplies. Where field decontamination is not practical or possible, it may be more practical to use dedicated devices or take a number of portable sampling devices into the

FIGURE 3.9
A bailer can be designated for use in a well, but not dedicated, because it must be removed from the well and exposed to atmospheric conditions during use. This designated bailer is hung from the well cap between sampling events.

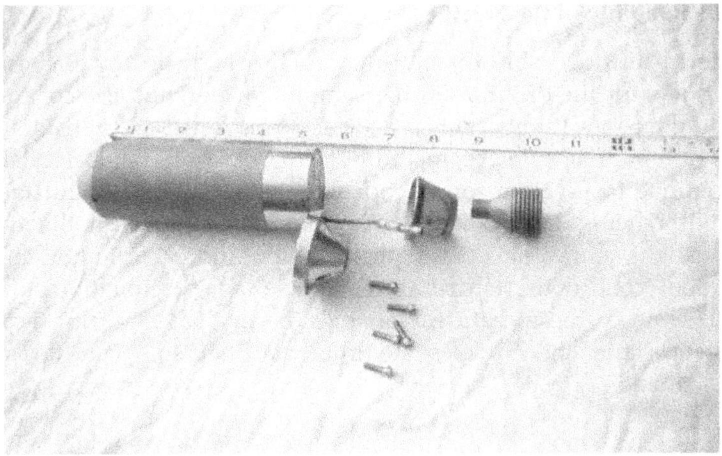

FIGURE 3.10
This electric submersible gear-drive pump is easily disassembled for cleaning, requiring removal of a threaded fitting to take off the inlet screen and five screws to access the internal gears, a process easily accomplished in a few minutes.

field and decontaminate them later at a more appropriate location. Following any cleaning procedure, equipment blanks should be collected to assess the effectiveness of the cleaning procedure. The reader is directed to Chapter 8 of this text for a more detailed discussion of field equipment cleaning procedures.

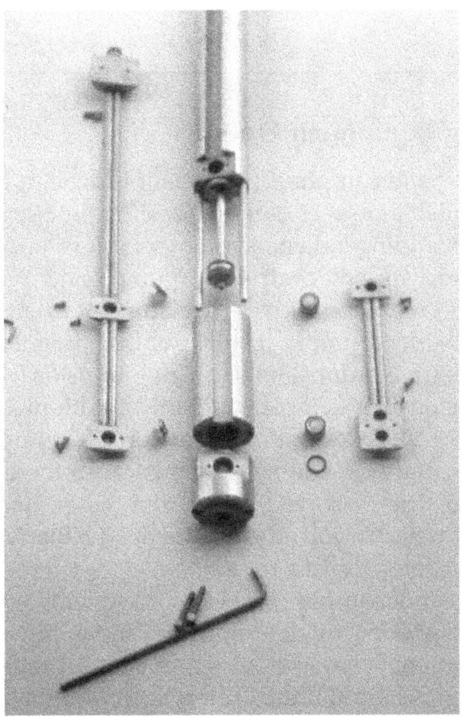

FIGURE 3.11
This double-acting piston pump is more difficult to disassemble for cleaning, as it has multiple valves, o-rings and discharge tubes, a pump cylinder and piston, and an inlet screen, most of which are held in place by multiple hex screws. Make sure you have spare parts, as the pump will not function without all of the pieces in place. (Photo courtesy of Bennett Sample Pumps Inc.)

Reliability and Durability of the Device

A sampling team must be able to rely on the chosen device to perform under field conditions not only for the duration of the sampling event, but also for the lifetime of a monitoring program. A number of practical issues must be considered related to reliability and durability of a device, including battery life, mechanical reliability (moving parts must continue to function for extended periods), ability to withstand chemically aggressive environments, ability to withstand rough handling in the field, ability to maintain water-tight connections, strength of tubing and cables, and sensitivity of the device to outdoor conditions (temperature, light, dust, precipitation, high humidity). Unfortunately, some of these features can only be evaluated during actual field use. Overly optimistic statements in sales literature may not hold true for all operational conditions in the field.

Purchase Price and Operating Costs

Both the initial capital cost and the operating cost (consumable supplies and maintenance costs) of the sampling device and accessory equipment are important considerations. However, cost considerations should not result in the selection of a device that compromises DQOs or sample accuracy and precision. Proper selection and use of purging and sampling devices will more than pay for the capital and operational costs by providing accurate and precise samples, resulting in cost savings from fewer false positive (or false negative) analytical results, fewer resampling events, and fewer problems in meeting regulatory or scientific goals and objectives.

Purging and Sampling Equipment Options

A wide variety of purging and sampling equipment is available for use in ground-water sampling programs. Available devices can be classified into six general categories: grab sampling mechanisms (including bailers, syringe samplers, and thief samplers), suction-lift mechanisms (including surface centrifugal pumps and peristaltic pumps), electric submersible centrifugal pumps, positive displacement mechanisms (including gas-displacement pumps, bladder pumps, piston pumps, progressing cavity pumps, and electric submersible gear-drive pumps), inertial-lift pumps and passive sampling devices (including passive diffusion bag samplers [PDBSs]). Although frequently used in the ground-water industry for well development, the air-lift pumping method is generally considered unsuitable for purging and sampling because the extensive mixing of drive gas and water is likely to strip dissolved gases from the ground water and significantly alter the concentration of other dissolved constituents (Gillham et al., 1983). This method is not discussed in this chapter for this reason.

Each of the purging and sampling devices described subsequently has specific operational characteristics that partly determine the suitability of each device for specific applications. These operational characteristics are listed in Table 3.2, which summarizes information derived from current manufacturers' specifications for the various devices.

Grab Samplers

Bailers, syringe samplers, and thief samplers (e.g., messenger samplers) are all examples of grab sampling devices used in traditional monitoring wells and direct-push sampling

TABLE 3.2

Operational Characteristics of Common Purging and Sampling Devices

Device	Operational Mechanism	Minimum Outside Diameter (in.)	Maximum Lift (ft)	Materials of Construction[a]	Dedicated or Portable	Power Source
Bailer	GS	0.375	Unlimited	Any	P	Manual or mechanical
Peristaltic pump	SL	0.375		Variable; tubing must be silicone or flexible PVC	P (pump); tubing may be designated or dedicated	Electric
Electric submersible centrifugal pump	CP	1.75	230	Stainless steel and PTFE	P or D	Electric
Double-acting piston pump	PD	0.625	1000	Stainless steel and PTFE	P or D	Pneumatic
Progressing cavity pump	PD	1.75	180	Stainless steel and EPDM or Viton	P or D	Electric
Electric submersible gear pump	PD	1.75	250	Stainless steel and PTFE	P or D	Electric
Bladder pump	PD	0.375	1000	Stainless steel, PTFE, PVC, and PVDF	P or D	Pneumatic or mechanical
Gas-displacement pump	PD	0.375	300	Stainless steel and PTFE	P or D	Pneumatic
Inertial-lift pump	IL	0.375	260	Acetal thermoplastic stainless steel HDPE and LDPE, PTFE, and PVC	P or D	Manual, mechanical, or electric

[a]For all pumps (except peristaltic), tubing may be composed of any flexible or rigid material.
GS = Grab sampler.
CP = Centrifugal pump.
SL = Suction lift pump.
PD = Positive displacement pump.
IL = Inertial lift pump.

points. These devices are lowered into the sampling point on a cable, rope, string, chain, or tubing to the desired sampling depth and then retrieved for purge water discharge, sample transfer, or direct transport of the device to the laboratory for sample transfer and analysis. Grab sampling devices are generally not limited to a maximum sampling depth, although use in very deep sampling points may be impractical. Because bailers can be manufactured in very small diameters (less than 0.5 in.), they are usually not limited in use to a particular diameter of sampling point; other types of grab samplers require inside diameters of at least 2 in. The rate at which water can be removed with a grab sampler will depend on the volumetric capacity of the device and the time required for lowering, filling, and retrieval and whether or not a bottom-emptying device is used to decant the sample.

Some grab samplers are prone to malfunction or damage by sediment in the well. Operational difficulty may be experienced in sandy or silty water due to check valve or seal leakage. When used portably, the ability to clean or decontaminate a grab sampler between wells will vary depending upon design. Bailers are generally easier to clean than other types of grab sampling devices and are widely available as disposable devices.

Bailers

The most commonly used grab samplers are bailers, in single check valve and dual check valve designs. Bailers are typically constructed of stainless steel, various plastics (e.g., PVC and PE), and fluorocarbon materials (e.g., PTFE) (Figure 3.12) and come in a variety of designs for portable, designated, and disposable applications. Bailers cannot be dedicated, but may be designated for use at a single sampling point.

The single check valve bailer is lowered into the sampling point. The act of lowering the bailer through the water column opens the check valve and water fills the bailer. Upon retrieval, the weight of the water inside the bailer closes the check valve as the bailer exits the water column. The water in the bailer is retained from the greatest depth to which the bailer was lowered. There is some potential for the contents of the bailer to mix with the surrounding water column during deployment or retrieval, depending on the design of the bailer top. A dual check valve bailer is intended to prevent mixing of the sample with the water column upon retrieval. Water passes through the bailer as it is lowered.

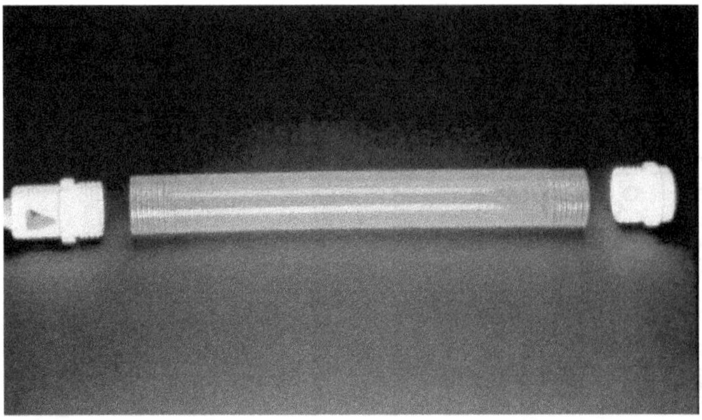

FIGURE 3.12
A bailer is a very simple device, consisting of a tubular body, a lifting bail at the top (left) and a check valve at the bottom (right). Bailers can be made of any material appropriate for contact with samples. This example is an all-PTFE bailer.

Upon retrieval, both check valves seat, ideally retaining a depth-discrete aliquot of water inside the bailer.

In the case of both single and dual check valve bailers, the sample water is decanted into a sample container following retrieval of the bailer. A bottom-discharge device with flow control may be used to provide improved control over the discharge of water from the bailer into the sample container. Figure 3.13 illustrates an example of this type of device. A bottom-discharge device may not work with a dual check valve bailer unless the bailer design allows for release of the upper check valve during sample decanting.

Bailers are commonly used for both purging and sampling in small diameter, shallow wells, primarily because of their convenience, ease of use, and low cost. While they have proven to be useful for collecting samples of LNAPLs from the top of the water column of water table wells, they cannot be reliably used to purge wells or to collect representative samples for most parameters, particularly those sensitive to agitation and oxidation.

Based on a large body of scientific evidence gathered in a number of field studies (e.g., Pohlmann and Hess, 1988; Yeskis et al., 1988; Pohlmann et al., 1991, 1994; Puls et al., 1991, 1992; Puls and Powell, 1992; Backhus et al., 1993; Heidlauf and Bartlett, 1993), it is exceedingly difficult to collect accurate or precise samples with bailers for a wide range of analytical parameters including VOCs, trace metals, colloid-associated analytes, and dissolved gases.

Muska et al. (1986), Imbrigotta et al. (1988), Yeskis et al. (1988), Tai et al. (1991), and Gibs et al. (1994) all concluded that bailers provided poor precision in analytical data for VOCs compared with several different types of pumps, and attributed the results to variability in operator technique. Heidlauf and Bartlett (1993) found that when wells were purged and sampled with bailers, ground-water samples consistently had high turbidity values

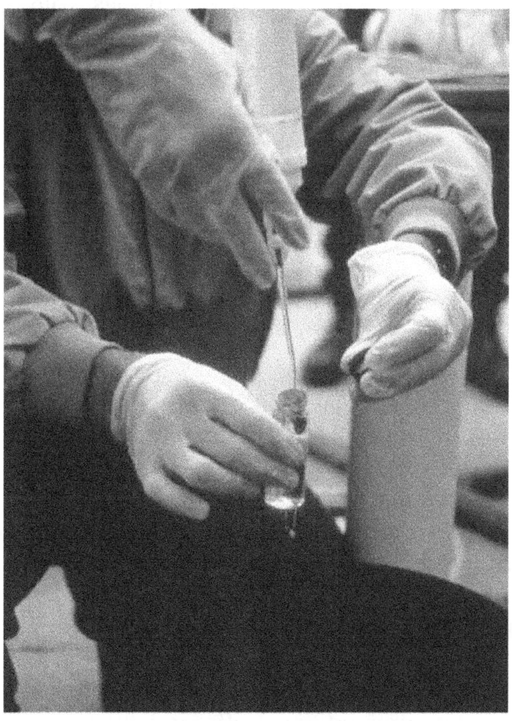

FIGURE 3.13
A bottom-discharge device, such as the one pictured here, makes control of sample discharge much easier than decanting from the top of the bailer.

(in excess of 100 NTUs) and that metal analyte concentrations of unfiltered samples were significantly higher than those of filtered samples. Puls et al. (1992) found that bailed samples had turbidities exceeding 200 NTUs, while samples collected by pumping at low flow rates had turbidities consistently less than 25 NTUs, and that As and Cr levels in bailed samples were five and two times higher, respectively, than in samples collected with a pump. Tests by Backhus et al. (1993) showed that bailing continued to produce samples with high turbidity even after 60 well volumes of purging. Further, they found that bailed and unfiltered samples contained up to 750 times greater concentrations of high molecular weight PAHs than pumped samples from the same wells at a coal tar contaminated site. These PAHs were determined to be sorbed primarily to the immobile aquifer solids and not part of the mobile contaminant load. Puls et al. (1990) found that Cr levels in bailed and unfiltered samples were two to three times higher than levels in samples collected by low-flow rate pumping. Use of data obtained from bailed samples at all of these sites could have led to substantial overestimation of mobile contaminant concentrations in ground water.

Pohlmann et al. (1994) found that particle size distributions of bailed samples were highly skewed toward particles greater than 5 μm in diameter (over 90% in some wells), while pumped samples contained a more uniform distribution of particle sizes. They also determined that bailers produced higher concentrations of particles that could be mobile in ground water (e.g., those in the range of 0.03 to 5.0 μm) than pumping, suggesting that concentrations of colloid-associated contaminants could be positively biased when bailing disturbs the sampling zone and elevates artifactual turbidity. Total particle concentrations in bailed samples were significantly higher (up to 20 times higher) than concentrations in pumped samples. Turbidity values obtained with bailers were as much as two orders of magnitude greater than pumped samples from the same wells. Backhus et al. (1993) also found concentrations the turbidity of bailed samples was several orders of magnitude higher than of pumped samples from the same wells and that there was an enormous difference in the size of particles in bailed samples versus pumped samples. Bailed samples contained particles in the range of 1 to 100 μm in diameter, whereas pumped samples mainly contained particles less than 5 μm in diameter. They determined that particles as large as 100 μm could not have been mobile at ambient ground-water flow rates and concluded that bailing collects particles and particle-associated contaminants that are not representative of *in situ* mobile contaminant loads.

To summarize, the primary drawbacks of using bailers to purge and sample wells include:

- The repeated insertion of the bailer into and withdrawal of the bailer from the water column in the well, even if it is done carefully, aerates the water column and creates a surging effect, which mixes and severely agitates the water column. The surging action that results from using bailers to purge a well creates two-directional flow within the well screen, resulting in continual development or over-development of the well (Puls and Powell, 1992) and grain flow within the filter pack (Pohlmann et al., 1994). As previously noted, bailing-induced agitation also mobilizes previously immobile aquifer matrix materials, creating substantial artifactual turbidity in water brought into the well. This may occur in wells installed in any type of formation, but is a particular problem in formations that are predominantly fine-grained (i.e., silts and clays). The surging effect also commonly results in resuspension of sediments that have accumulated and settled in the bottom of the well screen or sump, which also results in increased turbidity measurements. Ultimately, decreased life of the well can be expected to occur as a result of repeated bailing.

- Mixing the stagnant water column in the well casing with the dynamic water column in the well screen results in aeration of the entire water column and a composite averaging effect for all water in the well, which defeats the purpose of purging and makes it impossible to collect a sample representative of formation water chemistry. Puls and Powell (1992) and Pohlmann et al. (1994) determined that measuring representative values of DO and turbidity during purging with bailers was not possible and that equilibrium conditions for those parameters often could not be reached. DO concentrations in bailed samples are consistently higher than in pumped samples (in the study by Pohlmann et al. [1994], 10 to 20 times higher). Enrichment of DO during the sampling process not only produces samples that are unrepresentative with respect to oxygen content, but also results in oxidation and subsequent precipitation of reduced species, such as Fe^{2+}, and removal of other metal species through adsorption or co-precipitation (Stoltzenberg and Nichols, 1986). Alteration of dissolved constituents in this way during sampling may lead to erroneous conclusions about their concentration or speciation.
- Bailers are highly subject to uncontrolled operator variability, which leads to inconsistency in how wells are purged and samples are collected. This results in poor precision and accuracy (Imbrigotta et al., 1988; Yeskis et al., 1988; Tai et al., 1991; Muska et al., 1986). Operator variability can occur not only between different operators but also with the same operator (Yeskis et al., 1988).

Thief Samplers

Another type of grab sampler, called a thief sampler, employs a mechanical, electrical, or pneumatic trigger to actuate plugs or valves at either end of an open tube to open or close the chamber after lowering it to the desired sampling depth, thus sampling from a discrete interval within the well. Figure 3.14 is an example of this type of sampler called a Kemmerer sampler. This type of device is impractical to use for purging, but can be used to sample from any depth.

Syringe Samplers

The first syringe devices used for ground-water sampling were essentially commercially available medical syringes adapted to sample in a 2 in. diameter well (Gillham, 1982). Several manufacturers have since produced syringe devices specifically for ground-water sampling. The syringe sampler illustrated in Figure 3.15 is divided into two chambers by a moveable piston or float. The upper chamber is attached to a flexible air line that extends to the ground surface. The lower chamber is the sample chamber. The device is lowered into the sampling point and activated by applying suction to the upper chamber, thereby drawing the piston or float upward and allowing water to enter the lower chamber. In situations where the pressure exerted on the lower chamber by submergence is great enough to cause the piston or float to move upward prior to achieving the desired sampling depth, the upper chamber can be pressurized during placement in the well to prevent piston movement. The device is then activated by slowly releasing the pressure from the upper chamber, allowing water to fill the lower chamber under hydrostatic pressure. Syringe devices can be used to sample from any depth and for any parameter, but they are impractical to use for purging.

FIGURE 3.14
This Kemmerer sampler, typically used for surface-water sampling, has been adapted for use in a 2 in. diameter well. It is used in wells to sample from a discrete depth (e.g., from within the well screen), but is impractical to use for purging.

Other Grab-Sampling Devices

Several other grab-sampling devices have been developed in recent years to sample ground water, including the HydraSleeve^{TM}, the Discrete Interval Sampler, the Kabis sampler, the Snap Sampler, and the Pneumo-Bailer. The HydraSleeve is discussed in Chapter 4; only very limited information is available on the other devices and their applications and limitations. The interested reader is directed to Parker and Clark (2002), ITRC (2005) and Parsons (2005) for what information is available.

Suction-Lift Devices

Surface centrifugal pumps and peristaltic pumps are two common types of suction-lift pumps. These pumps, typically situated at or above ground level during purging and sampling, draw water to the surface by applying suction to an intake line through the use of impellers or rotors typically driven by an electric motor. In theory, suction-lift pumps are limited to lifting water approximately 34 ft, depending on altitude and barometric pressure. In practice, a lift of 15 to 25 ft is the typical upper limit. The diameter of sampling points to which these devices are applicable is limited only by the size of the intake tubing used; 0.25 or 0.375 in. tubing can be used to sample sampling points as small as 0.5 in. in diameter. Sediment has only a minor effect on suction-lift pumps, although large solids may plug the pump intake line.

The pressure decrease caused by suction-lift devices can result in a number of deleterious effects on samples. The most notable effects include loss of dissolved gases and VOCs and accompanying pH shifts that can affect dissolved metals concentrations.

(a) (b)

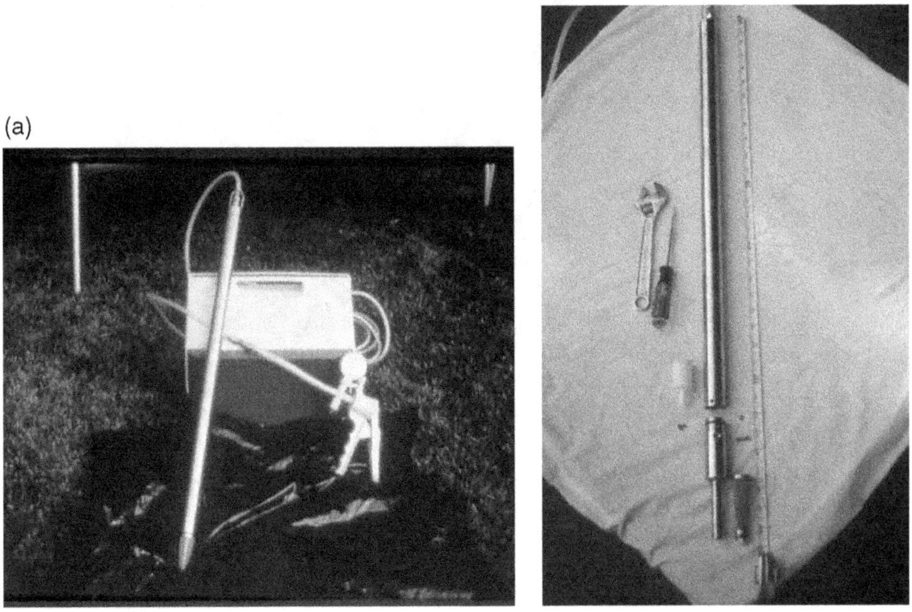

FIGURE 3.15
A syringe sampler collects discrete samples by exerting a slight negative pressure on a piston to draw the sample through a bottom check valve into the device. (a) Shows the device ready to install in a well (with a small hand pump to apply suction to the piston). (b) Shows the device disassembled for cleaning between uses. Syringe devices are impractical to use for purging.

Several researchers have noted losses ranging from as low as 4% to as high as 70% for a variety of VOCs (Ho, 1983; Barcelona et al., 1984; Devlin, 1987; Barker and Dickhout, 1988; Imbrigiotta et al., 1988) and losses from 7 to 17% for several trace metals including B, Ba, Sr, Hg, Mb, and Se (Houghton and Berger, 1984). Generally, losses of VOCs will be greater for compounds with high Henry's law constants. Suction-lift pumps are therefore best suited to collecting samples for parameters that are not sensitive to pressure decreases, such as the very stable major ions, and SVOCs with very low Henry's law constants, such as PCBs and pesticides.

Surface Centrifugal Pumps

Surface centrifugal pumps use impellers typically constructed of metal (brass or mild steel), plastic, or synthetic rubber, usually set within a cast iron or steel pump chamber. Before the pump can operate, it must be primed with water at the surface so it can create the suction necessary to lift water from a well. After priming, the pump can continue to operate without addition of more water. Figure 3.16 shows a representative design for this type of pump. These pumps can pump at rates of 2 to 40 gal/min, with 5 to 15 gal/min being more typical. Because surface centrifugal pumps operate by suction lift and because they are prone to cavitation, they are not appropriate for collection of samples to be analyzed for pressure-sensitive parameters, such as dissolved gases, VOCs, or trace metals. Because the pumped water contacts the pump mechanism, which is not constructed of inert materials, artifacts from sample contact with these materials should be considered when evaluating these pumps for sampling. In addition, these pumps can mix air from small leaks in the suction circuit into a sample, which can cause sample aeration. These pumps are typically difficult to adequately decontaminate between uses

FIGURE 3.16
Surface centrifugal pumps can be used to purge wells, but they are not appropriate for collecting samples because of their materials of construction and because of the strong negative pressure required to lift the water to the surface. (Photo courtesy of Todd Giddings.)

because they are difficult to disassemble. They may be useful for purging, but because of all of their limitations, they are not recommended for sampling.

Peristaltic Pumps

The peristaltic pump is the most common type of suction-lift device used for ground-water sampling in shallow sampling points. A peristaltic pump (Figure 3.17) consists of a rotor with rollers that squeeze flexible tubing as they revolve within a stator housing. This action generates a reduced pressure at one end of the tubing and an increased pressure at the other end. Peristaltic pumps operate at rates of less than 0.01 gal/min to more than

FIGURE 3.17
The pump head of a peristaltic pump contains a set of rollers and a rotor, driven by a pump motor, that squeeze flexible (usually silicone) tubing stretched over the rollers, to create negative pressure on one side of the pump head and positive pressure on the other. The negative pressure can bias samples collected for analysis of pressure-sensitive parameters including dissolved gases, VOCs, and trace metals.

FIGURE 3.18
To avoid contact between the flexible pump tubing and the sample, a sample transfer vessel can be installed between the well and the peristaltic pump. As suction is exerted on the transfer vessel, water is drawn into the vessel through the tubing extended down the well. This configuration results in a slight loss of pumping efficiency, decreasing the lift capability of the pump.

12 gal/min. Peristaltic pumps do not usually cause cavitation but, as in all suction-lift pumps, the application of a reduced pressure on the sample can bias the sample for pressure-sensitive parameters.

Several types of elastomeric material can be used for the pump tubing, although flexible PVC and silicone rubber are most common. The flexible tubing required for use in a peristaltic pump mechanism may cause sample bias. The plasticizers in flexible (Type II) PVC can contaminate samples with phthalate esters (Ho, 1983; Barcelona et al., 1985; Pearsall and Eckhardt, 1987; Barker and Dickout, 1988). The use of silicone rubber tubing, which contains no plasticizers, can obviate this problem. However, the potential for sample bias due to sorption or desorption exists with both materials (Barcelona et al., 1985). These pumps can be used with an intermediate transfer vessel, so the sample contacts only the intake tubing and vessel (which can be made of any appropriate material), avoiding contact with the pump tubing as illustrated in Figure 3.18. Alternatively, using a very short run of silicone rubber tubing at the pump head only can minimize this problem (Ho, 1983; Barker and Dickhout, 1988). Manufacturers' recommendations for tubing materials should be followed in conjunction with chemical compatibility charts.

Electric Submersible Centrifugal Pumps

An electric submersible centrifugal pump (Figure 3.19) consists of impellers housed within diffuser chambers that are attached to a sealed electric motor, which drives the

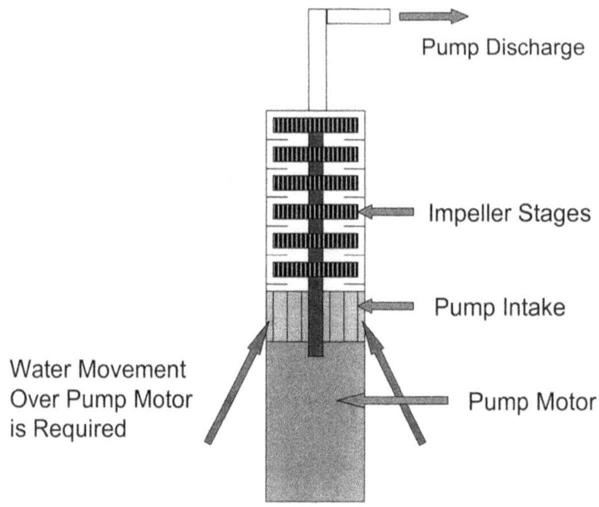

FIGURE 3.19
Schematic of an electric submersible centrifugal pump.

impellers through a shaft and seal arrangement. Water enters the pump by pressure of submergence, is pressurized by centrifugal force generated by the impellers, and discharged to the surface through tubing, hose, or pipe. An electric submersible centrifugal pump is suspended in a well by its discharge line or a support line. Electric power is supplied to the motor through a braided or flat multiple-conductor insulated cable. Figure 3.20 depicts an electric submersible centrifugal pump. These pumps are available in both fixed-speed and variable-speed configurations.

Electric submersible centrifugal pumps are driven by electric motors. Most designs require that water continually passes over the motor to cool it, while some designs can cool sufficiently by free convection in applications up to 86°F (30°C), provided that the pump motor is installed above the well intake zone. For designs that require flow for cooling, manufacturers of these pumps typically specify a minimum flow rate and velocity over the motor to prevent overheating. If the pump is located within the screened zone of the well, or if the well casing diameter is too large to provide sufficient flow over the motor, the use of a shroud may be required to achieve the necessary flow rate and velocity.

Flow rate and lift capability for pumps of this design are wide ranging. Fixed-speed pumps not specifically designed for ground-water sampling do not have controllable flow rates, and the flow rate is dependent on the horsepower rating of the pump, the number of impellers (stages) in the pump section, and the depth to static water level (i.e., lift). For variable-speed pumps designed for ground-water sampling, the discharge rate can be altered by regulating the frequency of the electrical power supply and controlling the motor speed to reduce or increase the flow rate. The rate is generally controllable between 0.03 and 8 gal/min. These pumps are capable of pumping at their highest rate at low lifts (10 ft); the rate decreases with increased lift, to a limit of about 230 ft.

With all pumps of this design, heat generated by the motor could increase sample temperature, which could result in loss of dissolved gases and VOCs from the sample and subsequent precipitation of trace metals. The temperature rise in samples is especially acute (between 5 and 7°C [Pohlmann et al., 1994; Oneacre and Figueras, 1996]) at low flow rates. Because the pump's impellers operate at very high speeds (between 2,500 and 3,500 r/min for fixed-speed designs; between 7,500 and 23,000 r/min for variable-speed

FIGURE 3.20
An electric submersible centrifugal pump mounted on a rig designed to make purging and sampling, and cleaning of the pump and discharge tubing, convenient for the sampling team.

designs), the accuracy of samples may also be affected by extreme agitation, which may include disaggregation of colloidal or other particulate material that may be entrained in the water. Electric submersible centrifugal pumps are considered acceptable for sampling major ions, radioactive constituents, and some dissolved metals (Pohlman et al., 1994), but they are generally not well-suited for sampling dissolved gases, VOCs, trace metals, or other parameters sensitive to temperature or agitation.

Electric submersible centrifugal pumps are only available in diameters that will fit into sampling points 2 in. or larger in diameter. These pumps can be damaged when used in silty or sandy water, requiring repair or replacement of pump components or the motor. If overheating occurs, there are three possible consequences. First, where the motor has internal water or oil in it for improved cooling, some of this liquid could be released into the sampling point, which could potentially contaminate the sampling point or samples collected from the sampling point. Because of this, motors that contain oil should not be used if the oil could interfere with the analytes of interest. Further, water used in motors should be of known chemistry and should be replaced between uses of the pump (i.e., between wells and between sampling events). Second, when this type of motor eventually cools, it can draw in water from the sampling point, which could cause future cross-contamination problems. Proper decontamination of the pump should include changing internal cooling fluid if the pump is to be used in nondedicated applications. As an alternative, dry sealed motors can be used to avoid these potential problems. Third, extensive or long-term overheating problems may result in motor failure, usually requiring replacement of the motor. Electric submersible centrifugal pumps should not be allowed to operate dry, or damage may occur to the pump seals or motor. Some pump designs may be difficult to disassemble in the field for cleaning or repair. For these pumps, if used portably, cleaning is usually performed by flushing the pump and discharge line and washing the exterior surfaces in accordance with ASTM Standard D 5088 (ASTM, 2006b).

Positive-Displacement Pumps

A number of different devices fall into the category of positive-displacement pumps: gas-displacement pumps, bladder pumps, piston pumps, progressing cavity pumps, and electric submersible gear-drive pumps. Until very recently, all of these devices were limited to use in 2 in. or larger diameter wells. However, some manufacturers of gas-displacement pumps and bladder pumps have developed smaller diameter pump models to fit into smaller direct-push wells and sampling tools.

Gas-Displacement Pumps

A gas-displacement pump (Figure 3.21) forces a discrete column of water to the surface via pressure-induced lift without the extensive mixing of drive gas and water produced by air-lift devices. The principle of operation of a gas-displacement pump is shown schematically in Figure 3.22. Hydrostatic pressure opens the inlet check valve and fills the pump chamber (fill cycle). The inlet check valve closes by gravity after the chamber is filled. Pressurized gas is then applied to the chamber, displacing the water up the discharge line (discharge cycle). After releasing the pressure, the cycle can be repeated. A check valve in the discharge line maintains the water in the line above the pump between discharge cycles. A pneumatic logic unit, or controller, is used to control the application and release of the drive-gas pressure. The lift capability of a gas-displacement pump is directly related to the pressure of the drive gas used (1 psi of gas pressure = 2.31 ft of lift) and is usually limited by the burst strength of the gas-supply tubing. PE and PTFE are the most commonly used materials for tubing.

(a) (b)

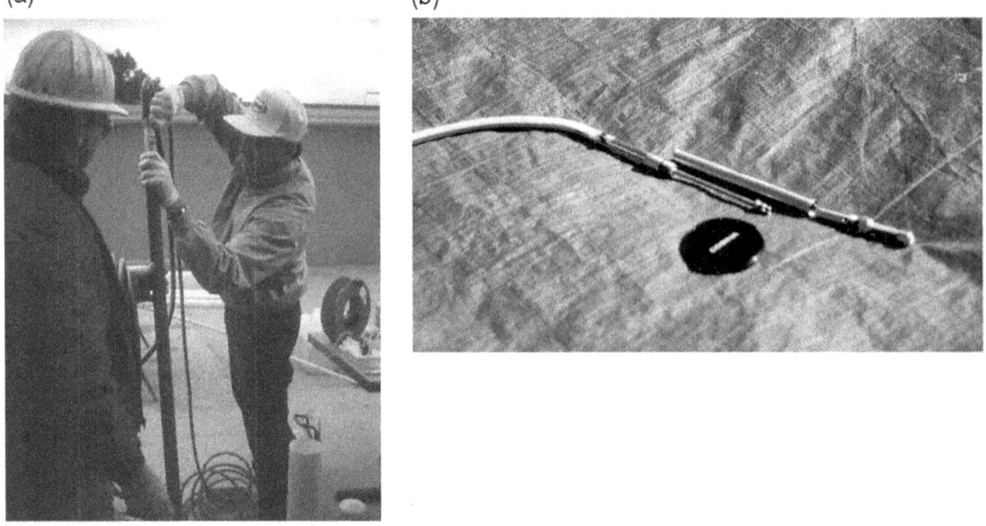

FIGURE 3.21
Gas-displacement pumps range from large, high-volume pumps (a) designed for high-flow-rate pumping to small-diameter, small-volume pumps (b) designed for small-diameter direct-push wells and multilevel monitoring systems. These pumps are easy to operate and easy to disassemble for cleaning.

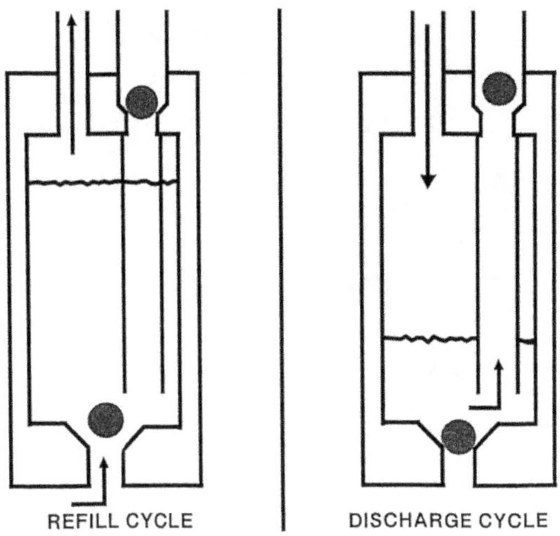

FIGURE 3.22
Schematic representation of a simple gas-displacement pump.

The maximum flow rate of a gas-displacement pump is based on the pump chamber volume, the pressure and volume of the drive-gas source, the cycling rate of the pump, the depth of the pump, and the submergence of the pump inlet. The flow rate can be controlled either by adjusting the pressure of the drive gas or the time allowed for the refill or discharge cycles to occur. Typical lifts for gas-displacement pumps rarely exceed 250 ft using single-stage compressors; greater lifts can be achieved using two-stage compressors or compressed-gas cylinders. Gas-displacement pumps are available for sampling points as small as 0.5 in. in diameter.

Gas-displacement pumps are generally not damaged by sediment, although sediment may reduce the maximum flow rate or temporarily clog the check valves and interrupt flow from the pump. These pumps are not damaged by pumping dry, so they are ideally suited for pumping in low-yield sampling points. They are typically easy to disassemble for cleaning, service, or repair.

There is a limited interface between the drive gas and the water in a gas-displacement pump. There is, however, a potential for loss of dissolved gases and VOCs across this interface (Barcelona et al., 1983; Gillham et al., 1983). This potential greatly increases if the pump is allowed to discharge completely, which would cause drive gas to be blown up the discharge line. Contamination of the sample may also result from impurities in the drive gas. It is highly recommended that an inert drive gas, such as nitrogen or helium, be used.

Bladder Pumps

Pneumatic bladder pumps, also known as gas-operated squeeze pumps, Middleburg-type pumps or diaphragm pumps, consist of a flexible membrane (bladder) enclosed by a rigid housing, with check valves on either side of the bladder (Figure 3.23). A schematic is shown in Figure 3.24. In the traditional pneumatic bladder pump, water enters the bladder under hydrostatic pressure through an inlet check valve at the pump bottom. The inlet check valve closes by gravity after the bladder is filled. Compressed gas is applied to the annular space between the outside of the bladder and pump housing, which squeezes the bladder. This action forces the water out of the bladder and up the discharge line to the surface; a check valve in the discharge line prevents discharged water from

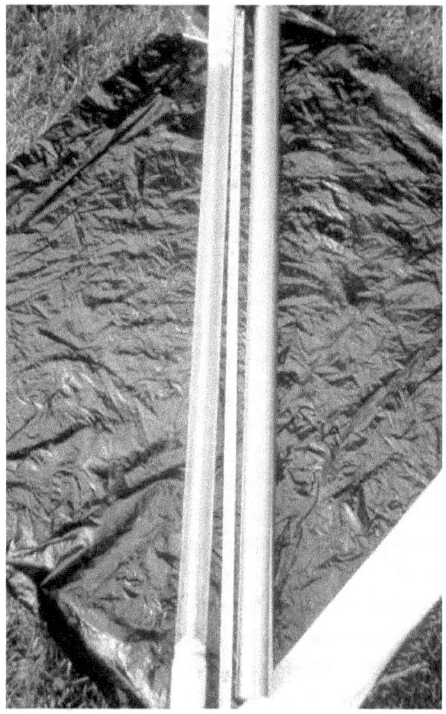

FIGURE 3.23
A dedicated pneumatic bladder pump disassembled to show the pump body (stainless steel) and the bladder (PTFE), with an inlet check valve at the bottom and a discharge check valve at the top (both PTFE). Dedicated pumps of this design are not meant to be disassembled.

re-entering the bladder. After releasing the gas pressure, this cycle can be repeated. In some bladder pump designs, the water and air chambers are reversed, with water entering the annular space between the pump housing and bladder; the bladder is then inflated to displace the water. A pneumatic logic controller controls the application and

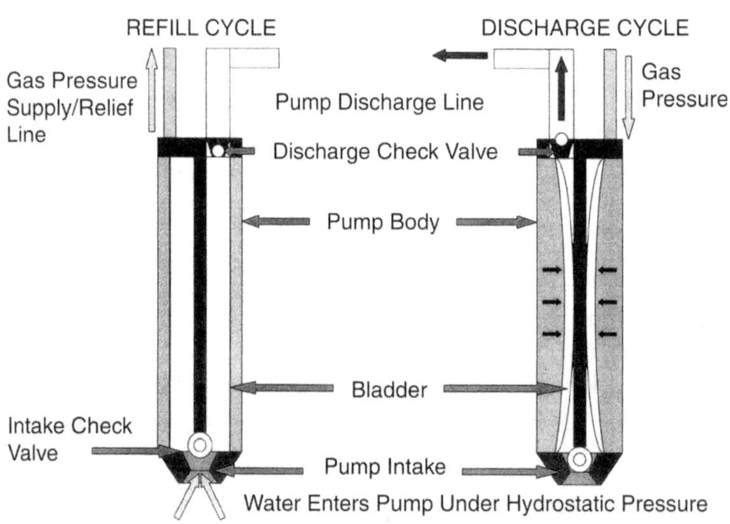

FIGURE 3.24
Schematic of a pneumatic bladder pump.

FIGURE 3.25
A portable pneumatic bladder pump designed to make it easy to disassemble for cleaning.

release of drive-gas pressure to the pump. The lift capability of bladder pumps is directly related to the pressure of the drive-gas source and is controlled to some degree by the burst strength of the tubing used and outer materials of pump construction (stainless steel has the greatest depth capability). Bladder pump flow rates are controlled by adjusting the drive-gas pressure or the discharge and refill cycle timing. Flow can be readily controlled, and discharge rates of between 50 ml/min and 4 l/min can be obtained in many applications. When dealing with high-volume sampling points, maximum flow rates from bladder pumps may be too low for high-flow-rate purging. In these situations, secondary purging pumps or packers can be used in conjunction with bladder sampling pumps to reduce purge time requirements or sampling protocols can be modified to include low-flow purging and sampling procedures. Bladder pumps designed to be used in dedicated applications may be difficult to clean if they are used portably. Some manufacturers have developed a portable version of their bladder pump (Figure 3.25) to permit easy disassembly and replacement of the bladders, which is difficult or impossible with the sealed, dedicated pump design. Pneumatic bladder pumps are also available for use in sampling points as small as 0.5 in. in diameter. These use the same principle of operation as the larger diameter pumps.

A small-diameter stainless steel mechanical bladder pump has been developed specifically for portable use in small-diameter direct-push wells and sampling tools (Figure 3.26). To operate this device, the pump, attached to an outer tubing of 0.43 in. diameter PE and an inner tubing of 0.25 in. diameter PE or Teflon, is lowered into the sampling point and is held in place by a clamp to prevent the pump from moving within the water column during operation. At the surface, a mechanical actuator employs either a circular stroke or a vertical stroke to retract the inner tubing on the intake stroke. As the inner tube is raised, the upper check ball seats and prevents water from flowing back into the Teflon bladder. The lower check ball opens and water is drawn from the well through the intake screen as the corrugated bladder expands. On the sampling stroke, the outer tubing is held stable and the inner tubing is lowered. As the inner tube is lowered, the upper check ball opens and water is pushed up the inner tube to the surface. The bladder

FIGURE 3.26
A mechanical bladder pump designed to operate in small-diameter direct-push rods or wells without reliance on a source of drive gas. (a) Shows the ball check valve on the intake of the pump. (b) Shows the pump being installed in a well. (c) Shows the configuration of the wellhead assembly. (d) Shows the manual cycling mechanism for the pump. (Photos courtesy of Geoprobe Systems.)

compresses as the inner tube is lowered and water is forced up and out of the bladder. As the bladder is compressed, the lower check ball is closed, preventing water from flowing out of the intake valve.

This mechanical bladder pump does not require the accessory equipment required by traditional bladder pumps, such as generators, compressors, pneumatic controllers, or compressed gases. It was evaluated by the U.S. EPA ETV Program (U.S. EPA, 2003), where it was determined that the pump could be used to collect VOC samples with low turbidity and that it was suitable for low-flow purging and sampling applications. The potential disadvantage of this device is that it is somewhat difficult to thoroughly decontaminate between sampling points, particularly the fine mesh of the inlet screen assembly.

All bladder pumps are susceptible to bladder damage or check valve malfunction caused by sediment. The use of inlet screens, often required by manufacturers as part of their pump warranty program, can minimize or eliminate these problems. Bladder pumps can be run dry without damage.

Bladder pumps provide representative samples under a wide range of field conditions. There is no contact between the drive gas and the water in a pneumatic bladder pump, eliminating the potential for stripping of dissolved gases and VOCs and the potential for sample contamination by the drive gas. Pressure gradients applied to the sample can be controlled by reducing the drive-gas pressure applied to the bladder, thus minimizing disturbance to sample chemistry. Bladder pumps are recommended for sampling all parameters under a wide variety of field conditions (Barcelona et al., 1983; Unwin and Maltby, 1988; Pohlmann et al., 1991, 1994; Tai et al., 1991; Kearl et al., 1992; Puls et al., 1992; Parker, 1994).

Piston Pumps

Piston pumps are a type of pneumatic positive displacement pump that use a drive-gas piston connected via a rod to a pump piston, which reciprocates within a stainless steel pump chamber. Movement of the pump piston draws water into inlet valves on one side of the piston, while simultaneously displacing water through discharge valves on the other side of the piston. Water displaced by the pump is driven to the surface through a discharge tube. The piston is cycled through the use of a pneumatic or mechanical actuator. Figure 3.27 provides a schematic of a piston pump illustrating the flow path of water through the pump.

Double-acting piston pumps (Figure 3.28) are available in several different diameters to fit 2 in. diameter and smaller sampling points. Currently, models are available in the following outside diameters: 0.625, 0.75, 0.875, 1, 1.4, and 1.8 in. The 1.4 and 1.8 in. diameter models are most widely used. The flow rate of a piston pump depends on the inside diameter of the pump cylinder and the stroke length and rate. The ability to control the minimum flow rate for sampling is dependent on the degree to which the stroke rate can be regulated by the controller unit at ground surface.

Owing to the complexity of the internal mechanisms of this pump, it is best suited to dedicated installations; however, it has been used successfully as a portable device. In a portable mode, accessory equipment can become cumbersome (i.e., large bottles of compressed gas or an oil-free compressor; large tubing bundles for deep applications) and will require a suitably sized field vehicle to transport equipment between sampling locations (Figure 3.29). In addition, the pump and tubing must be thoroughly cleaned between wells, which can be difficult because of the complex valving in the pump and the length of the tubing. Pump disassembly is time consuming, so most samplers simply flush the pump and tubing with cleaning solutions. If this is done, equipment blanks must be collected frequently as part of the QC program to confirm no cross-contamination of sampling points.

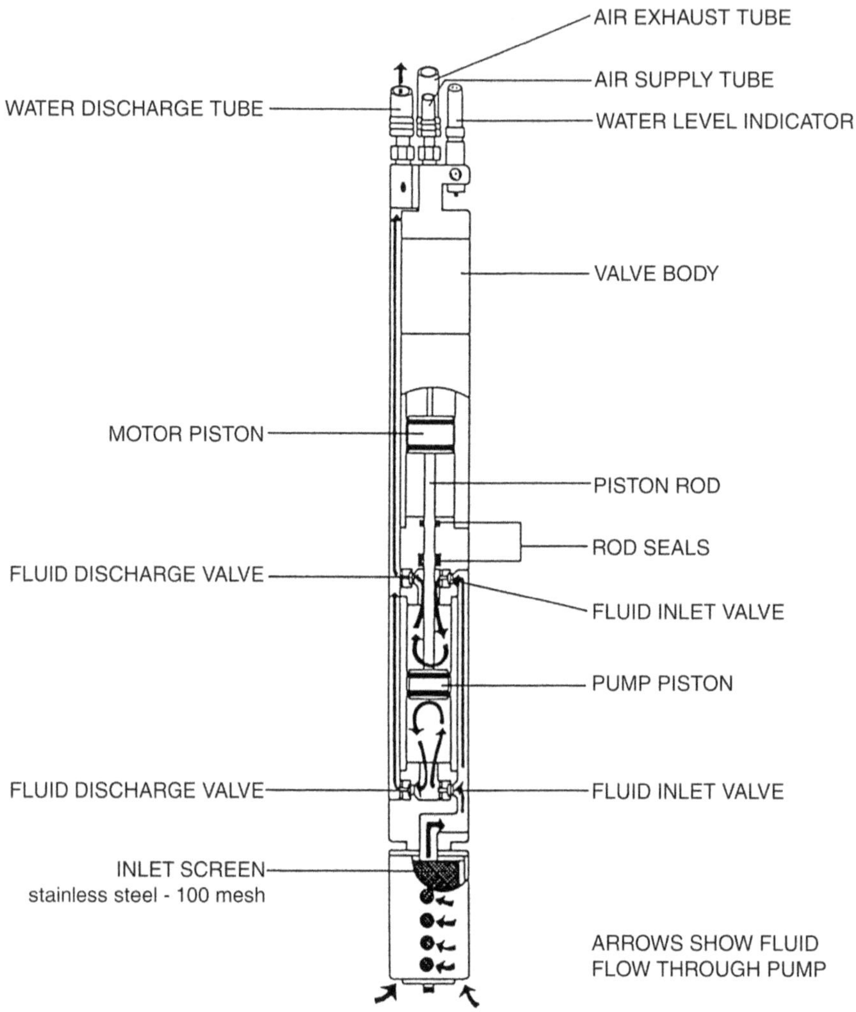

FIGURE 3.27
Schematic of a double-acting piston pump.

Piston pumps can provide representative samples for most parameters (Barcelona et al., 1983; Yeskis et al., 1988; Knobel and Mann, 1993), as they are constructed of inert materials, and can deliver samples at a controlled flow rate. Flow rates vary from less than 100 ml/min to up to 5 gal/min. These pumps can be used effectively in high-volume wells for low-flow purging. Dedicated installations of this pump indicate that the device is reliable in long-term monitoring programs. This pump has the greatest lift (over 1000 ft) of any small-diameter pump.

There are some concerns about what impact this device may have on sample chemistry due to the slight negative pressure produced during refill of the pump, although this effect is reduced as the pump cycling rate is decreased. Likewise, reducing the pump cycling rate also reduces the pressure applied to the sample, minimizing the potential for sample alteration. If a flow restrictor or valve is used to reduce the discharge rate, then the resultant pressure changes could alter sample chemistry (Barcelona et al., 1983; Gillham et al., 1983). Piston seals and inlet or discharge valves are subject to failure in highly turbid water; inlet screens can reduce or eliminate this damage. These pumps may also be damaged by running dry.

FIGURE 3.28
This double-acting piston pump is capable of pumping from lifts up to 1000 ft, at a rate of about a liter per minute. (Photo courtesy of Bennett Sample Pumps, Inc.)

FIGURE 3.29
Because the lift capability of the double-acting piston pump allows pumping from depths greater than 1000 ft, portable use often requires that a support vehicle carry the pump and support equipment. The tubing bundle (which consists of three separate lengths of tubing and, optionally, a cord for a pressure transducer) is cumbersome and large, as is the drive-gas source (in this case, a K-size cylinder of nitrogen). (Photo courtesy of Bennett Sample Pumps, Inc.)

Progressing Cavity or Helical Rotor Pumps

Progressing cavity pumps, also referred to as helical rotor or mono-type pumps, utilize a rotor driven by an electric motor against a stator assembly to displace water through a discharge line to ground surface. A schematic is shown in Figure 3.30. Rotation of the helical rotor causes the water-filled cavity between the rotor and stator to progress upward, thereby pushing water in a continuous flow upward through the discharge line. In some progressing cavity pumps, the discharge rate can be varied by adjusting the speed of the pump motor between 50 and 500 r/min. The progressing cavity pump (Figure 3.31) is typically suspended in a well by its discharge line or by a suspension cable. A two-conductor electric cable supplies power from a 12 V DC power supply and control box to the pump motor.

Progressing cavity pumps are commonly constructed of stainless steel with PTFE or PE materials used as seals. The rotors are generally constructed of stainless steel, while the stator material may consist of EPDM or Viton.

Progressing cavity pumps require a sampling point diameter of at least 2 in. and will operate at moderate to low flow rates up to a maximum lift of approximately 180 ft. The relatively low discharge rates attainable with most progressing cavity pumps make them most useful in applications where purging does not require removal of large volumes of water from monitoring wells. With variable flow rate progressing cavity pumps, once purging is complete, the discharge rate may be reduced before samples are collected. Owing to the sealed nature of the pump and the rotor/stator construction, these devices are well suited to dedicated installations, although they are commonly used portably because they are relatively easy to transport between sampling locations. It is highly recommended when the pump is run using 12 V batteries that a charging and backup battery system be available because the pump quickly drains batteries during use.

When not used frequently, these pumps are subject to locking where the rotor becomes stuck to the stator and is difficult or impossible to turn. This condition is difficult to overcome in the field. Additionally, the rotor and stator may be damaged when pumping silty or sandy water.

The operating principle of progressing cavity pumps makes them suitable for collection of samples for VOCs (Imbrigiotta et al., 1988). There is some evidence that these pumps

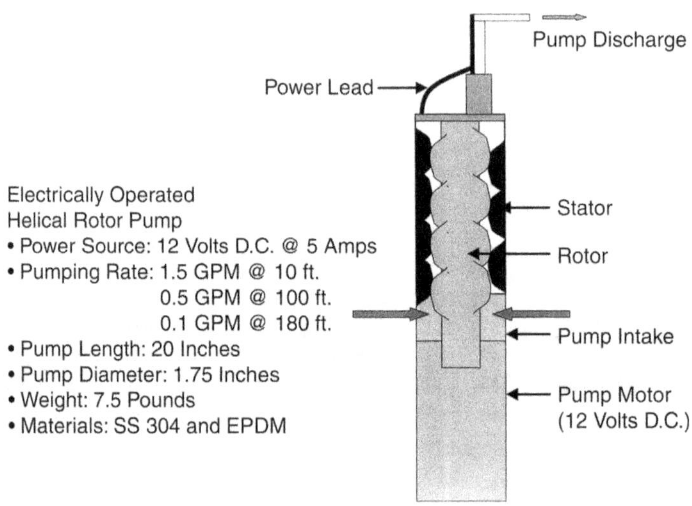

FIGURE 3.30
Schematic of a progressing cavity pump.

FIGURE 3.31
A progressing cavity pump is a relatively portable pump that operates on 12 V DC power supplied by a deep-cycle marine battery.

may not be suitable for sampling trace metals and other inorganic analytes at higher flow rates due to increased turbidity (Barcelona et al., 1983); to control turbidity, a variable speed pump controller can be used to reduce flow rate. The pressure applied to a sample is directly related to the motor speed and can be controlled in designs using variable-speed motor controls. Overheating of the motor may raise the temperature of the sample (Parker, 1994).

Electric Submersible Gear-Drive Pumps

Another type of positive displacement pump is the electric submersible gear-drive pump, shown schematically in Figure 3.32. In this type of pump, an electric motor drives a PTFE gear, which meshes with a second PTFE gear (Figure 3.33). As these gears rotate, their advancing teeth draw water into the pump through the pump intake port and push it through the gears in a continuous flow up the discharge line. The discharge rate can be varied using the pump controls to adjust the speed of the pump motor. As with many other submersible pumps, the gear-drive pump is usually suspended in a well by its discharge line. Electric power is supplied to the 36 V DC motor through a cable from the power source and control box at ground surface. The manufacturer provides either PE or Teflon-lined PE tubing with the pump, although other tubing can be made available if required.

This pump is commonly used portably due to its ease of transport and ease of decontamination between sampling locations (Figure 3.34). It can be installed as a dedicated pump, and a stand-alone control box and special dedicated installation well caps are available from the manufacturer.

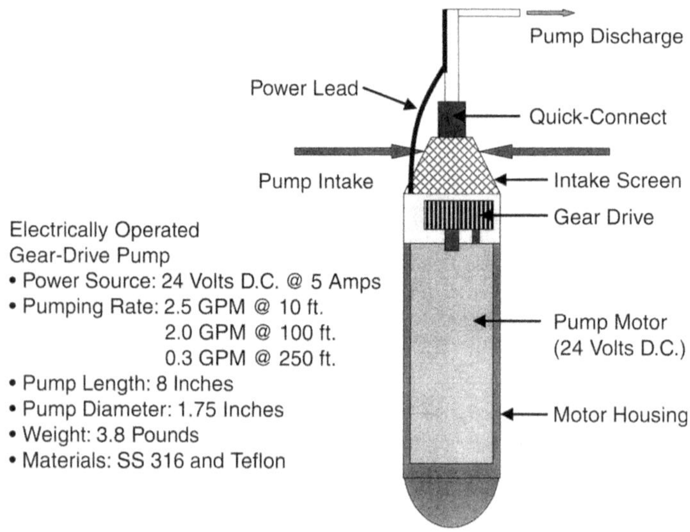

FIGURE 3.32
Schematic of an electric submersible gear-drive pump.

Electric submersible gear-drive pumps require a sampling point diameter of at least 2 in. Maximum discharge rates for gear-drive pumps range from more than 3 gal/min at lifts of less than 20 ft to 0.25 gal/min at lifts of 250 ft. Discharge rates are easily controlled using the flow control, which adjusts the power supplied to run the pump motor; pump discharge can be adjusted to less than 50 ml/min.

If electric submersible gear-drive pumps are used extensively for pumping water high in suspended solids, the PTFE gears may clog or wear, thereby reducing the discharge rate. Disassembly of the pump and replacement of the gears is a procedure easily accomplished in the field in a few minutes. Electric submersible gear-drive pumps are generally very easy to decontaminate when used in a portable mode.

FIGURE 3.33
The PTFE drive gear of a gear-drive pump (center) meshes with a second PTFE gear to displace water (which enters the pump under hydrostatic pressure) through the pump chamber into the discharge tubing to the surface.

FIGURE 3.34
Electric submersible gear-drive pumps are highly portable, requiring a 36 V DC gel cell power source, a 110 V AC source connected to a 36 V DC converter, or a 12 V DC marine battery connected to a 110 V AC inverter and a 36 V DC converter (shown here).

Electric submersible gear-drive pumps provide good sampling accuracy and precision for dissolved gases, VOCs, trace metals, and other inorganics and mobile colloids (Imbrigiotta et al., 1988; Backhus et al., 1993). Cavitation may occur if the pump is run at high speed, which could affect dissolved gases or VOCs. The potential for cavitation can be reduced or eliminated by controlling motor speed. The pressure applied by a gear-drive pump to a sample is directly related to the motor speed and can be controlled using the variable-speed motor controls. Electric submersible gear-drive pumps are constructed of materials acceptable for sampling sensitive ground-water parameters; the pump body is constructed entirely of stainless steel materials, while the gears are constructed of PTFE.

Inertial-Lift Pumps

Inertial-lift pumps (Figure 3.35), also known as tubing-check-valve pumps, consist of a discharge tubing (either flexible tubing or rigid pipe) with a ball-check foot valve attached to the lower end of the tubing. In operation, the tubing is lowered into a water column and cycled through reciprocating motion, either through manual action or through the use of a reciprocating mechanical arm mechanism driven by an electric motor or internal combustion engine, to discharge water. As the tubing is moved upward, water that has entered the tubing under hydrostatic pressure is lifted upward, held in the tubing by the seated foot valve. When the upward motion of the tubing is stopped, the inertia of the water column inside the tubing keeps it moving upward. As the tubing is pushed downward, the foot valve opens, allowing the tubing to refill, and the cycle is repeated to pump water from the sampling point.

Inertial-lift pumps can be constructed of any flexible or rigid tubing material that has sufficient strength to tolerate the pump cycling. Typically, these materials include rigid and flexible PVC, PE, PP, and PTFE. Tubing diameters of 0.25 or 0.375 in. can be used to

FIGURE 3.35
An inertial-lift pump simply consists of a length of rigid or flexible tubing with a foot valve on the bottom. (Photo courtesy of Solinst Canada Ltd.)

collect samples from sampling points as small as 0.5 in. in diameter. Inertial-lift pumps are commonly used in small-diameter direct-push sampling tools and wells.

The flow rate of an inertial-lift pump is directly related to the cycling rate. Flexing of the tubing in the sampling point can cause the flow rate to drop. To achieve discharge rates suitable for sample collection, it is necessary to insert a short length of small-diameter flexible tubing into the discharge line to divert a portion of the discharge stream into sample containers. To control pump discharge, the flexible tubing must be held stationary while the discharged water is directed into sample containers.

By nature of their simple design, inertial-lift pumps are not susceptible to damage by suspended solids or dry pumping, although check valve clogging will reduce the flow rate during operation. Some wear or damage may occur on the outer surface of the foot valve or tubing as it comes in contact with the well casing or screen or open borehole during cycling. These pumps are easily disassembled in the field for repairs if needed, although the mechanical cycling mechanisms may be difficult or impossible to repair in the field.

If inertial-lift pumps are cycled rapidly prior to or during sample collection, some loss of VOCs or dissolved gases could occur in the discharge stream. Inertial-lift pumps do not cause pressure changes in the sample. However, the cycling action of an inertial-lift pump in a sampling point can significantly increase sample turbidity and agitate and aerate the water column within the sampling point. This can result in alteration of concentrations of a wide variety of analytes, including dissolved gases, VOCs, and trace metals, and interference with analytical determinations in the laboratory because of the high suspended sediment content of samples.

Passive Diffusion Bag Samplers

A unique device for ground-water sample collection for selected VOCs without purging wells has been developed by the U.S. Geological Survey — the passive diffusion bag sampler (PDBS). A typical PDBS (Figure 3.36) consists of a 3 to 4 mil thick, 18 to 20 in. long, 1.25 in. diameter low-density polyethylene (LDPE) lay-flat tube sealed at both ends and filled with deionized water. The sampler is deployed in a well by attaching a weight to the bottom and a suspension cord to the top (Figure 3.37). It is positioned within the well screen so it is in contact with formation water and so VOCs that are present in formation water can diffuse through the PE bag into the deionized water contained within. The amount of time that the PDBS should be left in the well before retrieval depends on the time required for equilibration to occur and the time required for the well to recover from the disturbance caused by sampler deployment. Laboratory and field data suggest that 2 weeks of equilibration time is adequate for most applications, although in low hydraulic conductivity formations, longer equilibration times may be required (Vroblesky and Campbell, 2001; Vroblesky, 2001a).

Retrieval of a sample from a PDBS consists of pulling the sampler out of the well, puncturing the bag with a sharp rigid PE tube, and transferring the water from the bag into 40 ml sampling vials for later analysis. The concentrations of VOCs in the sample represent an integration of the chemical changes that occurred in the well over the most recent portion (approximately the last 48 to 166 h, depending on the water temperature and the individual compound) of the equilibration period (Vroblesky, 2001a). After samples are collected, the PDBS is discarded and another PDBS is installed in the well and left there until the next sampling event.

FIGURE 3.36
A PDBS prepared for deployment in a well. (Photo courtesy of EON Products.)

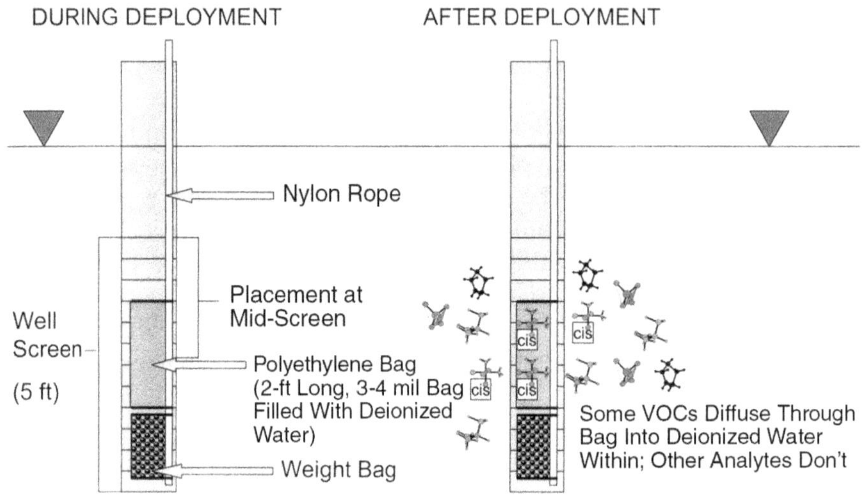

FIGURE 3.37
The PDBS is deployed in the well on a suspension cord, with a weight attached to the bottom and hung within the well screen. This sampler relies on diffusion of VOCs through the thin LDPE bag into the deionized water contained within. After a minimum 2-week equilibration period in the well, the bag is retrieved and the water within is decanted into VOC vials.

The effectiveness of PDBSs is dependent on the assumptions that there is horizontal flow through the well screen and that the water in the screen is representative of the water in the formation adjacent to the screen. PDBSs require sufficient horizontal flow to achieve chemical and hydraulic equilibrium with the formation. Some studies suggest that with sufficient flow through a properly constructed well, ground water in the well screen may be replaced in as little as 24 h (Robin and Gillham, 1987). For water in the PDBS to be representative of the VOCs in formation water, the rate of solute contribution from the aquifer to the well must equal or exceed the rate of in-well contaminant loss by processes such as volatilization or convection, which may not occur if ground-water velocities are very low or the well has a low yield resulting from low hydraulic conductivity formations or low hydraulic gradients (ITRC, 2004, 2005).

Several studies have determined that PDBSs are effective for determination of concentrations of low-solubility and low vapor pressure VOCs such as benzene, carbon tetrachloride, 1,2-dichlorothane, tetrachloroethylene, and vinyl chloride (Vroblesky, 2001a, b). They are not, however, effective for sampling to high solubility, high vapor pressure organic compounds (such as MTBE, methanol, or acetone), semi-volatile organic compounds (SVOCs), trace metals, or other inorganics that will not diffuse across the PE membrane (Vroblesky and Hyde, 1997; ITRC, 2004, 2005). Therefore, their applications are fairly limited.

These single-use bags are relatively inexpensive and are easy to deploy and retrieve from the well. One person can easily install the PDBS into a well and recover it during the next sampling event without any accessory equipment. Because the devices operate on the principle of diffusion, they are not affected by sediment or high turbidity within the water column. In situations where the objective of sampling is to monitor VOCs in more than one zone within a well, multiple samplers can be suspended in sequence to permit definition of VOC stratification in the water column. It is important to remember, however, that PDBSs are meant to be deployed in wells in which there is no vertical flow within the well. If there is vertical flow, the concentrations represented in the PDBS will

TABLE 3.3

Compounds Tested Under Laboratory Conditions for Use with PDBSs

Compounds showing good correlation (average differences in concentration of 11% or less versus control)			
Benzene	2-Chlorovinyl ether	*cis*-1,2-Dichloroethene	1,1,1-Trichloroethane
Bromodichloromethane	Dibromochloromethane	*trans*-1,2-Dichloroethene	1,1,2-Trichloroethane
Bromoform	Dibromomethane	1,2-Dichloropropane	Trichloroethane
Chlorobenzene	1,2-Dichlorobenzene	*cis*-Dichloropropene	Trichlorofluoromethane
Carbon tetrachloride	1,3-Dichlorobenzene	1,2-Dibromoethane	1,2,3-Trichloropropane
Chloroethane	1,4-Dichlorobenzene	*trans*-1,3-Dichloropropene	1,1,2,2-Tetrachloroethane
Chloroform	Dichlorodifluoromethane	Ethyl benzene	Tetrachloroethene
Chloromethane	1,2-Dichloroethane	Naphthalene	Vinyl chloride
	1,1-Dichloroethene	Toluene	Total xylenes
Compounds showing poor correlation (average differences in concentration of greater than 20% versus control)			
Acetone	Methyl-*tert*-butyl ether	Styrene	

Source: Vroblesky and Campbell, 2001. With permission.

be from water flowing vertically past the device and not from a discrete zone within the formation (ITRC, 2004).

PDBSs have a significant number of limitations and, as a result, are not widely accepted by regulatory agencies for use in ground-water sampling programs. The primary limitation is that they are suitable for only a relatively short list of VOCs (Table 3.3). They cannot be used for trace metals or other inorganics, SVOCs, pesticides, PCBs, or highly soluble VOCs. Therefore, they are only applicable to sites at which a limited number of low-solubility, low vapor pressure VOCs are of concern. If any other parameter must also be sampled at the site, a second sampling device must be used. PDBSs can introduce both negative and positive bias into samples. One source of negative bias is associated with the development of a biofilm on the surface of the bag during long-term deployment, which could reduce diffusion of some compounds through the bag. PDBSs can contribute phthalates to samples (Vroblesky, 2001a), and there is some concern that in highly contaminated ground water, the bags may degrade and contribute degradation products to samples.

PDBSs are available in two forms. One manufacturer sells PDBSs prefilled with deionized water of a known chemistry, while another sells the bags empty (to save costs associated with shipping) and requires the end user to provide the deionized water and seal the end of the device prior to deployment. This introduces a potential source of error associated with the variability in the chemistry of the deionized water used to fill the devices. This, in addition to variability in placement of the device within the water column between sampling events, can introduce imprecision when using PDBSs.

References

ASTM, Standard Guide for the Selection of Purging and Sampling Devices for Ground-Water Monitoring Wells, ASTM Standard D 6634, ASTM International, West Conshohocken, PA, 14 pp., 2006a.

ASTM, Standard Practice for Decontamination of Field Equipment Used at Nonradioactive Waste Sites, ASTM Standard D 5088, ASTM International, West Conshohocken, PA, 8 pp., 2006b.

ASTM, Standard Practice for Decontamination of Field Equipment Used at Low-Level Radioactive Waste Sites, ASTM Standard D 5608, ASTM International, West Conshohocken, PA, 8 pp., 2006c.

Backhus, D.A., J.N. Ryan, D.M. Groher, J.K. MacFarlane, and P.M. Gschwend, Sampling colloids and colloid-associated contaminants in ground water, *Ground Water*, 31(3), pp. 466–479, 1993.

Barcelona, M.J. and J.A. Helfrich, Well construction and purging effects on ground-water samples, *Environmental Science and Technology*, 20(11), pp. 1179–1184, 1986.

Barcelona, M.J., J.A. Helfrich, E.E. Garske, and J.P. Gibb, A laboratory evaluation of ground-water sampling mechanisms, *Ground-Water Monitoring Review*, 4(2), pp. 32–41, 1984.

Barcelona, M.J., J.P. Gibb, and R.A. Miller, A Guide to the Selection of Materials for Monitoring Well Construction and Ground-Water Sampling, Contract Report 327, Illinois State Water Survey, Champaign, IL, 78 pp., 1983.

Barcelona, M.J., J.P. Gibb, J.A. Helfrich, and E.E. Garske, Practical Guide to Ground-Water Sampling, Contract Report 374, Illinois State Geological Survey, Champaign, IL, 94 pp., 1985.

Barcelona, M.J., H.A. Wehrmann, and M.D. Varljen, Reproducible well-purging procedures and VOC stabilization criteria for ground-water sampling, *Ground Water*, 32(1), pp. 12–22, 1994.

Barker, J.F. and R. Dickout, An evaluation of some systems for sampling gas-charged ground water for volatile organic analysis, *Ground-Water Monitoring Review*, 8(4), pp. 112–120, 1988.

Devlin, J.F., Recommendations concerning materials and pumping systems used in the sampling of ground water contaminated with volatile organics, *Water Pollution Research Journal of Canada*, 22(1), pp. 65–72, 1987.

Gibs, J., T.E. Imbrigiotta, J.H. Ficken, J.F. Pankow, and M.E. Rosen, Effect of sample isolation and handling on the recovery of purgeable organic compounds, *Ground-Water Monitoring and Remediation*, 9(2), pp. 142–152, 1994.

Gillham, R.W., M.J.L. Robin, J.F. Barker, and J.A. Cherry, Ground-Water Monitoring and Sample Bias, American Petroleum Institute Publication, Environmental Affairs Department, Washington, DC, 206 pp., 1983.

Gillham, R.W. and S.F. O'Hannesin, Sorption of aromatic hydrocarbons by materials used in construction of ground-water sampling wells, Proceedings, ASTM Symposium on Standards Development for Ground Water and Vadose Zone Monitoring Investigations, Special Technical Publication 1053, D.M. Nielsen and A.J. Johnson, Eds., American Society for Testing and Materials, Philadelphia, PA, pp. 108–124, 1990.

Gillham, R.W., Syringe devices for ground-water sampling, *Ground-Water Monitoring Review*, 2(2), pp. 36–39, 1982.

Heidlauf, D.T. and T.R. Bartlett, Effects of monitoring well purge and sampling techniques on the concentration of metal analytes in unfiltered ground-water samples, Proceedings of the Seventh Outdoor Action Conference and Exposition, National Ground Water Association, Dublin, OH, pp. 437–450, 1993.

Ho, J.S.Y., Effect of sampling variables on recovery of volatile organics in water, *Journal of the American Water Works Association*, 75(11), pp. 583–386, 1983.

Holm, T.R., G.K. George, and M.J. Barcelona, Oxygen transfer through flexible tubing and its effects on ground-water sampling results, *Ground-Water Monitoring Review*, 8(3), pp. 83–89, 1988.

Houghton, R.L. and M.E. Berger, Effects of well-casing composition and sampling method on apparent quality of ground water, Proceedings of the Fourth National Symposium on Aquifer Restoration and Ground Water Monitoring, National Water Well Association, Worthington, OH, pp. 203–213, 1984.

Imbrigiotta, T.E., J. Gibs, T.V. Fusillo, G.R. Kish, and J.J. Hochreiter, Field evaluation of seven sampling devices for purgeable organic compounds in ground water, in *Ground-Water Contamination: Field Methods*, A.G. Collins and A.I. Johnson, Eds., Special Technical Publication 963, American Society for Testing and Materials, Philadelphia, PA, pp. 258–273, 1988.

ITRC, Technical and Regulatory Guidance for Using Polyethylene Diffusion Bag Samplers to Monitor VOCs in Ground Water, ITRC Publication, Interstate Technology and Regulatory Council, Washington, DC, 2004.

ITRC, Technical Overview of Passive Sampler Technologies, ITRC Publication, Interstate Technology and Regulatory Council, Washington, DC, 2005.

Kearl, P.M., N.E. Korte, M. Stites, and J. Baker, Field Comparison of micropurging vs. traditional ground-water sampling, *Ground-Water Monitoring and Remediation*, 14(4), pp. 183–190, 1994.

Kearl, P.M., N.E. Korte, and T.A. Cronk, Suggested modifications to ground-water sampling procedures based on observations from the colloidal borescope, *Ground-Water Monitoring Review*, 12(2), pp. 155–161, 1992.

Knobel, L.L. and L.J. Mann, Sampling for purgeable organic compounds using positive-displacement piston and centrifugal pumps: a comparative study, *Ground-Water Monitoring and Remediation*, 13(2), pp. 142–148, 1993.

Muska, C.F., W.P. Colven, V.D. Jones, J.T. Scogin, B.B. Looney, and V. Price, Jr., Field Evaluation of ground-water sampling devices for volatile organic compounds, Proceedings of the Sixth National Symposium and Exposition on Aquifer Restoration and Ground Water Monitoring, National Water Well Association, Worthington, OH, pp. 235–246, 1986.

Nielsen, D.M. and G.L. Yeates, A comparison of sampling mechanisms available for small-diameter ground-water monitoring wells, *Ground-Water Monitoring Review*, 5(2), pp. 83–99, 1985.

Nielsen, D.M. and R. Schalla, Design and Installation of Ground-Water Monitoring Wells; Chapter 10 in *The Practical Handbook of Environmental Site Characterization and Ground-Water Monitoring*, Second Edition, D.M. Nielsen, Editor, CRC Press/Taylor and Francis Group, Boca Raton, FL, pp. 639–806, 2006.

NWWA/PPI, *Manual on the Selection and Installation of Thermoplastic Water Well Casing*, National Water Well Association and Plastic Pipe Institute, Worthington, OH, 64 pp., 1981.

Oneacre, J. and D. Figueras, Ground-water variability at sanitary landfills: causes and solutions, *Proceedings, ASCE Conference on Uncertainty in the Geologic Environment*, Madison, WI, July 31–August 3, pp. 965–987, 1996.

Parker, L., Suggested guidelines for the use of PTFE, PVC and stainless steel in samplers and well casings, in *Current Practices in Ground Water and Vadose Zone Investigations*, D.M. Nielsen and M.N. Sara, Eds., Special Technical Publication 1118, American Society for Testing and Materials, Philadelphia, PA, pp. 217–229, 1991.

Parker, L., The effects of ground-water sampling devices on water quality: a literature review, *Ground-Water Monitoring and Remediation*, 14(2), pp. 130–141, 1994.

Parker, L.V. and C.H. Clark, Study of Five Discrete Interval Type Ground-Water Sampling Devices, ERDC/CRREL TR-02-12, U.S. Army Corps of Engineers, Cold Regions Research and Engineering Laboratory, Hanover, NH, 49 pp., 2002.

Parsons, Results Report for the Demonstration of No-Purge Ground-Water Sampling Devices at Former McClellan Air Force Base, California; Contract F44650-99-D-0005, Delivery Order DKO1, U.S. Army Corps of Engineers (Omaha District), Air Force Center for Environmental Excellence, and Air Force Real Property Agency, 2005.

Pearsall, K.A. and D.A. Eckhardt, Effects of selected sampling equipment and procedures on the concentrations of trichloroethylene and related compounds in ground-water samples, *Ground-Water Monitoring Review*, 7(2), pp. 64–73, 1987.

Pohlmann, K.F., G.A. Icopini, R.D. McArthur, and C.G. Rosal, Evaluation of Sampling and Field-Filtration Methods for Analysis of Trace Metals in Ground Water, EPA/600/R-94/119, U.S. Environmental Protection Agency, Office of Research and Development, Las Vegas, NV, 79 pp., 1994.

Pohlmann, K.F. and J.W. Hess, Generalized ground-water sampling device matrix, *Ground-Water Monitoring Review*, 8(4), pp. 82–83, 1988.

Pohlmann, K.F., R.P. Blegen, and J.W. Hess, Field Comparison of Ground-Water Sampling Devices for Hazardous Waste Sites: An Evaluation Using Volatile Organic Compounds, EPA/600/4–90/028, U.S. Environmental Protection Agency, Office of Research and Development, Las Vegas, NV, 102 pp., 1991.

Puls, R.W. and R.M. Powell, Acquisition of representative ground-water quality samples for metals, *Ground-Water Monitoring Review*, 12(3), pp. 167–176, 1992.

Puls, R.W., J.H. Eychaner, and R.M. Powell, Facilitated Transport of Inorganic Contaminants in Ground Water: Part I. Sampling Considerations, EPA/600/M-90/023, U.S. Environmental Protection Agency, Robert S. Kerr Laboratory, Ada, Oklahoma, 12 pp., 1990.

Puls, R.W., D.A. Clark, B. Bledsoe, R.M. Powell, and C.J. Paul, Metals in Ground Water: Sampling Artifacts and Reproducibility, *Hazardous Waste and Hazardous Materials*, 9(9), pp. 149–162, 1992.

Puls, R.W., R.M. Powell, D.A. Clark, and C.J. Paul, Facilitated Transport of Inorganic Contaminants in Ground Water: Part II, Colloidal Transport, EPA/600/M-91/040, U.S. Environmental Protection Agency, Robert S. Kerr Environmental Research Laboratory, Ada, OK, 1991.

Reynolds, G.W. and R.W. Gillham, Absorption of Halogenated Organic Compounds by Polymer Materials Commonly Used in Ground Water Monitoring, Proceedings of the Second Canadian/American Conference on Hydrogeology, National Water Well Association, Dublin, OH, pp. 125–132, 1985.

Robin, M.J.L. and R.W. Gillham, Field evaluation of well purging procedures, *Ground-Water Monitoring Review*, 7(4), pp. 85–93, 1987.

Stolzenburg, T.R. and D.G. Nichols, Effects of Filtration Method and Sampling on Inorganic Chemistry of Sampled Well Water, Proceedings of the Sixth National Symposium and Exposition on Aquifer Restoration and Ground Water Monitoring, National Water Well Association, Dublin, OH, pp. 216–234, 1986.

Tai, D.Y., K.S. Turner, and L.A. Garcia, The use of a standpipe to evaluate ground-water samplers, *Ground-Water Monitoring Review*, 11(1), pp. 125–132, 1991.

U.S. EPA, Environmental Investigations Standard Operating Procedures and Quality Assurance Manual, U.S. Environmental Protection Agency Region 4, Athens, GA, 1997.

U.S. EPA, Environmental Technology Verification Report: Geoprobe Inc., Mechanical Bladder Pump, Model MB470, EPA/600R-03/086, U.S. Environmental Protection Agency, Office of Research and Development, Washington, DC, 2003.

Unwin, J.P. and V. Maltby, Investigations of Techniques for Purging Ground-Water Monitoring Wells and Sampling Ground Water for Volatile Organic Compounds, Special Technical Publication 963, American Society for Testing and Materials, Philadelphia, PA, pp. 240–252, 1988.

Vroblesky, D.A., User's Guide for Polyethylene-Based Passive Diffusion Bag Samplers to Obtain Volatile Organic Compound Concentrations in Wells: Part 1, Development, Recovery, Data Interpretation and Quality Control and Assurance, Water Resources Investigations Report, 01-4060, U.S. Geological Survey, Reston, VA, 2001a.

Vroblesky, D.A., User's Guide for Polyethylene-Based Passive Diffusion Bag Samplers to Obtain Volatile Organic Compound Concentrations in Wells: Part 2, Field Tests, Water Resources Investigations Report, 01-4061, U.S. Geological Survey, Reston, VA, 2001b.

Vroblesky, D.A. and T.R. Campbell, Equilibration times, stability and compound selectivity of diffusion samplers for collection of ground water VOC concentrations, *Advanced Environmental Restoration*, 15(1), pp. 1–12, 2001.

Vroblesky, D.A. and W.T. Hyde, Diffusion samplers as an inexpensive approach to monitoring VOCs in ground water, *Ground-Water Monitoring and Remediation*, 17(3), pp. 177–184, 1997.

Yeskis, D., K. Chiu, S. Meyers, J. Weiss, and T. Bloom, A Field Study of Various Sampling Devices and Their Effects on Volatile Organic Contaminants, Proceedings of the Second National Conference on Aquifer Restoration, Ground Water Monitoring and Geophysical Methods, National Water Well Association, Dublin, OH, pp. 471–479, 1988.

4

Preparing Sampling Points for Sampling: Purging Methods

Gillian L. Nielsen and David M. Nielsen

CONTENTS

Objectives of Purging

Most traditional approaches to ground-water sampling are based upon the assumption that all water that resides in sampling points between sampling events is stagnant (i.e., does not interact with formation water) and does not represent the chemistry of water in the formation. Thus, the historical means of meeting the objectives of sampling programs has been to remove all of the water from the well and to induce fresh formation water to enter the well so it can be collected as a sample. This process is referred to as well purging. Several purging strategies, described in detail in the following sections, are in common use.

The SAP must specify which purging strategy or strategies will be used at a site for wells with high yield, as well as for wells with low yield. The SOPs for purging must provide step-by-step procedures that include specifying: water-level measurement requirements and methods; well-depth measurement requirements and methods; the equation used to calculate a well volume (if required); the device used for purging and guidance on placement of the device within the water column, operation of the device, and cleaning of the device if used portably; when and how to measure any required field parameters during purging; and how stabilization of field parameter measurements is defined, if required as part of the purging protocol.

It is apparent from research conducted over the last 20 years that the way in which a well is purged has one of the most significant impacts on sample quality. For example, a study conducted by Barcelona and Helfrich (1986) concluded that variations in water chemistry attributed to well purging were generally greater than errors associated with either sampling mechanism, tubing, or apparent well casing materials effects. This has, therefore, been a well known limitation of ground-water sampling practices for nearly two decades, yet little has been done to improve field methods that have persisted over the years. Traditional purging strategies have encouraged the use of portable devices, particularly grab sampling devices such as bailers, and high-speed submersible pumps.

The difficulty in collecting representative samples using traditional purging methods has been in accessing the water within the well screen (which, as discussed earlier, is representative of water in the formation screened by the well) without disturbing or mixing the water column in the well. Rather than focusing on methods that could be used to access the water in the screen directly, most purging strategies have focused on methods for removing large volumes of water from the well. Although it is important to purge some water from most wells before collecting a sample, purging too much water or purging at too high rate can cause mixing of water from zones of different quality and, potentially, contamination of noncontaminated zones by previously localized or stratified pollutants (Wilson and Rouse, 1983).

Purging strategies, as described in ASTM Standard D 6452 (ASTM, 2006a), are divided into two general categories: those appropriate for high-yield sampling points (points that do not go dry during purging and sampling) and those appropriate for low-yield sampling points (points that may go dry during purging and sampling). Low yield in a well can occur as a result of the low hydraulic conductivity of the formation screened by the sampling point, poor well construction, poor well maintenance, or insufficient sampling point volume. It is imperative that the most appropriate purging strategy is implemented on not just a site-specific basis but, in some instances, on a well-by-well basis.

Purging Strategies for High-Yield Wells

Traditional Strategies

There are three "traditional" strategies for purging high-yield sampling points: (1) removal of a fixed number of well volumes; (2) purging to stabilization of a predetermined list of field parameters; and (3) in the case of large-diameter wells or wells with large volumes of water, use of inflatable packers to physically reduce the volume of water that must be purged.

Fixed Well-Volume Purging

The most commonly applied purging strategy is to remove a fixed volume of water from the well using a bailer or a pump. Commonly, regulatory guidance mandates that three to five or four to six well volumes of water be removed prior to sample collection. A well volume may have different definitions, but it usually refers to either the total volume of water in the casing and the screen or that volume plus the volume of water contained in the filter pack.

Early research on ground-water sampling (Gibb et al., 1981) seemed to indicate that the removal of three to five well volumes of water from a well was necessary to ensure that

a representative sample of formation quality water could be collected. In hindsight, the primary reason for the need to remove this much water during the early studies was the effect that the insertion and use of the purging or sampling device had with respect to mixing of the water column in the well and the release of significant amounts of suspended sediment (turbidity). It often took the removal of three to five (or more) well volumes to negate the effects induced by the method used (Powell and Puls, 1993; Barcelona et al., 1994). As observed by Barcelona et al. (1994), there is no single number of well volumes that should be removed during purging that is best applied to all sites or that is suited to all hydrogeologic conditions. While existing rules requiring removal of a fixed volume of water are administratively convenient and easy to adhere to in theory, their field implementation often results in collection of samples of less than optimum quality that are not easily reproduced and that may not accurately represent formation water chemistry. This is because, with this approach, there is minimal concern about the effect that purging and sampling practices have on three-dimensional well hydraulics and aquifer chemistry, even though the stated objective is to collect a "representative" sample. As discussed earlier, this objective is exceedingly difficult to meet using traditional methods and, as a result, DQOs can rarely be satisfied. Imprecision in well purging and sample collection methods are common issues that call into question the validity of field and lab data generated by sampling events. Without confidence in data quality, decisions based on these data may not be scientifically or legally defensible.

Bailers or portable pumps are the devices used most often for traditional purging and sampling programs. Lowering these devices through the water column during purging causes significant mixing (Gillham et al., 1985; Robin and Gillham, 1987; Keely and Boateng, 1987a, b; Martin-Hayden et al., 1991) and release of considerable sediment (Kearl et al., 1992; Puls et al., 1992). The effects of mixing and the time for re-equilibration are directly related to purging rate, location of the pump intake, and extent of disturbance of the stagnant water column in the well (Puls and Powell, 1992). The way in which these devices have typically been used (bailing rapidly or pumping at a rate higher than the well will yield without appreciable drawdown) cause turbulent flow in the well and the surrounding formation materials and promote mixing of the water column (Figure 4.1 and Figure 4.2). The work of Puls et al. (1992) and Powell and Puls (1993) has clearly demonstrated that bailing and high-flow-rate pumping (in excess of the natural flow rate

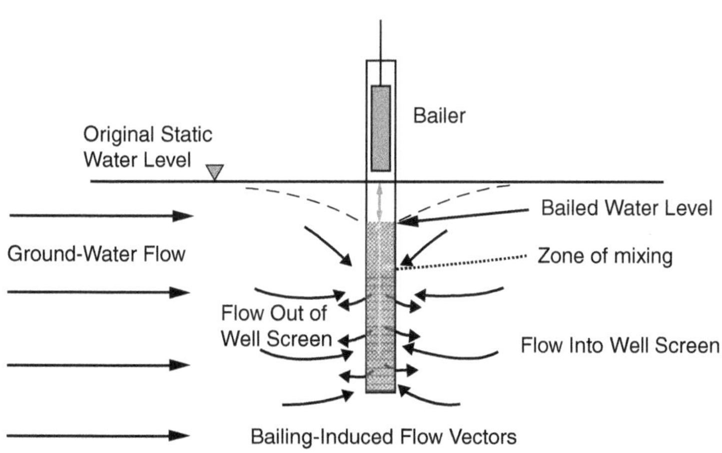

FIGURE 4.1
Conditions that occur in a well during bailing.

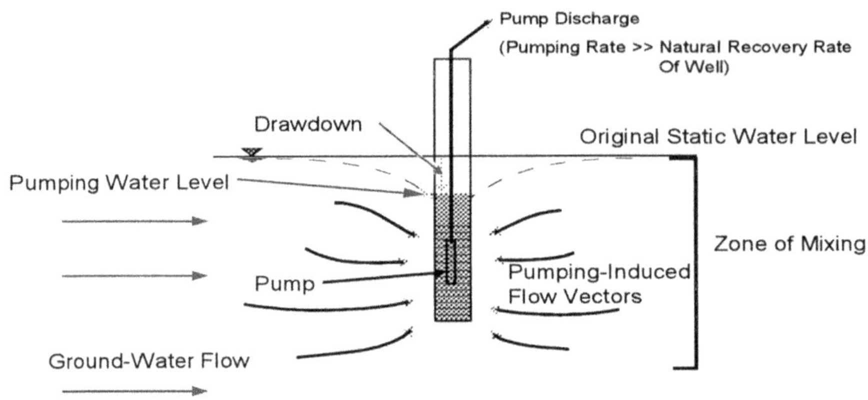

FIGURE 4.2
Conditions that occur in a well during high-flow-rate pumping.

through the screen) results in significant deleterious effects to the water sample and to the well. These problems include:

- Mixing of water from the zone of the formation targeted by the well screen with water from zones above and below the screen (or zones well beyond the borehole, in the case of fractured or solution-channeled rock [McCarthy and Shevenell, 1998]), resulting in dilution of samples or inclusion of water with constituents that are not the focus of the sampling program, and making data interpretation difficult (Robin and Gillham, 1987; Robbins and Martin-Hayden, 1991).
- Mobilization of fine-grained solids from the material surrounding the well, including the formation and the filter pack, and solids settled in the bottom of the well between sampling rounds, resulting in increased sample turbidity and gross overestimation of certain analytes of interest, particularly metals (Kearl et al., 1992; Puls and Paul, 1995). Many of these solids are artifactual (introduced to the subsurface during well construction) or immobile under natural flow conditions and mobilized from the aquifer material by shearing caused by bailing or pumping at high rates (Puls et al., 1992; Backhus et al., 1993).
- Possible dewatering of part of the screen, causing agitation, aeration, and oxidation of formation water and possibly causing precipitation of solids due to shifts in chemical equilibria, resulting in nonrepresentative samples.
- Potential for damage to the monitoring well filter pack (Barcelona et al., 1985, 1994; Pohlman et al., 1994) and disruption of the filter pack and aquifer matrix around the well that exposes fresh, reactive mineral surfaces and increases the possibility of sorption-desorption reactions (Palmer et al., 1987; Powell and Puls, 1993) along with increasing turbidity.
- Generating large amounts of purge water, especially from large diameter or deep wells, that may require management as hazardous waste, resulting in unnecessary expense and the potential for transfer of contaminants from one site to another.

The following discussion, taken from Puls and Paul (1995), illustrates a significant problem with sample accuracy resulting from the use of traditional well-volume purging methods.

> The volume of water purged, the rate at which it is withdrawn, and the location of the sampling device intake all contribute to how well the sample represents the water in the formation. The evacuation of 3 to 5 well volumes in a 200-foot deep well with a 20-foot screen in which the static water level is 20 feet below ground surface removes a large volume of water and averages a large volume of the aquifer in the water sample. Because many contaminant plumes can be narrow or thin, mixing with water from clean portions of the aquifer can provide misleading data concerning contaminant presence and concentration. If the sampling objective is a large, volume-averaged concentration of the water-bearing zone, then a consistently large sample volume, removed in a consistent manner, will generally provide reproducible values. However, if the sampling objective is to provide accurate spatial and temporal plume delineation, then alternative methods of sample collection are necessary.

Following removal of a fixed purge volume, samples are typically collected using either a bailer or a pump operated at a slow discharge rate. The device may or may not be the same device used for purging. Commonly, samples must be filtered to remove the high levels of turbidity produced by purging.

Using a fixed well-volume purging strategy, depending on what type of device is used, where it is placed in the well screen, and how it is used, there is no guarantee that all stagnant water will be removed from the well or that samples uncontaminated by the stagnant water will be obtained. Purging may be insufficient in some cases and much more than necessary in others. One common scenario observed by the authors at dozens of sites across the USA involves purging the well of three to five well volumes at a high rate with a submersible pump set within the well screen, then sampling the water at the top of the water column with a bailer. Considering the hydraulics of the well during pumping, this common practice is not likely to produce samples representative of the formation screened by the well. Most of the water pumped from the well during purging will come from the formation and, unless the water level in the well is drawn down to the intake of the pump (which can only be confirmed by measuring the water level in the well during pumping, which is rarely done), some stagnant water will remain in the well — this is very likely the water sampled by the bailer. Thus, while samplers using this method may adhere to the "letter of the law," they are not collecting samples that are representative of formation water. Only in the case in which the pump is set at the top of the water column at the start of purging and follows the water level down during purging to the point at which the water level stabilizes (and at least one well volume of water is removed from the well), could the pump truly remove all stagnant water from the well to prepare it for subsequent sampling. However, the same concerns regarding pumping at high rates and bailing noted earlier apply here.

In situations in which the static water level in the well is within the screen, removal of multiple well volumes of water is clearly unnecessary, because, as noted earlier, the water within the screen is representative of formation water. Because the top of the water column is in contact with atmospheric conditions (due to the existence of a headspace within the well), some water from the top of the water column should be purged from the well, as it may be chemically different from formation water. This can be confirmed by measurement of indicator parameters either downhole or in a flow-through cell during purging (discussed subsequently). However, application of a strict fixed-volume purging strategy in this type of situation is clearly excessive.

TABLE 4.1

Limitations of the Fixed Well-Volume Purging Strategy

No consideration of well-specific hydraulics or site-specific hydrogeology or geochemistry
No specifications on the rate of removal of water during purging, commonly resulting in hydraulically over-purging the well
No standardized definition of "well volume" can lead to errors in calculation of well volume
No measurement of water chemistry to determine, on a chemical basis, when fresh formation water has entered the well
Commonly results in removal of more water than is necessary, leading to decreased efficiency in sampling (increased time) and increased costs (labor and management and disposal of purge water)
Encourages removal of water at rates exceeding natural flow rates for the formation and well screen, which often results in high turbidity samples
Allows use of grab sampling devices and inertial-lift devices, which can severely agitate the water column and release significant turbidity

Source: Nielsen and Nielsen, 2002. With permission.

It is clear that the fixed well-volume purging approach has a number of shortcomings, as summarized in Table 4.1. Because of these limitations, the authors recommend that in most cases, use of this purging strategy should be discontinued, and other more appropriate methods (as discussed subsequently) should be used.

Purging to Stabilization of Indicator Parameters

A second approach to purging high-yield sampling points, which is less commonly applied in ground-water sampling programs, is to continuously monitor selected field water-quality indicator parameters during removal of water from the well. The indicator parameters to be measured and the frequency of measurements should be specified in the SAP. The most commonly measured parameters include (but are not limited to) pH, conductivity (or specific conductance), temperature, DO, and ORP. Parameters measured for any given sampling program should be selected based on knowledge of site-specific water chemistry, analytes of interest (and their relationship to indicator parameters), and any specific regulatory requirements. Frequency of measurement should be based on purging rate and method used for measurement (i.e., hand-held single- or multiparameter instruments versus sensors in a flow-through cell). The acceptable variation of parameter values to define stabilization and the minimum number of consecutive stable readings within the prescribed variation for each indicator parameter should be defined in the SAP. When the selected parameters stabilize, regardless of the volume of water removed, it is presumed that all stagnant water has been removed from the well and that fresh formation water is available for sampling. As in the fixed-volume strategy described earlier, this strategy permits use of grab samplers and portable pumps, in addition to dedicated pumping systems. The limitations of this purging strategy are summarized in Table 4.2.

Ideally, this purging method is implemented using a dedicated or portable pumping device in conjunction with a flow-through cell. A flow-through cell is of particular importance if any of the parameters monitored to determine water chemistry stabilization are sensitive to contact with ambient air, such as DO or ORP. Using a pumping device, rather than a grab sampling device, will ensure less disturbance of the water column and reduced turbidity, meaning that stabilization will be reached more quickly.

Use of a Packer During Purging

The third traditional method of purging high-yield wells is a strategy in which the volume to be purged from the well is reduced using a device referred to as a packer. A packer is an

TABLE 4.2

Limitations of Purging to Stabilization of Field Indicator Parameters

Purging rates are not controlled by, nor are they specific to, formation or well hydraulic conditions
Stabilization of selected field parameters does not necessarily reflect stabilization of contaminant chemistry (Gibs and Imbrigiotta, 1990)
Temperature and pH are not always reliable indicator parameters
Many potential sources of error and bias are associated with field parameter measurement (i.e., measuring DO or ORP in an open container)
Inconsistency in defining stabilization criteria
Method permits the use of grab samplers and inertial-lift devices for purging, which can make it very difficult to achieve stabilization for parameters sensitive to aeration and agitation of the water column in the well

Source: Nielsen and Nielsen, 2002. With permission.

expandable device placed within the well casing or competent rock just above the top of the well screen or open zone that, when deployed, physically isolates the water above the packer from the water below the packer. This negates the need to remove the water from above the packer and the need to purge multiple volumes of water from the well prior to sampling. When in use, a pump or pump intake is suspended below the packer (within the screen), to purge and sample the water within the screen without having to remove the overlying stagnant water in the casing (Figure 4.3). Packers are ineffective in isolating sampling zones when installed in the well screen or open zone. Water levels for use in piezometric surface mapping must be taken prior to placement of the pump or packer assembly to avoid bias. Water levels may also be taken in the water column above the packer during pumping to check for leakage of water from within the well casing past the packer; a measurable water-level drop during pumping indicates leakage.

Packers can be used in conjunction with any pumping device (Figure 4.4) in any well in which the static water level is above the top of the well screen, but are most efficient in large-diameter or deep wells in which the water level is significantly above the top of the well screen. Because the packer physically isolates the water within the well intake from

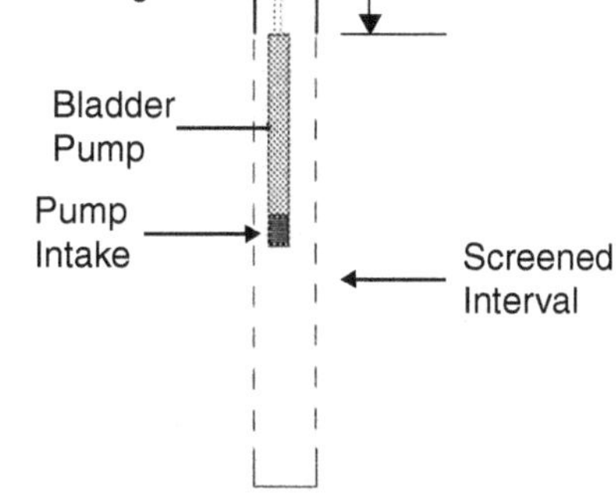

FIGURE 4.3
Placement of a packer to physically isolate the water column in the casing from the water column in the screen.

FIGURE 4.4
A packer used in conjunction with a progressing cavity pump, prior to installation.

the stagnant water in the casing, it is effective in minimizing purge volume and reducing the costs associated with management of purge water. Bailers cannot be used to sample when packers are used, and packers are not practical for use in low-yield wells, in which the open interval will dewater during purging.

Low-Flow Purging and Sampling

Researchers have evaluated the shortcomings of traditional purging methods and determined that each method imparts some type of error and bias, which can influence the representative nature of samples. An improved method of purging wells evolved out of research conducted to improve sample accuracy and precision and to encourage the collection of more representative samples. This method of purging is known as low-flow purging and sampling (also referred to as micropurging, low-stress purging, low-impact purging, or minimal-drawdown purging).

Unlike traditional purging methods, low-flow purging and sampling does not require the removal of large volumes of water from the well. The amount of water purged will vary with well diameter, but is typically less than one half of a well volume (Barcelona et al., 1994) and, in the authors' experience, often less than one third of a well volume. Some investigators have successfully sampled wells after purging as little as one to two times the volume of the sampling system, which includes the pump and discharge tubing (Shanklin et al., 1995). The actual volume purged prior to sampling depends on the time required for water-level stabilization and indicator parameter stabilization to occur and the pumping rate used over that time period, which is specific to each well.

Low-flow purging differs from traditional methods of purging in that its use is based on the observations of many researchers (cited in Chapter 1) that water moving through the formation also moves through the well screen. Thus, the water in the screen is representative of the formation water surrounding the screen. This assumes that the well has been properly designed, constructed, and developed as described in ASTM Standards D 5092 and D 5521 (ASTM, 2006b, 2006c). In wells in which the flow through the screen or intake zone is limited by hydraulic conductivity contrasts (e.g., borehole smearing, residual filter cake, filter pack grain size, or well screen open area), the head difference induced by low-flow pumping provides an exchange of water between the formation and the well. Samples collected during low-flow purging and sampling represent the water from the entire screened zone, and the chemistry of the sample represents a weighted average of the concentrations of solutes in the water in the screened interval (Martin-Hayden and Robbins, 1997; Puls and Paul, 1997). The effects of heterogeneities in geologic material screened by the well may change the contributions of various zones to the average, but do not change the overall effect of concentration averaging.

Low-flow purging involves removing water directly from the screened interval without physically or hydraulically disturbing the stagnant water column above the screen. This is done using a dedicated pump (or by very carefully installing a portable pump) with the pump intake set at or near the middle of the screen and pumping the well at a low enough flow rate to maintain a stabilized water level in the well as determined through water-level measurement during pumping. Typically, flow rates on the order of 0.1 to 0.5 l/min are used; however, this is dependent on site-specific and well-specific factors (Puls and Barcelona, 1996). Some very coarse-textured formations have been successfully purged and sampled in this manner at flow rates up to 1 l/min. Pumping water levels in the well and water-quality indicator parameters (such as pH, temperature, specific conductance, DO, and redox potential) are monitored during pumping, with stabilization indicating that purging is completed and sampling can begin (Figure 4.5 and Figure 4.6). Purging at a rate that minimizes drawdown will generally reduce purge volumes and time required to reach stabilization of indicator parameters.

"Low-flow" refers to the velocity that is imparted during pumping to the formation pore water adjacent to the well screen. This velocity must be minimized to avoid turbulent flow through the well screen and to preclude the entrainment of artifactual particulate matter in the water to be collected as a sample. Low-flow does not necessarily refer to the flow rate of water discharged by a pump at the surface, which can be affected by valves, restrictions in the tubing, or flow regulators. "Low stress" or "low impact" refers to the impact of pumping the well on formation hydraulics. Water-level drawdown

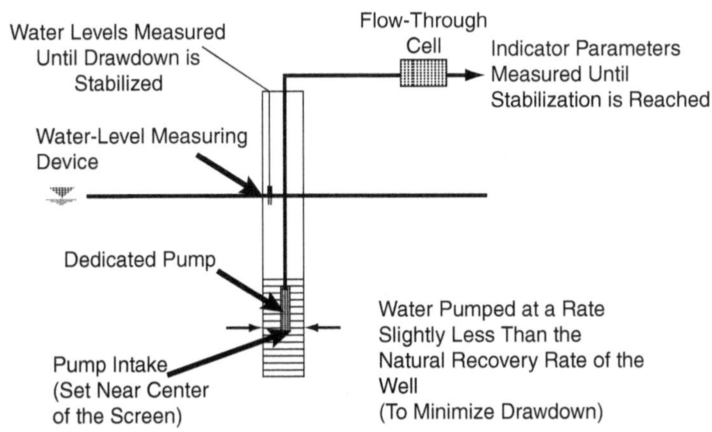

FIGURE 4.5
Illustration of the key components of low-flow purging and sampling.

FIGURE 4.6
The equipment setup for low-flow purging and sampling includes a pump (in this case, a dedicated bladder pump); a pump controller with drive gas (left); a flow-through cell (center) with sensors for measuring pH, temperature, specific conductance, DO and ORP, and a water-level gauge (right) for measuring drawdown during pumping.

provides a measurable indication of the stress or impact on a given formation imparted by a pumping device operated at a given flow rate. The objective of low-flow purging is to pump in a manner that minimizes stress or disturbance to the ground-water system to the extent practical. A stabilized water level in a well (not necessarily achieving a particular drawdown value) indicates that water subsequently pumped from the well is derived directly from the formation. Because the flow rate used for purging is, in many cases, the same as or only slightly higher than the flow rate used for sampling and because purging and sampling are conducted as one continuous operation in the field, the process is referred to as low-flow purging and sampling.

The most critical aspects of low-flow purging and sampling are summarized in Table 4.3. Low-flow purging requires the use of a pump. Grab samplers, such as bailers, and inertial-lift devices disturb the water column in the well and the formation and cannot reliably provide a representative sample. While dedicated pumps are preferred because they minimize disturbance to the well, portable pumps can be used with some precautions. Portable pumps must be installed carefully and lowered slowly into the screened zone to minimize disturbance of the water column. Even if done with the utmost care, the installation of a portable pump may result in some mixing of the water column above the well screen with that within the screened interval and the release of substantial suspended material (Kearl et al., 1992, 1994; Puls et al., 1992). This usually requires pumping for a longer period of time to achieve stabilization of water-quality indicator parameters and turbidity. Ideally, the pump should remain in place until any turbidity resulting from pump installation has settled and until horizontal flow through the well screen has been reestablished. Carefully lowering the pump into the well, then completing preparation of other equipment and materials to be used in the sampling event, often allows sufficient time for reduction of initial turbidity to acceptable levels. If, after the pump is started, initial turbidity readings are high (>100 NTU), it may be

TABLE 4.3

Summary of the Critical Aspects of Low-Flow Purging and Sampling

Requires the use of dedicated or portable pumps capable of operating at low speeds and low discharge rates (less than 1 l/min for most applications)
– Dedicated pumps are preferred. Portable pumps are appropriate for use, but must be used with some precautions
Bailers, inertial-lift pumps, and high-flow-rate pumps cannot be used for low-flow purging and sampling
Requires knowledge of well construction details, particularly information related to well screen placement and length
Pump intake should be set near the middle of the well screen or adjacent to the zone of highest hydraulic conductivity screened by the well (if it can be identified)
Pumping rate during purging must be equal to or less than the natural recovery rate of the well (minimizing drawdown in the well and allowing drawdown to stabilize prior to sampling)
Requires monitoring of water levels (to determine when the pumping water level has stabilized)
Requires monitoring of field water-quality parameters (to determine when formation water is being sampled)

Source: Nielsen and Nielsen, 2002. With permission.

necessary to stop the pump and allow turbidity to settle. The time required for turbidity to settle is well-specific and should be determined on a well-by-well basis.

In low-flow purging, the pump intake should be set either near the middle of the well screen (Figure 4.5), adjacent to the zone within the screen that has the highest hydraulic conductivity, or adjacent to the zone within the screen that has the highest level of contamination (if either of these are known). The well is pumped at a low flow rate, equal to or less than the natural recovery rate of the formation. The pumping flow rate used for any given well is dependent on the hydraulic performance of the well and the site-specific hydrogeology. A volume measuring device (e.g., a graduated cylinder or a container of known volume) and a time piece capable of measuring in seconds is necessary to calculate the flow rate from the discharge tube from the pump.

During low-flow purging and sampling, the water level in the well is continuously monitored using any water-level measurement equipment that does not disturb the water column in the well and that has an accuracy required by the sampling program (generally $\pm$0.01 ft). Devices such as downhole pressure transducers or bubblers (for continuous measurements) or electronic water-level gauges (for periodic measurements) can be used effectively. Water-level measurements should be taken every 1 to 2 min to the point at which the water level in the well has stabilized or at which drawdown ceases. The purging rate is adjusted so drawdown in the well is minimized and stabilized (does not increase over time). This, in effect, hydraulically isolates the column of stagnant water in the well casing from the water in the well screen and negates the need for its removal (Barcelona et al., 1985; Gillham et al., 1985; Robin and Gillham, 1987; Maltby and Unwin, 1992; Powell and Puls, 1993; Nielsen, 1996). After drawdown stabilizes, all water drawn into the pump must come from the formation and water-level measurement can be discontinued. In situations where the sampling rate is less than the purging rate, the well should be recharged during sampling (i.e., drawdown will be decreasing with time), further ensuring that the water sampled is from the formation.

Several researchers have proposed limits on the amount of drawdown that should be allowed before water-level stabilization occurs. In all cases, the proposed limits are arbitrary and no scientific rationale is provided for their adoption. For example, Puls and Barcelona (1996) proposed a limit of less than 0.1 m (0.33 ft or about 4 in.) drawdown for all wells, conceding that this goal may be difficult to achieve under some conditions due to geologic heterogeneities within the screened interval and may require adjustment based on site-specific conditions and personal experience. It should be emphasized that it is far more important to achieve a stabilized water level in a well during purging than to achieve a particular drawdown value, as each well is in a hydrogeologically unique position and thus will respond differently to pumping. In practical terms, to avoid the possibility of drawing stagnant water from the well casing into the pump intake, drawdown should not exceed the distance between the top of the well screen and the pump intake.

In addition to continuous water-level monitoring, low-flow purging and sampling requires continuous or periodic measurement of selected water-quality indicator parameters. The well is considered purged and ready for sampling when the chosen chemical and physical indicator parameters (commonly pH, temperature, specific conductance, and dissolved oxygen, sometimes also including redox potential [ORP] and turbidity) have stabilized, confirming that formation water is being pumped. Continuous monitoring in a closed flow-through cell of known volume generally provides the most consistent and reliable results, especially for DO and ORP, and is the preferred method of measuring indicator parameters. However, individual hand-held instruments designed to measure the most common water-quality indicator parameters (temperature, pH, and conductivity or specific conductance) or turbidity may also be used. DO and ORP measurements made after the purged water is exposed to atmospheric conditions, however, will not accurately reflect *in situ* conditions. All instruments used to measure indicator parameters and turbidity should be properly calibrated and maintained in accordance with manufacturers' instructions at the wellhead at the start of each day of sampling, and calibration should be checked periodically throughout the sampling event. Additional information on field parameter measurement follows later in this chapter.

Low-flow purging and sampling offers a number of benefits over traditional methods, including (ASTM, 2006d):

- Improved sample quality and reduced (or eliminated) need for sample filtration, through minimized disturbance of the well and the formation, which result in greatly reduced artifactual sample turbidity and minimization of false positives for analytes associated with particulate matter
- Improved sample data accuracy and precision and greatly reduced sample variability as a result of reduced stress on the formation, reduced mixing of the water column in the well and dilution of analytes, and reduced potential for sample agitation, aeration, and degassing or volatilization
- Samples represent a smaller section or volume of the formation, representing a significant improvement in the ability to detect and resolve contaminant distributions, which may vary greatly over small distances in three-dimensional space (Puls and McCarthy, 1995)
- Overall, improved sample reproducibility, especially when using dedicated pumps (Karklins, 1996)
- Improved ability to directly quantify the total mobile contaminant load (including mobile colloid-sized particulate matter) without the need for sample filtration (Puls and Barcelona, 1989; Puls et al., 1991, 1992; Puls and Powell, 1992a)
- Increased well life through reduced pumping stress on the well and formation, resulting in greatly reduced movement of fine sediment into the filter pack and well screen

- Greatly reduced purge-water volume (often 90 to 95%), resulting in significant savings of cost related to purge water handling and disposal or treatment, and reduced exposure of field personnel to potentially contaminated purge water
- Reduced purging and sampling time (much reduced at sites using dedicated pumps), resulting in savings of labor cost, depending on the time required for water-quality indicator parameters to stabilize

In addition, this purging technique can be applied to any well that can be pumped at a constant, low flow rate without continuous drawdown of the water level and it can be used in wells where the water table is above or within the well screen. It can also be effectively applied to bedrock well completions.

Although the application of low-flow purging and sampling will improve sampling results and produce significant technical and cost benefits at most sites, not all sites, and not all individual wells within a site, are well suited to this approach. It cannot be applied properly without consideration of site-specific hydrogeology and well-specific hydraulic performance. On a practical basis, low-flow purging and sampling is generally not suitable for use in very low-yield wells (e.g., those that will not yield sufficient water without continued drawdown with pumping over time). As discussed previously, low-flow purging cannot be performed using bailers, or inertial-lift devices, which severely agitate the water column in the well, resulting in significant mixing of the water column and release of considerable sediment, which shows up as increased turbidity in samples.

Low-flow purging and sampling is appropriate for collection of ground-water samples for all categories of aqueous phase contaminants and naturally occurring analytes. This includes VOCs and SVOCs, metals and other inorganics, pesticides, PCBs, other organic compounds, radionuclides, and microbiological constituents. It is particularly well-suited for use where it is desirable to sample aqueous-phase constituents that may sorb or partition to particulate matter. This method is not applicable to the collection of either LNAPLs or DNAPLs.

Four peer-reviewed field studies have been conducted in which analysis of samples collected by low-flow methods and carefully employed conventional methods (well-volume purging) were statistically compared (Powell and Puls, 1993; Kearl et al., 1994; Shanklin et al., 1995; Serlin and Kaplan, 1996). In each case, the authors reported no significant differences for a variety of analytes, including several VOCs, radionuclides, metals, trace metals, and other inorganics. This indicates that at the very least, the methods produce equivalent results.

In other peer-reviewed field studies, ground-water samples collected using low-flow purging and sampling have demonstrated dramatically lower concentrations of a variety of analytes associated with turbidity. Backhus et al. (1993) and Groher et al. (1990) reported levels of PAHs two to three orders of magnitude lower in low-flow samples than in bailed samples from the same wells. Bangsund et al. (1994) reported levels of some metals (Al and Fe) one to three orders of magnitude lower and dioxins or furans one to three orders of magnitude lower in samples collected using low-flow sampling techniques than in samples collected using traditional purging and sampling (well-volume purging and sampling with a bailer). Several researchers (Puls and Barcelona, 1989; Puls et al., 1992; Hurley and Whitehouse, 1995; Puls and Paul, 1995; McCarthy and Shevenell, 1998) have observed a direct relationship between flow rate (or bailing), turbidity, and metals content in unfiltered samples and a strong inverse correlation between turbidity and sample representativeness. In all instances, wells that had previously produced samples containing significant turbidity (>500 NTUs) using traditional purging and sampling protocols (bailing or high-flow-rate pumping) and produced samples with very low turbidity (typically <10 NTUs) using low-flow purging and sampling. In some cases, suspended solids levels two to three orders of magnitude lower were noted in low-flow samples (Puls et al., 1992; Backhus et al., 1993). This consistently results in samples with much lower concentrations of metals that more accurately reflect true ground-water conditions.

Some ground-water samples collected using low-flow methods may contain higher concentrations of some dissolved species than samples collected using traditional purging and sampling methods, primarily because sampling is more targeted to a specific zone, and samples do not exhibit strong effects of mixing and dilution. The natural tendency for higher horizontal hydraulic conductivity than vertical hydraulic conductivity in most geologic materials tends to reduce the amount of interzonal mixing if a lower pumping rate is used (Wilson and Rouse, 1983). Graham and Goudlin (1996) documented levels of several VOCs (including benzene, toluene, ethylbenzene, and xylenes [BTEX]; 1,1-dichloroethane; 1,2-dichloropropane; and 1,1,2-trichloroethane) that were 2 to 14 times higher in samples collected using low-flow methods than in samples collected using well-volume purging and sampling with bailers. This was apparently due to the lack of mixing between the water in the well screen (which represents formation-quality water) and the water in the well casing (which resulted in dilution when sampling with a bailer). Low-flow sampling in this instance provided more representative samples because it allowed collection of water from only the screened interval, which represented the higher levels of contamination that were the original targets of well screen placement and sampling.

The potential changes in concentrations of target analytes that could result from applying low-flow purging and sampling to wells that were previously sampled via conventional methods may cause some difficulties in data comparison and interpretation of temporal trends. With recognition of the possible changes in contaminant concentration noted earlier, these difficulties are easily overcome. However, anticipation of these difficulties has been offered by some samplers as a reason to continue sampling using traditional methods. Continuing the use of inappropriate sampling methods for the sake of maintaining consistency is not a valid argument; if questionable or bad data were collected before, collection of good data becomes even more important. The cost of making incorrect decisions based on poorly collected samples is much greater than the cost of sampling correctly. It must be recognized that all decisions based on sample data of questionable value will themselves be questionable. The extra time and cost of carefully collecting samples that provide estimates of contaminant concentrations must be weighed against the cost of making a bad decision based on inaccurate estimates. All efforts in sampling ground water should be directed toward collecting the highest quality samples possible, to ensure that decisions based on sampling results are scientifically and legally defensible.

Purging and Sampling Strategies for Low-Yield Wells

At some facilities, such as landfill sites, it is necessary to install sampling points in fine-grained formations, such as silts and clays, which have inherently low hydraulic conductivity. To date, monitoring well construction and development technologies and standards have not been developed to overcome the problems of sampling from low-yield formations (refer to Nielsen and Schalla [2006] for additional information). The approach to purging and sampling wells installed in these formations must be different than approaches used for high-yield wells, as these wells typically run dry prior to removal of a specified number of well volumes. In sampling points with very low yield, even low-flow purging and sampling may not be appropriate because excessive water-level drawdown may result in increased soil particle or colloidal transport into the well (Sevee et al., 2000). As a consequence, in relatively low hydraulic conductivity materials, a sample collected directly from the well screen, without excessively pumping and dewatering the sampling point, will provide the most representative sample of formation-quality water possible.

Regulatory requirements for purging and sampling low-yield monitoring wells are highly variable, but most involve removal of all water from the well during purging, then sampling upon recovery of the well, because these wells will not yield multiple well volumes in a reasonable time. To assist in determining which purging and sampling strategy is most appropriate to use in low-yield wells, some guidance defines "low yield" and "high yield." For example, in current ground-water sampling guidance for San Diego County, California, definitions are provided for both of these terms as follows:

- *Slow Recharging Well [Low-Yield Well]*
  "A well is considered to be slow recharging if recovery to 80% of its static condition takes longer than two hours."
- *Fast Recharging Well [High-Yield Well]*
  "A well is considered to be fast recharging if recovery to 80% or more of its static condition occurs within two hours."

Traditional Strategies

Two traditional purging and sampling strategies have been used in low-yield wells: (1) purging to dryness and sampling during well recovery and (2) purging to the top of the well screen and sampling the water in the well screen. Each of these strategies imparts bias on samples collected following purging; however, they continue to be the most widely used purging strategies implemented in the field for low-yield wells.

Purging to Dryness

The most commonly prescribed strategy for purging and sampling low-yield wells is to purge the well to dryness and then sample during or following well recovery. Examples of sampling strategies following well dewatering are summarized in Table 4.4. The most significant problem encountered in low-yield wells is that dewatering of the well screen (Figure 4.7) results in potentially significant chemical alteration of the water eventually collected as a sample, including the following:

- The time required for sufficient recovery of the well may be excessive, affecting sample chemistry through prolonged exposure of the water in the formation surrounding the well to atmospheric conditions. As Puls et al. (1991) point out, the oxidation reactions that occur under these conditions can dramatically alter sample results for a variety of inorganic analytes. Perhaps the most significant effect is the loss of VOCs from samples collected after recovery of the water level, in some cases, exceeding 70% in only a few hours (McAlary and Barker, 1987; Herzog et al., 1988).
- Purging the well dry causes a significant increase in the hydraulic gradient in formation materials surrounding the well (Figure 4.8 and Figure 4.9), increasing the flow velocity toward the well and resulting in turbulent flow in the formation and filter pack immediately adjacent to the well and mobilization of formation

TABLE 4.4

Traditional Strategies for Collecting Samples Following Well Dewatering

Sample after a defined period of time (e.g., 2, 4, 24 h)
Sample after a defined recovery of the water level to a percentage of the initial static water level (e.g., 80% of static)
Sample after a defined volume of water has re-entered the well (e.g. enough water has recharged to permit filling all sample containers required)

Source: Nielsen and Nielsen, 2002. With permission.

fines, and creating increased turbidity in water entering the well (Giddings, 1983). When this turbid sample is acidified (following standard chemical preservation techniques for metals), metallic ions that were previously adsorbed onto or contained within the suspended clays and silts are released, causing elevated concentrations of metals in the sample. This renders the sample useless for detection of some metals and may also bias other general chemistry parameters.

- Purging the well dry may cause cascading of water as the well recovers (Figure 4.9), resulting in a change in dissolved gases and redox state, and ultimately affecting the concentration of the analytes of interest through the oxidation of dissolved metals and loss of VOCs.
- Draining water from the filter pack surrounding the screen can result in air being trapped in the pore spaces, with lingering effects on dissolved gas levels and redox state.
- In some cases, the well may not recover sufficiently to produce the sample volume required within a reasonable time period.

Thus, while rules requiring complete removal of water from the well during purging are administratively convenient and relatively easy to adhere to, it is clear that this approach results in the collection of samples that are not representative of formation water chemistry. Some state regulatory agencies have recognized this and suggest

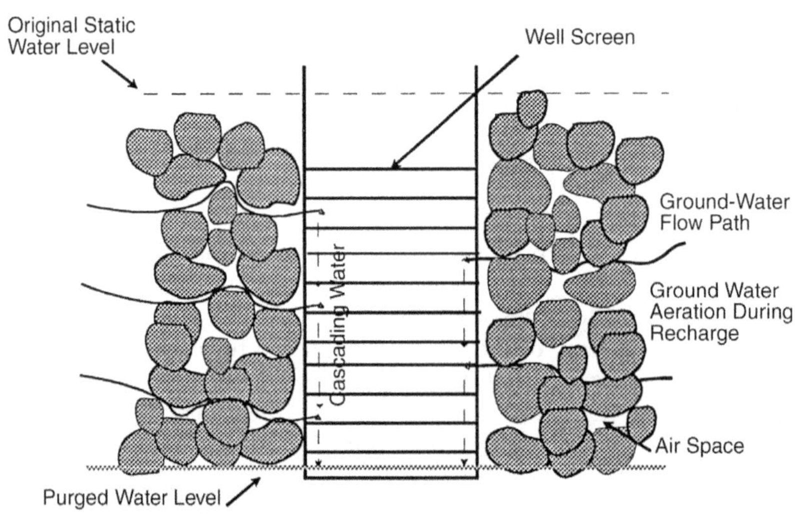

FIGURE 4.7
Well-screen dewatering can cause significant changes in water chemistry for a number of parameters including those associated with artifactual turbidity (i.e., hydrophobic organics, and trace metals) and those affected by exposure to atmospheric air (i.e., VOCs, dissolved gases, and trace metals).

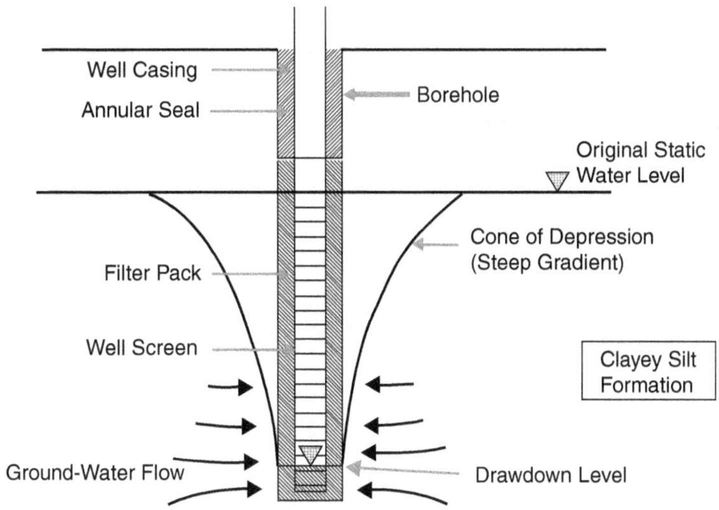

FIGURE 4.8
Purging a low-yield well (in this case, installed in an unconsolidated formation material) often dewaters the well and creates a very steep gradient toward the well. This causes turbulent flow in the formation (at the pore-space level), which results in the mobilization of sediment from the formation that shows up as turbidity in samples.

avoiding pumping wells to dryness, because of the many deleterious effects that this approach has on sensitive analytical parameters (Thurnblad, 1995).

Another problem in sampling low-yield wells is that the approach to determining the volume of water to be removed from a low-yield well prior to sampling is not consistent between regulatory jurisdictions. In some locations, samplers are still required to remove three to five well volumes prior to sample collection. This represents a misapplication of a purging strategy intended for high-yield wells and can result in a tremendous increase in

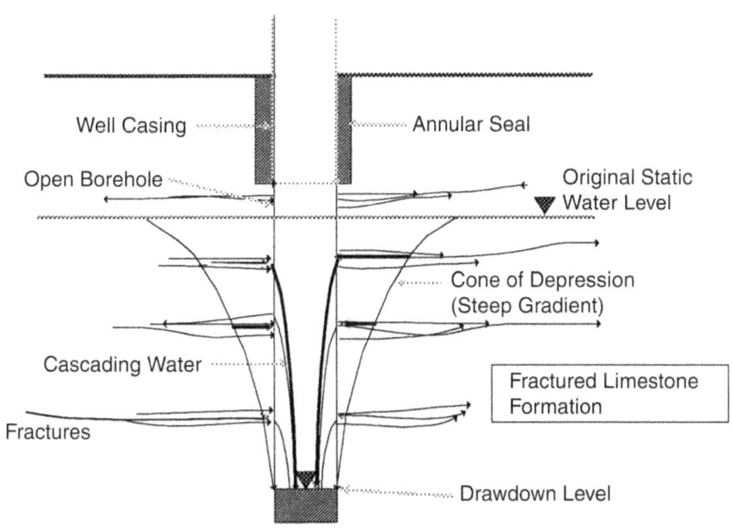

FIGURE 4.9
Purging a low-yield well (in this case, installed in a fractured bedrock) often dewaters the well and causes a very steep gradient toward the well. Discrete flow through fractures can carry sediment into the well and can also cause cascading of water into the well (in this case, an open bedrock borehole), which results in significant aeration and agitation as the water streams down the open borehole.

time and expense, with no benefit with respect to ensuring collection of representative samples. It can often require several days to sample a single well using this approach.

Purging Water from the Casing Only

To avoid the problems associated with dewatering the well screen, an alternative approach to purging low-yield wells that is less commonly applied is to remove only that water in storage in the well casing (i.e., only the water above the well screen), then begin sampling after this stagnant water has been removed. This requires the sampler to know the depth to and length of the well screen, to use a purging device that permits accurate and precise placement of the intake relative to the top of the well screen, to measure water levels in the well during purging (to assess drawdown), and to ensure that the water column in the casing and screen is not agitated or mixed during purging and sampling. This precludes the use of bailers or inertial-lift pumps. Using this purging method, the water in storage in the casing is typically removed by pumping from the top of the water column and following the water level down to the top of the screen. Once water in storage is removed, samples can be collected directly from the screened portion of the well, usually using the same device used for purging. The water in the screen comprises the most representative sample that can be collected from a low-yield well.

Other Methods for Purging and Sampling Low-Yield Wells

There are several better alternatives to using these traditional approaches to purge and sample low-yield wells. Two alternative methods that should be considered for use in these wells include (1) minimum-purge sampling (also known as passive sampling) and (2) use of a device to "core" the water column in the well screen without disturbing the water column.

Minimum-Purge Sampling

Minimum-purge sampling or passive sampling, as described in Puls and Barcelona (1996), Powell and Puls (1993), and Nielsen and Nielsen (2002), appears to offer one of the best alternatives for collecting samples from low-yield wells with minimal alteration of formation water chemistry (Figure 4.10). Minimum-purge sampling generally requires

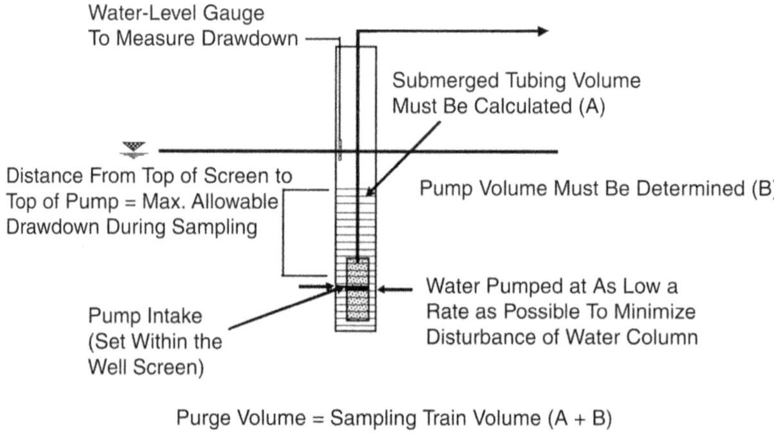

FIGURE 4.10
Illustration of the concepts behind minimum-purge or passive sampling.

the use of a dedicated pump with the pump intake installed near or below the middle of the well screen. It involves the removal of the smallest possible volume of water prior to sample collection. This volume is generally limited to the volume of water in the sampling system, including the volume of water in the dedicated pump chamber and the volume of water in the discharge tubing submerged below the static water level in the well. Immediately following removal of this small volume, samples are collected, with the expectation that the water pumped immediately after evacuation of the sampling system (i.e., the water from the screened zone) represents formation water quality. Flow rates used for minimum-purge sampling are about the same as or slightly lower than those used for low-flow purging and sampling — generally 100 ml/min or less. Because very low hydraulic conductivity formations do not yield sufficient water to satisfy the demands of the pump even at these low flow rates, drawdown cannot be avoided. Thus, to determine the volume of water available for sampling, it is necessary to calculate the volume of water within the well screen above the pump intake. Only this volume should be collected as the sample, and sampling must be discontinued once drawdown has reached a prescribed level in the well relative to the pump intake.

The minimum-purge sampling method consists of collecting only the water in the screened zone, thus avoiding the pitfalls of complete evacuation of the well and providing a better opportunity to collect samples that reflect *in situ* water quality. Although the low flow rate of water through the well screen provides only a limited exchange of water with the formation, avoiding the sources of error and bias associated with dewatering the well screen and disturbing the water column is of greater importance.

Minimum-purge sampling can be applied to any well in which there is sufficient water to ensure submergence of the pump intake throughout purging and sample collection, although it is most often applied to wells installed in very low hydraulic conductivity formations. Minimum-purge sampling is appropriate for collection of samples for all categories of naturally occurring analytes and aqueous-phase contaminants including VOCs, SVOCs, metals and other inorganics, pesticides, PCBs, other organic compounds, radionuclides, and microbiological constituents. However, it cannot be used to collect samples of either LNAPLs or DNAPLs.

Minimum-purge sampling differs from low-flow purging and sampling in that while low-speed portable pumps can be used for low-flow purging, they are impractical in most situations where minimum-purge sampling could be applied. Because disturbance of the water column can result in significant mixing of the stagnant water column in the casing with the water column in the well screen, as well as the release of significant turbidity, if a portable pump is used, the pump must be set in the well long before sampling is planned. Generally, the pump must be placed far enough ahead of the time of sampling so that the effect of pump installation has completely dissipated prior to sample collection. The time required will vary from well to well, but may be in excess of 48 h (Kearl et al., 1992; Puls and Barcelona, 1996). For this reason, dedicated pumps are the most practical devices used for nearly all minimum-purge sampling applications. Under no circumstances can grab sampling devices, such as bailers or inertial-lift pumps, be used for minimum-purge sampling.

In minimum-purge sampling, placement of the pump intake depends more on the sample volume required to satisfy the objectives of the sampling program than on other factors. The pump intake should be set within the well screen, but not too close to the bottom of the screen (to avoid drawing in sediment that may have settled in the bottom of the well) or too close to the top of the screen (to avoid incorporating stagnant water from the casing in the sample). Two to three feet above the bottom of the screen is generally sufficient. Because the volume that can be collected as a sample is, for practical purposes, limited to the volume available within the well screen (minus the volume displaced by the pump), it is first necessary to calculate the volume available for pumping and compare that to the volume required for samples — this will help determine the optimum pump setting. To provide a safety factor against including stagnant water from the casing in the sample, drawdown during sampling should be limited to about 2 ft less than the distance between the top of the screen and the pump intake as illustrated in the following example:

Example

In a 2 in. nominal diameter well with a 10 ft long well screen (in which the static water level is above the top of the well screen), if the pump intake (at the bottom of the pump) is set at 2 ft above the bottom of the screen, and the pump is 1 ft long and 1.75 in. O.D., the allowable drawdown is 6 ft, and the effective volume of water available for removal is that within 5 ft of the screen (accounting for the volume displaced by the pump). For a 2 in. well, that volume is 0.16 gal/ft × 5 ft or 0.80 gal (3 l).

In this example, as long as volume requirements for the sampling program are equal to or less than 0.8 gal (3 l), the pump setting will be satisfactory. If additional sample volume appears necessary, the pump should not be set lower in the screen, to avoid the possibility of drawing in sediment that may have settled in the bottom of the screen. Rather, the sampling team should ask the laboratory if they can get by with less volume for each parameter (which is nearly always the case) to ensure that the volume available is sufficient to satisfy sampling and analytical requirements.

With minimum-purge sampling, there is no requirement to monitor indicator parameters to determine chemical stabilization of the well during purging due to the low volume of water collected. Therefore, equipment such as flow-through cells and field parameter instrumentation is not necessary to implement minimum-purge sampling, although some field parameter measurements may still be required by regulatory agencies for other purposes.

"Coring" the Water Column in the Well Screen

A device known as the HydraSleeve (Figure 4.11), a low-cost, disposable grab-sampling device, allows samplers to capture a "core" of water from any discrete interval within the screened portion of the sampling point without purging the well prior to sample collection. During use of this device, there is no change in water level and minimal disturbance to the water column in the well. The device consists of a flexible, lay-flat, PE sleeve at the top of which is a flexible top-loading check valve. Although the standard size of the device is 1.75 in. in diameter and 30 in. in length (holding a volume of about 1 l when full), it can be made in a variety of different lengths and diameters to meet project-specific and sampling-point-specific requirements. To ensure that the device is placed at the bottom of the well screen (so it is in position to core the water in the screen at a later date), a reusable stainless steel weight is attached to the sealed bottom of the flexible sleeve. An optional weight can be attached to the top of the device to compress the sampler in the bottom of the well, if desired. A collar at the top of the device allows attachment of a suspension cable or cord to permit retrieval of the device for sample collection (Figure 4.12). In cases where vertical stratification of contaminants in the well screen is of concern, it is possible to stack several HydraSleeves on the suspension cable to permit vertical profiling of the water column.

FIGURE 4.11
The HydraSleeve is a lightweight disposable sampler, ideally suited for sampling low-yield wells. This photo shows a HydraSleeve after retrieval from a newly installed well that has not yet been developed (thus the high sediment content). In practice, these samplers are capable of collecting sediment-free samples from low-yield wells, by coring the water column in the screen.

During deployment, the HydraSleeve is carefully lowered into the sampling point and positioned at or near the bottom of the well. While it is lowered into the sampling point, hydrostatic pressure keeps the sleeve collapsed and the check valve closed, preventing water from entering the sampler. The slim cross-section of the collapsed sleeve minimizes disturbance of the water column during placement, helping to reduce the time required for the sampling point to return to chemical and hydraulic equilibrium. During equilibration, the sleeve remains collapsed with the check valve closed due to the water pressure difference between the outside and inside of the device. Once the device is situated at the desired sampling interval, the suspension cable is attached to the top of the well (e.g., to the underside of the well cap) to allow samplers to retrieve it after the sampling point has equilibrated (generally between 48 h and 1 week).

To collect a sample after the sampling point has equilibrated, the HydraSleeve is removed from the screened interval by pulling up on the suspension cord at a rate of 1 ft/sec or faster. The upward motion can be accomplished using one long continuous pull, several short rapid strokes, or any combination that moves the device the required distance. When this upward motion is initiated, the check valve at the top of the device opens and the device fills with water from within the screened interval. The total upward distance the device must be pulled to complete the filling process is about one to one and a half times the length of the sampler. As the device fills with water, there is no change in water level in the well and there is no agitation of the water column or the sample. After the device is full, the check valve closes, thus preventing additional water from overlying portions of the water column from entering the device during retrieval. At ground

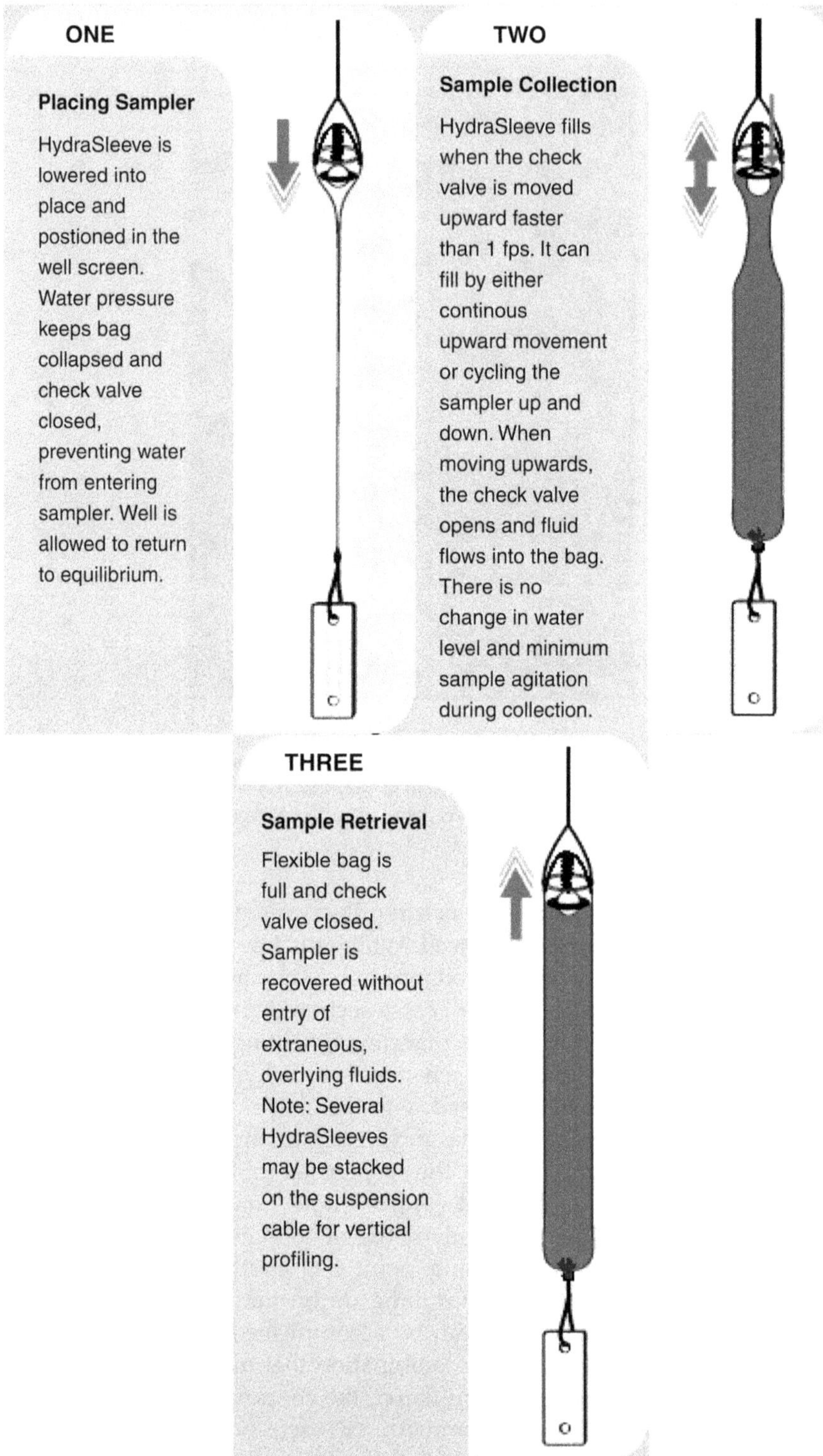

FIGURE 4.12
Illustration of the process of collecting samples using a HydraSleeve. (Diagram courtesy of GeoInsight, Inc.)

surface, the sample is removed from the HydraSleeve by puncturing the sleeve with a sharp, small-diameter discharge tube at the base of the check valve. By raising and lowering the bottom of the sampler or by pinching the sample sleeve just below the discharge tube, the sample discharge flow rate can be controlled. The sample is discharged directly into any parameter-specific sample container. At the conclusion of sample collection, a new HydraSleeve can be lowered into the sampling point and left in place for the next sampling event. This saves time and improves sampling precision and accuracy by ensuring that the formation has reached equilibrium prior to the subsequent sampling event.

The HydraSleeve can be used to collect representative samples for all chemical parameters due to its construction with inert materials, lack of aeration and agitation of the water column and samples (and resulting low turbidity of samples), and its single use (i.e., disposable) design. Studies to date have shown that because samples are collected at *in situ* pressure with no aeration or agitation of the water column, there are no problems related to loss of volatile constituents or oxidation of sensitive parameters such as trace metals (Parker and Clark, 2002; ITRC, 2005; Parsons, 2005) that are common with other grab-sampling devices such as bailers (Parker and Clark, 2002).

Field Measurement of Water-Quality Indicator Parameters and Turbidity

As discussed earlier in this chapter, monitoring of selected chemical and physical indicator parameters in ground water in the field is an integral component of some strategies used to purge high-yield wells. Chemical parameters most commonly measured include pH, specific conductance (or conductivity) and, more recently, DO and ORP (redox potential, also measured as Eh). In addition, some sampling programs include the measurement of temperature, and some include turbidity, a physical parameter that is an indicator of the disturbance caused to the water column by the purging and sampling method used.

The SAP must detail exactly which indicator parameters are to be measured in the field, and where, when (in the context of well purging and sample collection), and how (related to the type of equipment used) those parameters are to be measured. For applications involving measurement to stabilization, the SAP must also provide an accurate definition of "stabilization" for each parameter, that is specific to the instrumentation used. It is the experience of the authors that this is a common deficiency of many SAPs, which results in a great deal of error in field parameter measurement. Table 4.5 summarizes the common errors in field parameter measurement that lead to inaccuracy and imprecision of data generated in the field.

Some indicator parameter measurements can be made in an open container (Figure 4.13) (e.g., pH, temperature, and specific conductance) if a device other than a pump is used for purging (e.g., a bailer). Other parameters, such as DO and ORP, should never be measured in an open container where there is an air–water interface. In these circumstances, readings will be affected by exposure to atmospheric air and may, in fact, never be accurate and never truly stabilize. DO and ORP should always be measured either downhole (with an *in situ* probe) (Figure 4.14 and Figure 4.15) or in a flow-through cell at the surface (Figure 4.16). The latter option requires that a pump be used for purging and sampling. Several researchers (Puls et al., 1991, 1992; Barcelona et al., 1994; Puls and McCarthy, 1995) have shown temperature and pH (and, in some instances, specific

TABLE 4.5

Common Errors in Field Parameter Measurement

No instrument-specific definition of stabilization
No calibration or incorrect calibration of instrumentation under field conditions
Use of expired or incorrect calibration standards
Poor equipment cleaning and maintenance practices
No training of field personnel on the proper use of specified instrumentation
Failure to understand operating ranges, accuracy, resolution, and operational features of individual parameter measurement probes or test methods
Failure to recognize errors in field parameter measurements
Failure to record units of measure and "+" or "−" values for parameters such as ORP
Measurement of DO and ORP in open containers
Taking too long to measure parameters that are temperature sensitive
Errors in collection and handling of subsamples analyzed for turbidity

conductance) to be the least sensitive indicators of equilibrated conditions, while ORP and DO are more sensitive indicators.

During purging, the purpose of monitoring field parameters is to determine when formation-quality water is available for sample collection. This is interpreted to occur when the selected indicator parameters have stabilized. The term stabilization must be clearly understood by all samplers to ensure precision between sampling teams and to avoid overpurging or underpurging the sampling point. In the context of low-flow purging and sampling, indicator parameters are considered stable when three consecutive readings made several minutes apart fall within the ranges presented in Table 4.6

FIGURE 4.13
Only a few indicator parameters (specific conductance, pH, and temperature) can be reliably measured in an open container at the surface, but only if it is done quickly. Other parameters (DO and ORP) are strongly affected by exposure to atmospheric conditions, and should not be measured in open containers.

FIGURE 4.14
A multiparameter sonde designed for taking *in situ* measurements of indicator parameters in a 2 in. diameter well. This is the best method for collecting representative data for parameters strongly affected by exposure to atmospheric conditions. This sonde is capable of measuring *in situ* pH, specific conductance, temperature, DO, ORP, and salinity. (Photo courtesy of YSI, Inc.)

FIGURE 4.15
Using a multiparameter sonde designed for collecting *in situ* measurements of indicator parameters allows chemical profiling in the well. The differences in chemistry in the water column are often significant enough to allow accurate location of the top of the well screen or location of contaminated zones within the screen.

FIGURE 4.16
A flow-through cell can also be used to collect representative data on indicator parameters that are affected by exposure to atmospheric conditions. In this case, the same type of equipment shown in Figure 4.14 and Figure 4.15 (with the addition of turbidity as a measurement option) has been adapted for use in a small-volume (250 ml) cell, with the inlet at the bottom and the outlet at the top. This configuration is preferred to achieve flow over the measurement sensors and accurate indicator parameter measurement.

(Nielsen and Nielsen, 2002). In the context of traditional purging to stabilization of indicator parameters, stabilization is often defined as the point at which measured values are within ±10% for all parameters for three consecutive readings taken 3 min apart (Nielsen and Nielsen, 2002).

While the criteria in Table 4.6 are reasonable criteria for many hydrogeochemical situations, it should be recognized that firm criteria for indicator parameter stabilization may not be appropriate for some situations because of a variety of factors, including variability in aquifer properties, monitoring well hydraulics (Robbins and Martin-Hayden, 1991; Barcelona et al., 1994; Robin and Gillham, 1987), and natural spatial and temporal variability in ground-water chemistry and contaminant distribution (Keely and Boateng, 1987a; Huntzinger and Stullken, 1988; Clark and Baxter, 1989). Therefore, the criteria in Table 4.6 should be compared to well-specific measurements to determine whether the site-specific criteria need to be adjusted. Additionally, these criteria should be evaluated to select those that are most important and relevant to meeting the sampling objectives for the specific site. Not all criteria need to be met for all sites. Shanklin et al. (1995) and Puls and McCarthy (1995) point out that stabilization criteria that are too stringent may unnecessarily lead to overpurging of the sampling point and the generation of large amounts of contaminated purge water, without providing the benefit of ensuring

TABLE 4.6

Example Criteria for Defining Stabilization of Water-Quality Indicator Parameters

Temperature	±0.2°C
pH	±0.2 pH units[a]
Conductivity	±3% of the reading
DO	±10% of the reading or ±0.2 mg/l, whichever is greater[a]
Eh or ORP	±20 mV[a]

[a]Related to the measurement accuracy of commonly available field instruments.

that the samples are any more representative. This commonly occurs when stabilization criteria are set at levels that exceed the measurement capability or accuracy of the instrumentation. It is important for sampling team members to understand that just because an instrument displays a measurement to two decimal places (resolution), it does not mean that the displayed number is accurate to two decimal places. ORP is a good example of a parameter for which this problem is commonly seen. For ORP, samplers need to remember that the recommended stabilization criterion is ± 20 mV.

If stabilization is not achieved, sampling should be done at the discretion of the field sampling team (or in consultation with the project manager). Factors that commonly affect field parameter stabilization and that should be checked prior to sampling without achieving stabilization include improper instrument calibration, failure to allow instrument sensors to warm up and stabilize prior to use, poor condition of instrument sensors, air leaks in the pump tubing or flow-through cell connectors, and too low a flow rate through the flow-through cell. If samples are collected without reaching water-quality indicator parameter stabilization, the conditions under which the samples are collected should be documented.

For in-line flow-through cells, the frequency of the measurements for indicator parameters should be documented in the SAP. The first measurement should generally be made after the volume of the pump and tubing has passed through the flow-through cell. Subsequent measurements should be made based on the volume of the cell and the time required to completely evacuate one volume of the cell at the flow rate used for purging, to ensure that independent measurements are made. For example, a 500 ml flow-through cell in a system pumped at a rate of 250 ml/min will be evacuated in 2 min, so measurements should be made at least 2 min apart. It is important, therefore, that the sampling team establish the following volumes and rates in the field prior to the sampling event:

- Volume of the pump and discharge tubing
- Optimum pump discharge rate
- Volume of the flow-through cell corrected for the displacement volume of the field parameter measurement instrumentation installed inside the flow-through cell

It is also important to know the manufacturers' recommendations for the amount of time required to allow individual sensors (e.g., DO) used to measure field parameters to stabilize to ensure that representative data are collected.

For wells in which dedicated pumps are used, indicator parameters should stabilize shortly after the volume of the pump and tubing has been removed. In a sampling program in which bladder pumps were used for low-flow purging and sampling, Shanklin et al. (1995) found that they achieved indicator parameter stabilization and that they could collect representative samples after purging twice the calculated volume of the pump and discharge line. For wells in which portable pumps are used, the effects of pump installation on the water column in the well usually result in the need to remove significantly more water before chemical indicator parameters (and, as noted subsequently, turbidity) reach stabilization. Owing to the agitation and aeration of the water column caused by a bailer, which results in increased levels of DO and increased amounts of artifactual particulate matter suspended in the water, the volume of water required to be purged and the time required to reach stabilization will be greater than for a scenario in which a pump is used. For some parameters (particularly DO), stabilization may never be reached in bailed samples because of the disturbance caused to the water column in the well (Pohlmann et al., 1994).

Although not a chemical parameter, and not indicative of when formation-quality water is being pumped, turbidity may also be a useful parameter to measure (Figure 4.17). Turbidity is a physical parameter that provides a measure of the suspended particulate matter in the water removed from the well. Turbidity is indicative of pumping stress on the formation or disturbance to the water column in the well during installation of portable pumps. Sources of turbidity in sampling points can include:

- Naturally occurring colloid-sized or larger solids that may be in transit through the formation
- Artifactual solids from well drilling and installation (e.g., drilling fluids, filter pack, and grout) that have not been effectively removed by well development
- Naturally occurring fine-grained formation materials that are mobilized by agitation of the water column (i.e., by bailing, by installation of a portable pump, or by over-pumping the well)
- Microbial growth that often occurs in monitoring wells in the presence of certain types of contaminants (i.e., petroleum hydrocarbons)
- Precipitation caused by different redox conditions in the sampling point than in the formation

Turbidity levels elevated above the natural formation condition can result in biased analytical results for many parameters, including trace metals, hydrophobic organic

FIGURE 4.17
Field turbidity measurements can provide a useful indication of the degree of disturbance of the water column in the well caused by purging and sampling. In this case, turbidity is being measured using a stand-alone turbidimeter. Some multi-parameter sondes used downhole or in flow-through cells (e.g., the one depicted in Figure 4.16) are also capable of measuring turbidity.

compounds, and other strongly sorbed chemical constituents. Naturally occurring turbidity in some ground water can exceed 10 NTU (Puls and Barcelona, 1996) and may be unavoidable. Turbidity in a properly designed, constructed, and developed well is most often a result of significant disturbance of the water column or excessive stress placed on the formation by overpumping.

The primary reason for minimizing turbidity during purging and sampling is that turbidity can affect the ability to accurately determine the aqueous concentration of some analytes. Analysis of organic analytes can be hampered by the physical presence of suspended solids, and accurate analysis of aqueous inorganics can be affected by stripping of cations, particularly metal species, from the surfaces of suspended clay particles by the sample preservation process (i.e., acidification).

To avoid artifacts in sample analysis, turbidity should be as low as possible when samples are collected. Turbidity measurements should be taken at the same time that chemical indicator parameter measurements are made or, at a minimum, once when pumping is initiated and again just prior to sample collection, after chemical indicator parameters have stabilized. In the context of low-flow purging and sampling, the stabilization criterion for turbidity is $\pm 10\%$ of the prior reading or ± 1.0 NTU, whichever is greater (ASTM, 20046d). If turbidity values are persistently high, the pumping rate should be lowered until turbidity decreases. If high turbidity persists even after lowering the pumping rate, the pump may have to be stopped for a period of time until turbidity settles and the purging process restarted. If this fails to solve the problem, well maintenance or redevelopment may be necessary. In the context of low-flow purging and sampling, difficulties with high turbidity should be identified during pilot tests prior to implementing low-flow purging or during the initial low-flow sampling event, and contingencies should be established to minimize the problem of elevated turbidity.

References

ASTM, Standard Guide to Purging Methods for Wells Used for Ground-Water Quality Investigations, ASTM Standard D 6452, ASTM International, West Conshohocken, PA, 6 pp., 2006a.

ASTM, Standard Practice for Design and Installation of Ground-Water Monitoring Wells in Aquifers, ASTM Standard D 5092, ASTM International, West Conshohocken, PA, 20 pp., 2006b.

ASTM, Standard Guide to Development of Ground-Water Monitoring Wells in Granular Aquifers, ASTM Standard D 5521, ASTM International, West Conshohocken, PA, 15 pp., 2006c.

ASTM, Standard Practice for Low-Flow Purging and Sampling for Wells and Devices Used for Ground-Water Quality Investigations, ASTM Standard D 6771, ASTM International, West Conshohocken, PA, 7 pp., 2006d.

Backhus, D.A., J.N. Ryan, D.M. Groher, J.K. MacFarlane, and P.M. Gschwend, Sampling colloids and colloid-associated contaminants in ground water, *Ground Water,* 31(3), pp. 466–479, 1993.

Bangsund, W.J., C.G. Peng, and W.R. Mattsfield, Investigation of contaminant migration by low flow rate sampling techniques, *Proceedings of the Eighth Annual Outdoor Action Conference,* National Ground Water Association, Dublin, OH, pp. 311–326, 1994.

Barcelona, M.J. and J.A. Helfrich, Well construction and purging effects on ground-water samples, *Environmental Science and Technology,* 20(11), pp. 1179–1184, 1986.

Barcelona, M.J., J.P. Gibb, J.A. Helfrich, and E.E. Garske, Practical Guide to Ground-Water Sampling, Contract Report 374, Illinois State Geological Survey, Champaign, IL, 94 pp., 1985.

Barcelona, M.J., H.A. Wehrmann, and M.D. Varljen, Reproducible well-purging procedures and VOC stabilization criteria for ground-water sampling, *Ground Water,* 32(1), pp. 12–22, 1994.

Clark, L. and K.M. Baxter, Ground-water sampling techniques for organic micropollutants: UK experience, *Quarterly Journal of Engineering Geology,* 22, pp. 159–168, 1989.

Gibb, J.P., R.M. Schuller, and R.A. Griffin, Procedures for the Collection of Representative Water Quality Data from Monitoring Wells, Cooperative Ground Water Report 7, Illinois State Water and Geological Surveys, Champaign, IL, 61 pp., 1981.

Giddings, T., Bore-volume purging to improve monitoring well performance: an often-mandated myth, *Proceedings of the Third National Symposium on Aquifer Restoration and Ground Water Monitoring*, National Water Well Association, Dublin, OH, pp. 253–256, 1983.

Gillham, R.W., M.J.L. Robin, J.F. Barker, and J.A. Cherry, Ground-Water Monitoring and Sample Bias, American Petroleum Institute Publication, Environmental Affairs Department, Washington, DC, 206 pp., 1983.

Graham, B.S. and T.C. Goudlin, Comparison of purge-and-bail sampling to low-stress sampling at an NPL site, *Proceedings of the Tenth Annual Outdoor Action Conference*, National Ground Water Association, Columbus, OH, pp. 605–619, 1996.

Groher, D., P.M. Gschwend, D. Backhus, and J. MacFarlane, Colloids and sampling ground water to determine subsurface mobile loads, *Proceedings of the Environmental Research Conference on Ground-Water Quality and Waste Disposal*, I.P. Murarka and S. Cordle, Eds., EPRI EN-6749, Electric Power Research Institute, Palo Alto, CA, pp. 19.1–19.10, 1990.

Herzog, B.L., S.F.J. Chou, J.R. Valkenburg, and R.A. Griffin, Changes in volatile organic chemical concentrations after purging slowly recovering wells, *Ground-Water Monitoring Review*, 8(4), pp. 93–99, 1988.

Huntzinger, T.L. and L.E. Stullken, An Experiment in Representative Ground-Water Sampling for Water-Quality Analysis, Water Resources Investigations Report 88-4178, U.S. Geological Survey, Reston, VA, 1988.

Hurley, D.F. and J.M. Whitehouse, ACL monitoring using a low-flow sampling technique: a case study, *Proceedings of the Ninth Annual Outdoor Action Conference*, National Ground Water Association, Columbus, OH, pp. 603–616, 1995.

ITRC, Technical Overview of Passive Sampler Technologies, ITRC Publication, Interstate Technology and Regulatory Council, Washington, DC, 2005.

Karklins, S., Ground Water Sampling Desk Reference, Publication PUBL-DG-037 96, Wisconsin Department of Natural Resources, Bureau of Drinking Water and Ground Water, Madison, WI, 165 pp., 1996.

Kearl, P.M., N.E. Korte, M. Stites, and J. Baker, Field Comparison of micropurging vs. traditional ground-water sampling, *Ground-Water Monitoring and Remediation*, 14(4), pp. 183–190, 1994.

Kearl, P.M., N.E. Korte, and T.A. Cronk, Suggested modifications to ground-water sampling procedures based on observations from the colloidal borescope, *Ground-Water Monitoring Review*, 12(2), pp. 155–161, 1992.

Keely, J.F. and K. Boateng, Monitoring well installation, purging, and sampling techniques — Part 1: Conceptualization, *Ground Water*, 25(3), pp. 300–313, 1987a.

Keely, J.F. and K. Boateng, Monitoring well installation, purging, and sampling techniques — Part 2: Case histories, *Ground Water*, 25(4), pp. 427–439, 1987b.

Maltby, V. and J.P. Unwin, A field investigation of ground-water monitoring well purging techniques, in *Current Practices in Ground Water and Vadose Zone Investigations*, Special Technical Publication 1118, D.M. Nielsen and M.N. Sara, Eds., American Society for Testing and Materials, Philadelphia, PA, pp. 281–299, 1992.

Martin-Hayden, J.M. and G.A. Robbins, Plume distortion and apparent attenuation due to concentration averaging in monitoring wells, *Ground Water*, 35(2), pp. 339–346, 1997.

Martin-Hayden, J.M., G.A. Robbins, and R.D. Bristol, Mass balance evaluation of monitoring well purging — Part II: Field tests at a gasoline contamination site, *Journal of Contaminant Hydrology*, 8(1), pp. 225–241, 1991.

McAlary, T.A. and J.F. Barker, Volatilization losses of organics during ground-water sampling from low permeability materials, *Ground-Water Monitoring Review*, 7(4), pp. 63–68, 1987.

McCarthy, J. and L. Shevenell, Obtaining representative ground-water samples in a fractured and karstic formation, *Ground Water*, 36(2), pp. 251–260, 1998.

Nielsen, G.L., Ground-Water Sampling: Field Practices for Regulatory Compliance, Workshop Notebook, Tenth Annual Outdoor Action Conference, National Ground Water Association, Columbus, OH, pp. 89–97, 1996.

Nielsen, D.M. and R. Schalla, Design and Installation of Ground-Water Monitoring Wells; Chapter 10 in *The Practical Hand book of Environmental Site Characterization and Ground-Water Monitoring*, Second Edition, D.M. Nielson, Editor, CRC Press/Taylor and Francis Group, Boca Raton, FL, pp. 639–806, 2006.

Nielsen, D.M. and G.L. Nielsen, *Technical Guidance on Low-Flow Purging and Sampling and Minimum-Purge Sampling*, NEFS-TG001-02, Nielsen Environmental Field School, Galena, OH, 59 pp., 2002.

Palmer, C.D., J.F. Keely, and W. Fish, Potential for solute retardation on monitoring well sand packs and its effect on purging requirements for ground-water sampling, *Ground-Water Monitoring Review*, 7(2), pp. 40–47, 1987.

Parker, L.V. and C.H. Clark, Study of Five Discrete Interval Type Ground-Water Sampling Devices, ERDC/CRREL TR-02-12, U.S. Army Corps of Engineers, Cold Regions Research and Engineering Laboratory, Hanover, NH, 49 pp., 2002.

Parsons, Results Report for the Demonstration of No-Purge Ground-Water Sampling Devices at Former McClellan Air Force Base, California; Contract F44650-99-D-0005, Delivery order DKO1, U.S. Army Corps of Engineers (Omaha District), Air Force Center for Environmental Excellence, and Air Force Real Property Agency, 2005.

Pohlmann, K.F., G.A. Icopini, R.D. McArthur, and C.G. Rosal, Evaluation of Sampling and Field-Filtration Methods for Analysis of Trace Metals in Ground Water, EPA/600/R-94/119, U.S. Environmental Protection Agency, Office of Research and Development, Las Vegas, NV, 79 pp., 1994.

Powell, R.M. and R.W. Puls, Passive sampling of ground-water monitoring wells without purging: multilevel well chemistry and tracer disappearance, *Journal of Contaminant Hydrology*, 12, pp. 51–77, 1993.

Puls, R.W. and R.M. Powell, Acquisition of representative ground-water quality samples for metals, *Ground-Water Monitoring Review*, 12(3), pp. 167–176, 1992.

Puls, R.W. and C.J. Paul, Low-flow purging and sampling of ground-water monitoring wells with dedicated systems, *Ground-Water Monitoring and Remediation*, 15(1), pp. 116–123, 1995.

Puls, R.W. and C.J. Paul, Multi-layer sampling in conventional monitoring wells for improved estimation of vertical contaminant distributions and mass, *Journal of Contaminant Hydrology*, 25, pp. 85–111, 1997.

Puls, R.W. and J.F. McCarthy, Well Purging and Sampling (Workshop Group Summary), *Ground Water Sampling — A Workshop Summary*, EPA/600/R-94/205, U.S. Environmental Protection Agency, Office of Research and Development, Washington, DC, pp. 82–87, 1995.

Puls, R.W. and M.J. Barcelona, Ground-Water Sampling for Metals Analysis, EPA/540/4-89/001, Superfund Ground Water Issue, U.S. Environmental Protection Agency, Office of Solid Waste and Emergency Response, Washington, DC, 6 pp., 1989.

Puls, R.W. and M.J. Barcelona, Low-Flow (Minimal-Drawdown) Ground-Water Sampling Procedures, EPA/540/5-95/504, Ground Water Issue, U.S. Environmental Protection Agency, Office of Solid Waste and Emergency Response, Washington, DC, 12 pp., 1996.

Puls, R.W., D.A. Clark, B. Bledsoe, R.M. Powell, and C.J. Paul, Metals in Ground Water: Sampling Artifacts and Reproducibility, *Hazardous Waste and Hazardous Materials*, 9(9), pp. 149–162, 1992.

Puls, R.W., R.M. Powell, D.A. Clark, and C.J. Paul, Facilitated Transport of Inorganic Contaminants in Ground Water: Part II, Colloidal Transport, EPA/600/M-91/040, U.S. Environmental Protection Agency, Robert S. Kerr Environmental Research Laboratory, Ada, OK, 1991.

Robbins, G.A. and J.M. Martin-Hayden, Mass balance evaluation of monitoring well purging, Part 1. Theoretical models and implications for representative sampling, *Journal of Contaminant Hydrology*, 8, pp. 203–224, 1991.

Robin, M.J.L. and R.W. Gillham, Field evaluation of well purging procedures, *Ground-Water Monitoring Review*, 7(4), pp. 85–93, 1987.

Serlin, C.L. and L.M. Kaplan, Field Comparison of Micropurge and Traditional Ground-Water Sampling for Volatile Organic Compounds, *Proceedings of the Conference on Petroleum Hydrocarbons and Organic Chemicals in Ground Water*, National Ground Water Association, Westerville, OH, pp. 177–190, 1996.

Sevee, J.E., C.A. White, and D.J. Maher, An analysis of low-flow ground-water sampling methodology, *Ground-Water Monitoring and Remediation*, 20(2), pp. 87–93, 2000.

Shanklin, D.E., W.C. Sidle, and M.E. Ferguson, Micropurge low-flow sampling of uranium-contaminated ground water at the Fernald environmental management project, *Ground-Water Monitoring and Remediation*, 15(3), pp. 168–176, 1995.

Thurnblad, T., Minnesota Pollution Control Agency Ground Water Sampling Guidance: Development of Sampling Plans, Protocols and Reports, Minnesota Pollution Control Agency, St. Paul, MN, 1995.

Wilson, L.C. and J.V. Rouse, Variations in water quality during initial pumping of monitoring wells, *Ground-Water Monitoring Review*, 3(1), pp. 103–109, 1983.

5

Ground-Water Sample Pretreatment: Filtration and Preservation

Gillian L. Nielsen and David M. Nielsen

CONTENTS

Sample Pretreatment Options

Another group of parameter-specific field protocols that must be evaluated and included in the SAP are methods for sample pretreatment, including sample filtration and physical and chemical preservation. Sample pretreatment must be performed at the wellhead at the time of sample collection to ensure that physical and chemical changes do not occur in the samples during the time that the sample is collected and after the sample container has been filled and capped. ASTM International has published Standard Guides that address both types of sample pretreatment. ASTM Standard D 6564 (ASTM, 2006a) provides a detailed guide for field filtration of ground-water samples, and ASTM Standard D 6517 (ASTM, 2006b) discusses physical and chemical preservation methods for ground-water samples. Each type of sample pretreatment is discussed subsequently.

Sample Filtration

Ground-water sample filtration is a sample pretreatment process implemented in the field for some constituents, when it is necessary to determine whether a constituent is truly "dissolved" in ground water. Filtration involves passing a raw or bulk ground-water sample directly through a filter medium of a prescribed filter pore size either under negative pressure (vacuum) or under positive pressure. Particulates finer than the filter pore size pass through the filter along with the water to form the filtrate, which is submitted to the laboratory for analysis. Particulate matter larger than the filter pore size is retained by the filter medium. In the case of most ground-water monitoring programs, this material is rarely analyzed, although it may be possible to analyze the retained fraction for trace metals or for some strongly hydrophobic analytes such as PCBs or PAHs. Figure 5.1 illustrates a common vacuum filtration setup, and Figure 5.2 illustrates one form of positive pressure filtration.

What Filter Pore Size to Use

The most common method for distinguishing between the dissolved and particulate fractions of a sample has historically been filtration with a 0.45 μm filter (see, e.g., U.S. EPA, 1991). The water that passes through a filter of this pore size has, by default, become the operational definition of the dissolved fraction, even though this pore size does not accurately separate dissolved from colloidal matter (Kennedy et al., 1974; Wagemann and Brunskill, 1975; Gibb et al., 1981; Laxen and Chandler, 1982). Some colloidal matter is small enough to pass through this pore size, but this matter cannot be considered dissolved. For this reason, Puls and Barcelona (1989) reported that the use of a 0.45 μm filter was not useful, appropriate, or reproducible in providing information on metals solubility in ground-water systems and that this filter size was not appropriate for determining truly dissolved constituents in ground water.

The boundary between the dissolved phase and the colloidal state is transitional. There is no expressed lower bound for particulate matter and no clear cutoff point to allow selection of the optimum filter pore size to meet the objective of excluding colloidal

FIGURE 5.1
A vacuum filtration system used for ground-water samples. This practice is not encouraged.

FIGURE 5.2
A positive-pressure filtration system is a better option to use for ground-water samples. Note the removal of sediment achieved by the cartridge filter.

particles from the sample. The best available evidence indicates that the dissolved phase includes matter that is less than 0.01 μm in diameter (Smith and Hem, 1972; Hem, 1985), suggesting that a filter pore size of 0.01 μm is most appropriate. However, filters with such small pore sizes are subject to rapid plugging, especially if used in highly turbid water, and are not practical to use in the field. Kennedy et al. (1974) and Puls et al. (1991) provide a strong case for the use of a filter pore size of 0.1 μm for field filtration to allow better estimates of dissolved metal concentrations in samples. Puls et al. (1992) and Puls and Barcelona (1996) also recommend the use of 0.1 μm (or smaller) filters for determination of dissolved inorganic constituents in ground water. Such filters are considerably more effective than filters with larger pore sizes (e.g., 0.45 or 1.0 μm) in terms of removing fine particulate matter. These filters are widely available and practical for use in the field for most situations, although in some highly turbid water, filter plugging may make the filtration process difficult and protracted. All factors considered, 0.1 μm field filtration, although it is a compromise, appears to offer the best opportunity for collecting samples that best represent the dissolved fraction.

Yao et al. (1971) indicate that colloids larger than several microns in diameter are probably not mobile in aquifers under natural ground-water flow conditions due to gravitational settling. Puls et al. (1991) also suggest that colloidal materials up to 2 μm are mobile in ground water systems. With the upper bound for colloidal matter described by many investigators as being between 1.0 and 10 μm, it seems reasonable to suggest that a filter pore size of 10 μm would include all potentially mobile colloidal material and exclude the larger, clearly nonmobile artifactual fraction. However, it should be noted that using this filter pore size, artifactual colloidal material that is finer than 10 μm in diameter will be included in the sample. Although this filter pore size is a compromise, it will lead to conservative estimates of total mobile contaminant load while excluding at least a portion of the particulate matter that is artifactual in nature. The collection and analysis of both filtered and unfiltered samples is sometimes suggested as a means of discriminating between natural and artifactual colloidal material or between dissolved and colloidal contaminant concentrations.

Functions of Filtration

Historically, filtration of ground-water samples has served several important functions in ground-water sampling programs. Filtration helps minimize the problem of data bounce, which commonly results from variable levels of suspended particulate matter in samples between sampling events and individual samples, making trend analysis and statistical evaluation of data more reliable. In addition, by reducing suspended particle levels, filtration makes it easier for laboratories to accurately quantify metals concentrations in samples. Perhaps most importantly, filtration of samples makes it possible to determine actual concentrations of dissolved metals in ground water that have not been artificially elevated as a result of sample preservation (acidification), which can leach metals from the surfaces of artifactual or colloidal particles (Nielsen, 1996). The assumption that the separation of suspended particulates from water samples to be analyzed eliminates only matrix-associated (artifactual) constituents may often be incorrect (EPRI, 1985a; Feld et al., 1987), as at least some potentially mobile natural colloidal material will be retained on most commonly used filter pore sizes.

Filtration is often performed as a post-sampling "fix" to exclude from samples any particulate matter that may be an artifact of poor well design or construction, inappropriate sampling methods (use of bailers, inertial-lift pumps, or high-speed, high-flow-rate pumps), or poor sampling techniques (agitating the water column in the well). Filtration may be considered particularly important where turbid conditions caused by high particulate loading might lead to significant positive bias through inclusion of large quantities of matrix metals in the samples (Pohlmann et al., 1994). Alternatively, as discussed earlier, the presence of artifactual particulate matter in samples may also negatively bias analytical results through removal of metal ions from solution during sample shipment and storage as a result of interactions with particle surfaces. However, filtration is not always a valid means of alleviating problems associated with artifactual turbidity, as it often cannot be accomplished without affecting the integrity of the sample in one way or another.

Which Parameters to Filter

During the planning phase of a ground-water sampling program, each parameter to be analyzed in ground-water samples should be evaluated to determine its suitability for field filtration and the most suitable filtration medium. As a general rule of thumb, parameters that are sensitive to the following effects of filtration should not be filtered in the field:

- Pressure changes that would result in degassing or loss of volatile constituents
- Temperature changes
- Aeration and agitation that may occur during filtration processes

Table 5.1 presents a summary of parameters for which filtration may be used and of parameters for which filtration should not be used in the field.

Samples to be analyzed for alkalinity must be field filtered if significant particulate calcium carbonate is suspected in samples, as this material is likely to impact alkalinity titration results (Puls and Barcelona, 1996). Care should be taken in this instance, however, as filtration may alter the CO_2 content of the sample and, therefore, affect the results.

Filtration is not always appropriate for ground-water sampling programs. If the intent of filtration is to determine truly dissolved constituent concentrations (e.g., for

TABLE 5.1

Analytical Parameter Filtration Recommendations

Examples of parameters that may be field filtered
Alkalinity
Trace metals
Major cations and anions
Examples of parameters that should not be filtered
VOCs
TOC
TOX
Dissolved gases (e.g., DO and CO2)
"Total" analyses (e.g., total arsenic)
Low molecular weight, highly soluble, and nonreactive constituents
Parameters for which "bulk matrix" determinations are required

Source: U.S. EPA, 1991.

geochemical modeling purposes), the inclusion of colloidal matter less than 0.45 μm in the filtrate will result in overestimated values (Wagemann and Brunskill, 1975; Bergseth, 1983; Kim et al., 1984). This result is often obtained with Fe and Al, where "dissolved" values are obtained which are thermodynamically impossible at the sample pH (Puls et al., 1991). Conversely, if the purpose of sampling is to estimate total mobile contaminant load, including both dissolved and naturally occurring colloid-associated constituents, significant underestimates may result from filtered samples, due to the removal of colloidal matter that is larger than 0.45 μm (Puls et al., 1991). A number of researchers have demonstrated that some metal analytes are associated with colloids that are greater than 0.45 μm in size (Gschwend and Reynolds, 1987; Enfield and Bengtsson, 1988; Ryan and Gschwend, 1990) and that these constituents would be removed by 0.45 μm filtration. Kim et al. (1984) found the majority of the concentrations of rare earth elements to be associated with colloidal species that passed through a 0.45 μm filter. Wagemann and Brunskill (1975) found more than twofold differences in total Fe and Al values between 0.05 and 0.45 μm filters of the same type. Some Al compounds, observed by Hem and Roberson (1967) to pass through a 0.45 μm filter, were retained on a 0.10 μm filter. Kennedy et al. (1974) found errors of an order of magnitude or more in the determination of dissolved concentrations of Al, Fe, Mn, and Ti using 0.45 μm filtration as an operational definition for "dissolved." Sources of error were attributed to passage of fine-grained clay particles through the filter.

Evidence from several field studies (Puls et al., 1992; Puls and Powell, 1992; Backhus et al., 1993; McCarthy and Shevenell, 1998) indicates that field filtration does not effectively remedy the problems associated with artifactual turbidity in samples. These and other studies indicate that filtration may cause concentrations of some analytes to decrease significantly, due to removal of colloidal particles that may be mobile under natural flow conditions. Puls and Powell (1992) noted that 0.45 μm filtered samples collected with a bailer had consistently lower As concentrations than samples obtained using low-flow-rate pumping. They suggested that the difference may have been due to filter clogging from excessive fines reducing the effective pore size of the filters or adsorption onto freshly exposed surfaces of materials brought into suspension by bailing. Puls et al. (1992) found that high-flow-rate pumping resulted in large differences in metals concentrations between filtered and unfiltered samples, with neither value being representative of values obtained using low-flow-rate sampling. Ambiguous sampling results found by McCarthy and Shevenell (1998) were attributed to analytical values for metals obtained using low-flow sampling that fell between filtered and unfiltered values from samples collected using bailing or high-flow-rate pumping. Discrepancies in analytical values for some metals (Al and Fe) exceeded an order of magnitude in this study. They determined that filtration of turbid samples may have occluded pores in filters, leading to removal of colloidal particles that may be representative of the load of mobile contaminants in ground water. Puls and Barcelona (1989) also point to the removal of potentially mobile species as an effect of filtration, indicating that filtration of ground-water samples for metals analysis will not provide accurate information concerning the mobility of metal contaminants.

If the objective of a ground-water sampling program is to determine the exposure risk of individuals who consume ground-water from private water supply wells, filtration of those samples would not produce meaningful results. To make this type of exposure risk determination, it is important to submit samples for analysis that are representative of water as it is consumed, and, because most people do not have 0.45 μm filters at their taps, unfiltered samples should be collected. In addition, it is important to remember that MCL and MCLG values set for drinking-water standards are based on unfiltered samples.

Sources of Error and Bias in Filtration

The very act of filtration can introduce significant sources of error and bias into the results obtained from analysis of sample filtrate (Braids et al., 1987). Some of these changes in sample chemistry result from pressure changes in the sample, as well as sample contact with filtration equipment and filter media. It is critical to evaluate the suitability of filtration on a parameter-specific basis and to carefully select filtration methods, equipment, and filtration media when developing site-specific filtration protocols to minimize sample bias caused by filtration. The following sources of negative and positive sample bias need to be considered:

- Potential for negative bias to occur due to adsorption of constituents from the sample (U.S. EPA, 1991; Horowitz et al., 1996). For example, Puls and Powell (1992) found that in-line polycarbonate filters adsorbed Cr onto the surface of the filter medium, resulting in an underestimation of Cr concentrations in the ground-water samples being filtered.
- Potential for positive bias to occur due to desorption or leaching of constituents into the sample (Jay, 1985; Puls and Barcelona, 1989; Puls and Powell, 1992; Horowitz et al., 1996). In the Puls and Powell (1992) study, K was observed to leach from nylon filters that were not adequately preconditioned prior to use.
- Removal of particulates smaller than the original filter pore size due to filter loading or clogging as filtered particles accumulate on the filter surface (Danielsson, 1982; Laxen and Chandler, 1982) or variable particle size retention characteristics (Sheldon, 1965; Sheldon and Sutcliffe, 1969).
- Removal of particulate matter with freshly exposed reactive surfaces, through particle detachment or disaggregation, that may have sorbed hydrophobic, weakly soluble, or strongly reactive contaminants from the dissolved phase (Puls and Powell, 1992). This material itself may have been immobile prior to initiation of sampling and mobilized by inappropriate sampling procedures.
- Removal of solids (metal oxides and hydroxides) that may have precipitated during sample collection (particularly where purging or sampling methods that may have agitated or aerated the water column are used) and any adsorbed species that may associate with the precipitates. Such precipitation reactions can occur within seconds or minutes (Reynolds, 1985; Grundl and Delwiche, 1992; Puls et al., 1992), and the resultant solid phase possesses extremely high reactivity

(high capacity and rapid kinetics) for many metal species (Puls and Powell, 1992). Most metal adsorption rates are extremely rapid (Sawhney, 1966; Posselt et al., 1968; Ferguson and Anderson, 1973; Anderson et al., 1975; Forbes et al., 1976; Sparks et al., 1980; Benjamin and Leckie, 1981; Puls, 1986; Barrow et al., 1989). Additionally, increased reaction rates are generally observed with increased sample agitation.

- Exposure of anoxic or suboxic ground water (in which elevated levels of Fe^{2+} are typically present) to atmospheric conditions during filtration can also lead to oxidation of samples, resulting in formation of colloidal precipitates and causing removal of previously dissolved species (NCASI, 1982; EPRI, 1987; Puls and Eychaner, 1990; Puls and Powell, 1992; Puls and Barcelona, 1996). The precipitation of ferric hydroxide can result in the loss of dissolved metals due to rapid adsorption or co-precipitation potentially affecting As, Cd, Cu, Pb, Ni, and Zn (Kinniburgh et al., 1976; Gillham et al., 1983; Stoltzenburg and Nichols, 1985; Kent and Payne, 1988).
- During sample filtration, care should be taken to minimize sample handling to the extent possible to minimize the potential for aeration. If sample transfer vessels are used, they should be filled slowly and filtration should be done carefully to minimize sample turbulence and agitation. Stoltzenburg and Nichols (1986) demonstrated that the use of sample transfer vessels during filtration imparted significant positive bias for DO and significant negative bias for dissolved metal concentrations. For this reason, the use of transfer vessels is discouraged. In-line filtration is preferred because of the very low potential it poses for sample chemical alteration.

Filtration Methods and Equipment

After a decision is made to field filter ground-water samples to meet DQOs for an investigation, decisions must be made regarding selection of the most appropriate field filtration method. The ground-water sample filtration process consists of several phases: (1) selection of a filtration method; (2) selection of filter media (materials of construction, surface area, and pore size); (3) filter preconditioning; and (4) implementation of field filtration procedures. Information on each part of the process must be presented in detail in the SAP to provide step-by-step guidance for sampling teams to implement in the field.

A wide variety of methods are available for field filtration of ground-water samples. In general, filtration equipment can be divided into positive-pressure filtration and vacuum (negative pressure) filtration methods, each with several different filtration medium configurations. As discussed previously, ground-water samples undergo pressure changes as they are brought from the saturated zone (where ground water is under pressure greater than atmospheric pressure) to the surface (where it is under atmospheric pressure), potentially resulting in changes in sample chemistry. The pressure change that occurs when the sample is brought to the surface may cause changes in sample chemistry, which include loss of dissolved gases and precipitation of dissolved constituents such as metals. When handling samples during filtration operations, additional turbulence and mixing of the sample with atmospheric air can cause aeration and oxidation of Fe^{2+} to Fe^{3+}. Fe^{3+} rapidly precipitates as amorphous iron hydroxide and can adsorb other dissolved trace metals (Stolzenburg and Nichols, 1986). Vacuum filtration methods further exacerbate pressure changes and changes due to sample oxidation. For this reason, positive-pressure filtration methods are preferred (Puls and Barcelona, 1989, 1996; U.S. EPA, 1991).

Table 5.2 presents equipment options available for positive pressure and vacuum filtration of ground-water samples.

When selecting a filtration method, the following criteria should be evaluated on a site-by-site basis:

- Possible effect on sample integrity, considering the potential for the following to occur:
  a. Sample aeration, which may result in sample chemical alteration
  b. Sample agitation, which may result in sample chemical alteration
  c. Change in partial pressure of sample constituents resulting from application of negative pressure to the sample during filtration
  d. Sorptive losses of components from the sample onto the filter medium or components of the filtration equipment (e.g., flasks, filter holders, etc.)
  e. Leaching of components from the filter medium or components of the filtration equipment into the sample
- Volume of sample to be filtered
- Chemical compatibility of the filter medium with ground-water sample chemistry
- Anticipated amount of suspended solids and the attendant effects of particulate loading (reduction in effective filter pore size)
- Time required to filter samples. Short filtration times are recommended to minimize the time available for chemical changes to occur in the sample
- Ease of use
- Availability of an appropriate medium in the desired filter pore size
- Filter surface area
- Use of disposable versus nondisposable equipment
- Ease of cleaning equipment if not disposable
- Potential for sample bias associated with ambient air contact during sample filtration
- Cost, evaluating the costs associated with equipment purchase price, expendable supplies and their disposal, time required for filtration, time required for decontamination of nondisposable equipment, and QC measures

TABLE 5.2

Examples of Equipment Options for Positive-Pressure and Vacuum Field Filtration of Ground-Water Samples

Positive-pressure filtration equipment
In-line capsules
– Attached directly to a pumping device discharge hose
– Attached to a pressurized transfer vessel
– Attached to a pressurized bailer
Free-standing disk filter holders
Syringe filters
Zero headspace extraction vessels
Vacuum filtration equipment
Glass funnel support assembly

The filtration method used for any given sampling program should be documented in the site-specific SAP and should be consistent throughout the life of the sampling program to permit comparison of data generated. If an improved method of filtration is determined to be appropriate for a sampling program, the SAP should be revised in lieu of continuing use of the existing filtration method. In this event, the effect on comparability of data needs to be examined and quantified to allow proper data analysis and interpretation. Statistical methods may need to be used to determine the significance of any changes in data resulting from a change in filtration method.

Filtration equipment and filter media are available in a wide variety of materials of construction. Materials of construction should be evaluated in conjunction with parameters of interest being filtered with particular regard to minimizing sources of sample bias, such as adsorption of metals from samples (negative bias) or desorption or leaching of constituents into samples (positive bias). Materials of construction of both the filter holder or support and the filter medium itself need to be carefully selected based on compatibility with the analytes of interest (Puls and Barcelona, 1989). Filter holders that are made of steel are subject to corrosion and may introduce artifactual metals into samples. Glass surfaces may adsorb metals from samples.

Table 5.3 presents a summary of the most commonly used filtration media available for field filtration of water samples. The potential for sample bias for these filter media materials is variable, therefore, filter manufacturers should be consulted to determine recommended applications for specific filtration media and for guidelines on the most effective preconditioning procedures.

Large-diameter filter media (>47 mm) are recommended for ground-water sample filtration (Puls and Barcelona, 1989). Because of the larger surface area of the filter, problems of filter clogging and filter pore size reduction are minimized. High-capacity in-line filters have relatively large filter media surface areas, which may exceed 750 cm^2. This can improve the efficiency of field sample filtration.

Filter Preconditioning

Filter media require proper preconditioning prior to sample filtration (Jay, 1985; U.S. EPA, 1995; Puls and Barcelona, 1996; ASTM, 2006a). The purposes of filter preconditioning are: (1) to minimize positive sample bias associated with residues that may exist on the filter surface or constituents that may leach from the filter, and (2) to create a uniform wetting front across the entire surface of the filter to prevent channel flow through the filter and increase the efficiency of the filter surface area. Preconditioning the filter medium may not completely prevent sorptive losses from the sample as it passes through the filter medium.

In most cases, filter preconditioning should be done at the wellhead immediately prior to use (Puls and Barcelona, 1989). In some cases, filter preconditioning must be done in a laboratory prior to use (e.g., GF/F filters must be baked prior to use). Some manufacturers "preclean" filters prior to sale. These filters are typically marked "precleaned" on filter packaging and provide directions for any additional field preconditioning required prior to filter use.

The procedure used to precondition the filter medium is determined by the following: (1) the design of the filter (i.e., filter capsules or disks); (2) the material of construction of the filter medium; (3) the configuration of the filtration equipment; and (4) the parameters of concern for sample analysis. Filtration medium manufacturers' instructions should be followed prior to implementing any filter preconditioning protocols in the field to ensure that proper methods are employed and to minimize potential bias of filtered samples.

TABLE 5.3

Examples of Common Filter Media Used in Ground-Water Sampling

	Filter Medium						
	Acrylic Copolymer	Glass Fiber	Mixed Cellulose Esters	Nylon	Polycarbonate	Polyethersulfone	Polypropylene
Analytes						X	X
Major ions	X	—	—	—	X	X	X
Minor ions	—	—	—	—	X	X	X
Trace metals	X	—	—	X	X	X	X
Nutrients	X	—	X	—	X	—	—
Organic compounds	—	X	—	—	—	X	—
Filter effective area (cm^2)							
17	X	X	X	X	X	X	—
20	X	X	X	X	X	X	—
64	—	X	—	—	—	X	—
158	X	X	X	—	—	X	—
250	—	—	—	—	—	X	—
600	—	—	—	X	—	X	—
700	X	—	—	—	—	—	—
770	—	—	—	X	—	X	X
Pore size (μm)							
0.1	—	—	X	—	X	X	—
0.2	—	—	X	—	X	X	—
0.45	X	—	X	X	X	X	—
1.0	X	X	X	X	—	X	X
5.0	X	—	X	X	—	X	X
Filter type							
Flat disk	X	X	X	X	X	X	—
Capsule	X	—	—	X	X	X	X
Syringe	X	X	—	X	—	X	—
Funnel	X	—	X	X	X	X	—

Source: ASTM, 2006a.

These instructions will specify filter-specific volumes of water or medium-specific aqueous solutions to be used for optimum filter preconditioning.

The volume of water used in filter preconditioning is dependent on the surface area of the filter and the medium's ability to absorb liquid. Many filter media become fragile when saturated and are highly subject to damage during handling. Therefore, saturated filter media should be handled carefully and are best preconditioned immediately prior to use in the field.

Disk filters (also known as plate filters) should be preconditioned as follows: (1) hold the edge of the filter with filter forceps constructed of materials that are appropriate for the analytes of interest; (2) saturate the entire filter disk with manufacturer-recommended, medium-specific water (e.g., distilled water, deionized water, or sample water) while holding the filter over a containment vessel (not the sample bottle or filter holder) to catch all run-off; (3) then place the saturated filter on the appropriate filter stand or holder in preparation for sample filtration; (4) complete assembly of the filtration apparatus; (5) pass the recommended volume of water through the filter to complete preconditioning; (6) discard preconditioning water; and (7) begin sample filtration using a clean filtration containment vessel or flask. When preconditioning disk filters, care should be taken not to perforate the filter. The filter medium should not be handled with anything other than filter forceps. Otherwise, there may be a reduction in the porosity and permeability of the filter medium. In addition, care should be taken to avoid exposure of the filter medium to airborne particulates to minimize introduction of contaminants onto the filter surface.

Preconditioning of capsule filters requires that liquid be passed through the filter prior to sample filtration and collection. A volume of manufacturer-recommended, medium-specific water (e.g., distilled water, deionized water, or sample water) should be passed through the filter while holding the capsule upright, prior to sample collection. In general, large-capacity capsule filters require that 1000 ml of water be passed through the filter prior to sample collection, while small-capacity filters require approximately 500 ml of water to be passed through the filter.

Sample Preservation

The second form of pretreatment of ground-water samples is physical and chemical preservation. As described in ASTM Standard D 6517 (ASTM, 2006b), ground-water samples are subject to unavoidable chemical, physical, and biological changes relative to *in situ* conditions when samples are brought to ground surface during sample collection. These changes result from exposure to ambient conditions, such as pressure, temperature, ultraviolet radiation, atmospheric oxygen, and atmospheric contaminants, in addition to any changes that may be imparted by the sampling device as discussed earlier in this chapter.

Objectives of Sample Preservation

The fundamental objective of physical and chemical preservation of samples is to minimize further changes in sample chemistry associated with sample collection and handling from the moment the sample is placed in the sample container to the time it is removed from the container for extraction or analysis in the laboratory. Sample preservation methods are determined on a parameter-specific basis and must be specified in the SAP prior to sample collection. Requirements for sample container type, storage and shipping temperature, and chemical preservatives are specified in the analytical method used for each individual parameter selected for analysis. Sampling team members are encouraged to speak with a laboratory representative prior to the sampling event to ensure that the correct types and numbers of sample containers (along with a few spares) and necessary chemical preservatives are shipped to the field site in sufficient time for the scheduled sampling event. Sampling team members must also learn from the laboratory what the parameter-specific holding times are (the amount of time that can transpire from the moment the sample container is filled to the time the sample is extracted or analyzed by the lab) for each parameter to ensure that samples are received by the laboratory in a timely fashion. This is particularly critical when sampling is conducted late in the work week for parameters that have a short holding time (e.g., 48 h).

Physical Preservation Methods

Physical preservation methods for ground-water samples include: (1) use of appropriate sample containers for each parameter being analyzed; (2) use of appropriate packing and packaging of samples to prevent damage during transport to the laboratory; and (3) temperature control of samples during delivery to the laboratory. Sample containers are specified on a parameter-specific basis by the chosen analytical method for the sampling program (ASTM, U.S. EPA SW846 [U.S. EPA, 1996], AWWA Standard Methods [APHA, AWWA, and WPCF, 1985]), as well as in Federal (40 CFR Part 136), state, and local regulatory guidelines on ground-water sample collection and preservation. Containers are specified with a number of design criteria in mind, to protect the integrity of the analytes of interest, including shape, volume, gas tightness, materials of construction, use of cap liners, and cap seal or thread design (Figure 5.3). Table 5.4 presents a summary of some of the more common ground-water sample containers used. These required containers are subject to change as methods are revised.

Sampling team members must also be aware that how they package sample containers for either hand delivery to the laboratory or commercial shipping is a critical aspect of physical preservation procedures. Field personnel should package and ship samples in compliance with applicable shipping regulations as discussed in ASTM Standard D 6911

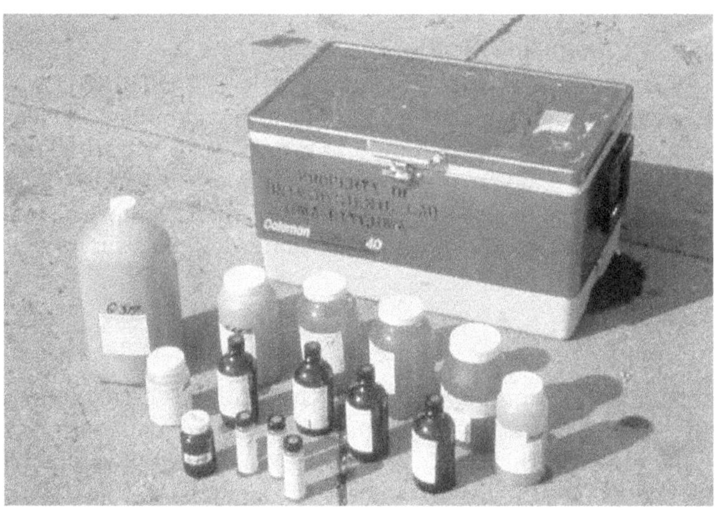

FIGURE 5.3
Sample containers vary in design based on the analytes to be measured in the sample. This example depicts containers used for a full suite of parameters included in the RCRA detection monitoring program as collected from one well.

TABLE 5.4

Examples of Frequently Used Containers for Ground-Water Samples

	Parameter of Interest	Container	Volume (ml)
Inorganic tests	Chloride	P	125
	Cyanide (total and amenable)	P	1000
	Nitrate	P	125
	Sulfate	P	250
	Sulfide	P	500
Metals	Cr^{6+}	P	500
	Mercury	P	500
	Metals except Cr^{6+} and Hg	P	1000
Organic tests	Acrolein and acrylonitrile	G, PTFE ls	40
	Benzidines	G am, PTFE lc	1000
	Chlorinated hydrocarbons	G am, PTFE lc	1000
	Dioxins and furans	G am, PTFE lc	1000
	Haloethers	G am, PTFE lc	1000
	Nitroaromatics and cyclic ketones	G am, PTFE lc	1000
	Nitrosamines	G am, PTFE lc	1000
	Oil and grease	G am, wm	1000
	Total organic carbon	G am PTFE ls	40
	Organochlorine pesticides	G am, PTFE lc	1000
	Organophosphorus pesticides	G am, PTFE lc	1000
	PCBs	G am, PTFE lc	1000
	Phenols	G am, PTFE lc	1000
	Phthalate esters	G am, PTFE lc	1000
	Polynuclear aromatic hydrocarbons	G am, PTFE lc	1000
	Purgeable aromatic hydrocarbons	G, PTFE ls	40
	Purgeable halocarbons	G, PTFE ls	40
	TOX	G am, PTFE lc	250
Radiological tests	Alpha, beta, and radium	P	1000

Notes: P, high density PE; G, glass; G am, amber glass; wm, wide mouth; PTFE, polytetrafluorethylene (Teflon®); lc, lined cap; ls, lined septum; TOX, total organic halides.
Source: U.S. EPA, 1996.

(ASTM, 2006c). Shipping regulations such as the U.S. Department of Transportation Title 49 Code of Federal Regulations Part 172 and the International Air Transport Association (IATA) regulations should be consulted by sampling team members prior to a sampling event where ground-water samples may be sufficiently contaminated to require classification as dangerous goods for shipping purposes or where concentrated chemical preservatives require shipment. These regulations will provide definitive instructions on the correct packaging of regulated samples for shipment to the laboratory. Sample containers should be shipped in a manner that will ensure that the samples are received intact by the laboratory at the appropriate temperature as soon as possible after sample collection, to permit sufficient time for the laboratory to perform the requested analyses within the prescribed holding time for each analyte. Care must be taken by sampling team members to ensure that sample containers are packed sufficiently tight within the outer shipping container and to prevent movement during shipment that could result in container breakage. It is also a good practice to avoid packing glass containers against glass containers whenever possible (plastic against glass is a better configuration). Special shock-absorbing sleeves and containers with a plastic coating have been designed to help reduce the incidence of container breakage during shipment and handling. Use of bubble wrap around containers can also minimize container breakage. Commercial carriers often recommend that absorbent pads be placed in the bottom of sample shipping containers and on the top of sample containers after the shipper is filled, to absorb shock during transit.

Another important consideration during handling of samples in the field, following collection and during transport to the laboratory, is temperature control. Many parameters require that samples be stored at 4°C in the field between sample collection locations, during sample shipment (or delivery if by hand), and upon arrival at the laboratory. (4°C is the temperature at which water is at its maximum density and is most chemically stable.) To accomplish this, sample temperatures should be lowered immediately after sample containers have been filled, labeled, and had any required security seals affixed to them.

Cooling can be accomplished using on-site refrigeration systems if they are available in the field or, more commonly, using wet (natural) ice. Wet ice is the preferred method to cool samples to 4°C. It is inexpensive, readily available, and will not get samples so cold that they will freeze. Wet ice will, however, require replenishment throughout the day to maintain sample temperatures, especially when sampling in warm ambient temperatures. Wet ice should be double bagged to prevent leakage into the shipping container as the ice melts.

Reusable chemical ice packs (also called blue ice packs) (Figure 5.4) are neither suitable for lowering sample temperatures to 4°C from *in situ* temperatures nor suitable for maintaining sample temperatures at 4°C during field handling and shipment. Thus, they are not recommended for use during ground-water sampling (Kent and Payne, 1988; ASTM, 2006b). There is also some concern about what chemicals may be released into the shipping container should a chemical ice pack be punctured or leak during sample shipment.

Dry ice is sometimes specified for use for sample cooling. Unfortunately, when dry ice is used, samples often become too cold and smaller volume containers commonly freeze, resulting in container breakage and sample loss. Dry ice is also relatively expensive and difficult to obtain in the field. It requires special handling procedures in the field and is a regulated substance under shipping regulations.

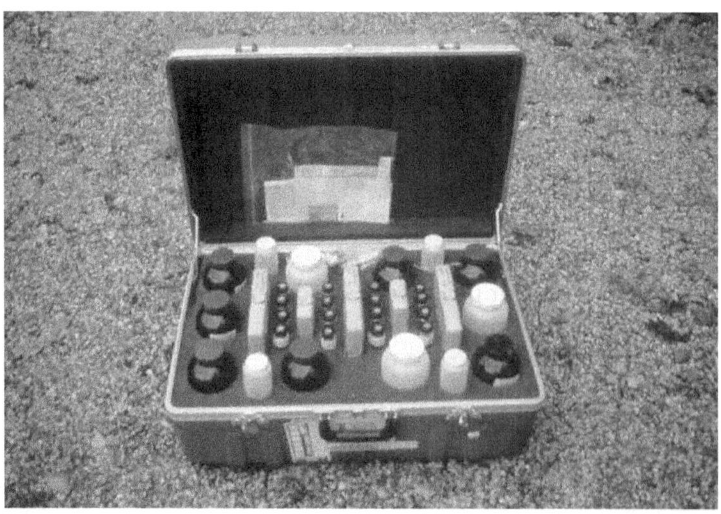

FIGURE 5.4
Reusable chemical ice packs, such as those used in this sample shuttle, are not recommended for ground-water sampling because they are incapable of achieving the desired sample temperature (4°C) in most cases and they may contribute contaminants to samples if they rupture or leak in transit to the laboratory.

To verify appropriate temperature control of samples, it is recommended that samplers include a QC sample referred to as a temperature blank (ASTM, 2006b) in the same shipping container as the ground-water samples. Upon receipt at the laboratory, a laboratory representative will check the temperature blank to determine whether samples were approximately preserved with respect to temperature. If the temperature blank is not at the required temperature (4 ± 2°C), the laboratory representative will contact the sampling team to notify them of the sample arrival temperature and to determine an appropriate course of action.

Chemical Preservation Methods

Chemical preservation is an important field procedure that samplers must implement to ensure that chemical change in samples is minimized during sample handling and shipment. Chemical preservation involves the addition of one or more chemicals (reagent grade or better) to the ground-water sample during sample collection. Chemicals can be used to adjust sample pH to keep constituents in solution or to inhibit microbial degradation of samples. Chemical preservatives are specified by each analytical method for each parameter and the preservatives are typically provided by the laboratory. Table 5.5 provides examples of common chemical preservatives used for ground-water samples.

Ground-water samples can be chemically preserved in one of several ways: (1) titration of pH-adjusting compounds (e.g., nitric acid) while monitoring pH change with a pH meter or narrow-range litmus paper; (2) addition of a fixed volume of liquid preservative (e.g., sulfuric acid contained in glass vials or ampoules) to the sample container; (3) addition of a fixed amount of pelletized preservative (e.g., sodium hydroxide) to the sample container; or (4) placement of preservatives in empty sample containers prior to shipment of the containers to the field (i.e., prepreserved sample containers). Titration methods for sample preservation, while theoretically a valid approach, are not always practical under field conditions where samplers are required to handle large volumes of concentrated preservatives and work with glass titration apparatus under less than ideal conditions (wind, rain, dust, etc.). A modified version of this method is to use calibrated dropper bottles of preservative rather than glass burettes for titration. This ensures that the correct preservative is titrated into the sample while monitoring pH changes, but in a safer fashion in the field.

Vials or ampoules of preservatives are commonly used for sample preservation in the field (Figure 5.5). The laboratory provides the vials or ampoules containing a fixed volume of the required preservative for each sample container requiring chemical preservation. The sampling team should be provided with directions on which preservative must be added to which container on a parameter-specific basis, as well as guidance on whether the preservative should be added to the container before or after filling. One common error made by sampling teams is to assume that the amount of preservative provided in vials or ampoules will always be sufficient to reach the required end pH for the analyte. This is not a safe assumption, especially in situations where ground-water pH may be abnormally high or low based on contaminant chemistry or the natural pH of formation-quality water. For this reason, it is essential that both an initial and final pH measurement be taken to check for pH anomalies and to ensure that the required end pH for the sample has been reached (ASTM, 2006b). To take these measurements, a small aliquot of sample should be decanted into another container (e.g., a clean empty VOC vial without preservative that will not be used for sample collection or a clean small-volume beaker) and the pH measured using either a calibrated pH probe or narrow-range litmus paper. If the sample pH is not at the required endpoint, additional

TABLE 5.5

Examples of Commonly Used Ground-Water Sample Chemical Preservatives and Holding Times

	Parameter of Interest	Preservative	Holding Time
Inorganic tests	Chloride	Cool to 4°C	ASAP
	Cyanide (total and amenable)	Cool to 4°C[a]; if oxidizing agents present, add 5 ml 0.1 *N* $NaAsO^2$/l or 0.06 g ascorbic acid/l; pH > 12 with 50% NaOH	14 days
	Nitrate	Cool to 4°C; boric acid for method 9210	ASAP
	Sulfate	Cool to 4°C	ASAP
	Sulfide	Cool to 4°C; pH > 9 with NaOH; Zn acetate; no headspace	7 days
Metals	$Chromium^{6+}$	Cool to 4°C	1 day
	Mercury	pH < 2 HNO3	28 days
	Metals Except Cr^{6+} and Hg	pH < 2 HNO3	6 months
Organic tests	Acrolien and acrylonitrile	Cool to 4°C; pH 4–5[a] Na2S2O3	14 days
	Benzidines	Cool to 4°C; 0.008% Na2S2O3[a]	7e/40ae
	Chlorinated hydrocarbons	Cool to 4°C; 0.008% Na2S2O3[a]	7e/40ae
	Dioxins and furans	Cool to 4°C; 0.008% Na2S2O3[a]	30e/45ae
	Haloethers	Cool to 4°C; 0.008% Na2S2O3[a]	7e/40ae
	Nitroaromatics and cyclic ketones	Cool to 4°C; 0.008% Na2S2O3[a] dark	7e/40ae
	Nitrosamines	Cool to 4°C; 0.008% Na2S2O3[a] dark	7e/40ae
	Oil and grease	Cool to 4°C; pH < 2 HCl	ASAP
	Organic carbon, total (TOC)	Cool to 4°C; pH < 2 H2SO4 or HCl; dark	ASAP
	Organochlorine pesticides	Cool to 4°C	7e/40ae
	Organophosphorous pesticides	Cool to 4°C; 0.008% Na2S2O3[a], pH 5–8	7e/40ae
	PCBs	Cool to 4°C; 0.008% Na2S2O3[a]	7e/40ae
	Phenols	Cool to 4°C; 0.008% Na2S2O3[a]	7e/40ae
	Phthalate esters	Cool to 4°C; 0.008% Na2S2O3[a]	7e/40ae
	Polynuclear aromatic hydrocarbons	Cool to 4°C; 0.008% Na2S2O3[a]	7e/40ae
	Purgeable aromatic hydrocarbons	Cool to 4°C; Na2S2O3 pH < 2 HCl or H2SO4 or NAHSO4	14 days
	Purgeable aromatic halocarbons	Cool to 4°C; Na2S2O3 pH < 2 HCl or H2SO4 or NAHSO4	14 days
	TOX	Cool to 4°C; pH < 2 H2SO4; dark; no headspace	28 days
Radiological tests	Alpha, beta, and radium	pH < 2 HNO3	6 months

[a]Only add a reducing agent if the sample contains free or combined chlorine. A field test kit needs to be used for this determination.
7e: sample extraction must be completed within 7 days of sample collection.
40ae: analysis must be completed within 40 days after sample extraction.
Note: ASAP, analysis should be performed as soon as possible. Samplers should discuss with the laboratory how ASAP is to be interpreted on a project-specific basis. In many cases, if the parameter can be analyzed with accuracy and precision in the field that is preferred. Otherwise, many laboratories use a 24 h holding time.
Source: U.S. EPA, 1996.

preservative must be added until it is reached. For this reason, sampling teams must ask the laboratory to provide additional preservative, preferably in a vial that can be resealed if only a few extra drops of preservative are required. Sampling teams must resist the temptation to double the preservative required by the method simply for the sake of convenience. This can result in the samples becoming a corrosive liquid for shipping

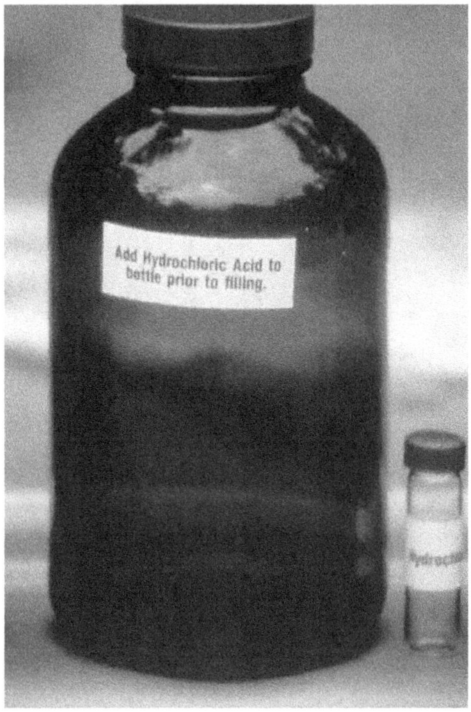

FIGURE 5.5
Chemical preservation using vials.

purposes as described in ASTM Standard D 6911 (ASTM, 2006c) and can detrimentally affect the chemistry of the sample.

It is generally accepted that the sample dilution attributed to the addition of chemical preservatives should be limited to a maximum of 0.5% (ASTM, 2006b). The pH of samples should be checked upon arrival at the laboratory to ensure that appropriate sample preservation procedures were implemented in the field. If the pH is not where it is required to be, the laboratory will consider the sample to be inappropriately preserved and the sampling team will be contacted to discuss an appropriate course of action.

The most convenient method of chemically preserving ground-water samples is to use prepreserved containers (Figure 5.6). Prepreserved containers are either purchased by the laboratory already prepared or they can be prepared by the laboratory. These containers hold a fixed volume of the parameter-specific preservative and are shipped to the sampling team with information about which preservatives have been added to which containers. The advantages of this method of sample preservation are: (1) the sampling team does not have to handle preservatives; (2) theoretically, no errors associated with adding an incorrect preservative to a sample can be made by field personnel; and (3) time savings. There are also several limitations to using prepreserved containers. As when using vials and ampoules to add preservative to samples, the volume of preservative is fixed. Thus, difficulties can arise in the field if field verification of end pH determines that the volume of preservative provided in the container is insufficient. In this situation, sampling teams must have available additional preservative (the same as that used to prepare the container), so the required end pH can be achieved prior to shipping the sample to the laboratory. This may be impossible for the laboratory to provide if they have purchased prepreserved containers from a supplier. From a practical perspective,

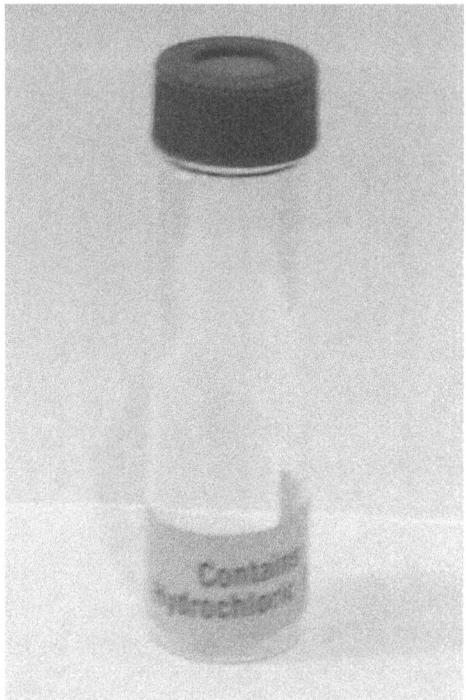

FIGURE 5.6
A prepreserved container used for chemical preservation of samples.

one common complaint of sampling team members using prepreserved containers is that it is easy to lose preservative from the container, either through accidentally knocking over the container during filling or through overfilling, especially when attempting to collect zero-headspace samples in 40 ml vials.

Another concern over prepreserved containers is related to the fact that concentrated preservatives may react with the empty container during storage prior to sample collection. For example, nitric acid, when in long-term storage in a high-density PE container, will chemically react with the container, resulting in alteration of the container walls (evidenced by orange staining inside the container) and, ultimately, as the authors have observed, failure of the container (the plastic container will crack linearly when squeezed). In glass containers stored in hot ambient conditions, acid preservatives will commonly evaporate to form an acid vapor, which is released to ambient air when the container is opened, leaving little preservative available to lower the pH of the sample and creating a breathing hazard for samplers. In addition, the vapor can deteriorate sample container lid threads — a problem that may not be detected until the container is taken into the field and the cap crumbles into pieces when it is removed for container filling. For these reasons, some sampling protocols do not permit the use of prepreserved containers. In other programs, prepreserved containers are allowed but with storage time restrictions that can vary from days to hours. It is recommended that sampling teams work with the laboratory and regulatory agencies during the planning phases of the ground-water sampling program to determine how long a prepreserved container can remain in storage prior to sample collection.

References

Anderson, M.A., J.F. Ferguson, and J. Gavis, Arsenate adsorption on amorphous aluminum hydroxide, *Journal of Colloid and Interface Science*, 54(3), pp. 391–399, 1975.

APHA, AWWA, and WPCF, *Standard Methods for the Examination of Water and Wastewater*, 17th ed, American Public Health Association, American Water Works Association and Water Pollution Control Federation, Washington, DC, 1989.

ASTM, Standard Guide for Field Filtration of Ground-Water Samples, ASTM Standard D 6564, ASTM International, West Conshohocken, PA, 5 pp., 2006a.

ASTM, Standard Guide for Field Preservation of Ground-Water Samples, ASTM Standard D 6517, ASTM International, West Conshohocken, PA, 6 pp., 2006b.

ASTM, Standard Guide for Packaging and Shipping Environmental Samples for Laboratory Analysis, ASTM Standard D 6911, ASTM International, West Conshohocken, PA, 6 pp., 2006c.

Backhus, D.A., J.N. Ryan, D.M. Groher, J.K. MacFarlane, and P.M. Gschwend, Sampling colloids and colloid-associated contaminants in ground water, *Ground Water*, 31(3), pp. 466–479, 1993.

Barrow, N.J., J. Gerth, and G.W. Brunner, Reaction kinetics of the adsorption and desorption of nickel, zinc, and cadmium by geothite II: modeling the extent and rate of reaction, *Journal of Soil Science*, 40, pp. 437–450, 1989.

Benjamin, M.M. and J.O. Leckie, Multiple-site adsorption of Cd, Cu, Zn, and Pb on amorphous iron oxyhydroxide, *Journal of Colloid and Interface Science*, 79(1), pp. 209–221, 1981.

Bergseth, H., Effect of filter type and filtrate treatment on the measured content of Al and other ions in ground water, *Acta Agriculturae Scandinavica*, 33, pp. 353–359, 1983.

Braids, O.C., R.M. Burger, and J.J. Trela, Should ground-water samples from monitoring wells be filtered before laboratory analysis? *Ground-Water Monitoring Review*, 7(2), pp. 58–67, 1987.

Danielsson, L.G., On the use of filters for distinguishing between dissolved and particulate fractions in natural waters, *Water Resources*, 16, pp. 179–182, 1982.

EPRI, Field Measurement Methods for Hydrogeologic Investigations: A Critical Review of the Literature, EPRI Report EA-4301, Research Project 2485-7, Electric Power Research Institute, Palo Alto, CA, 260 pp., 1985a.

EPRI, Sampling Guidelines for Ground Water Quality, EPRI Report EA-4952, Electric Power Research Institute, Palo Alto, CA, 47 pp., 1987.

Enfield, C.G. and G. Bengtsson, Macromolecular transport of hydrophobic contaminants in aqueous environments, *Ground Water*, 26(1), pp. 64–70, 1988.

Feld, J., J.P. Connelly, and D.E. Lindorff, Ground water sampling — addressing the turbulent inconsistencies, Proceedings of the First National Outdoor Action Conference on Aquifer Restoration, Ground Water Monitoring and Geophysical Methods, National Water Well Association, Worthington, OH, pp. 237–251, 1987.

Ferguson, J.F. and M.A. Anderson, Chemical forms of arsenic in water supplies and their removal, in, Chapter 7 *Chemistry of Water Supply, Treatment, and Distribution*, A.J. Rubin, Ed., Ann Arbor Science Publishers, Ann Arbor, MI, 1973.

Fetter, C.W., Potential sources of contamination in ground-water monitoring, *Ground-Water Monitoring Review*, 3(2), pp. 60–64, 1983.

Forbes, E.A., A.M. Posner, and J.P. Quirk, The specific adsorption of divalent Cd, Co, Cu, Pb, and Zn on geothite, *Journal of Soil Science*, 27, pp. 154–166, 1976.

Gibb, J.P., R.M. Schuller, and R.A. Griffin, Procedures for the Collection of Representative Water Quality Data from Monitoring Wells, Cooperative Ground Water Report 7, Illinois State Water and Geological Surveys, Champaign, IL, 61 pp., 1981.

Gillham, R.W., M.J.L. Robin, J.F. Barker, and J.A. Cherry, Ground-Water Monitoring and Sample Bias, American Petroleum Institute Publication, Environmental Affairs Department, Washington, DC, 206 pp., 1983.

Grundl, T. and J. Delwiche, Kinetics of ferric oxyhydroxide precipitation, *Geochimica et Cosmochimica Acta*, 56, 1992.

Gschwend, P.M. and M.D. Reynolds, Monodisperse ferrous phosphate colloids in an anoxic ground water plume, *Journal of Contaminant Hydrology*, 1, pp. 309–327, 1987.

Hem, J.D., Study and Interpretation of the Chemical Characteristics of Natural Water, Water Supply Paper 2254, U.S. Geological Survey, Reston, VA, 263 pp., 1985.

Hem, J.D. and C.E. Roberson, Form and Stability of Aluminum Hydroxide Complexes in Dilute Solution, Water Supply Paper 1827-A, U.S. Geological Survey, Reston, VA, 1967.

Horowitz, A.J., K.R. Lum, C. Lemieux, J.R. Garbarino, G.E. Hall, and C.R. Demas, Problems associated with using filtration to define trace element concentrations in natural water samples, *Environmental Science and Technology*, 30(3), pp. 954–963, 1996.

Jay, P.C., Anion contamination of environmental samples introduced by filter media, *Analytical Chemistry*, 57, pp. 780–782, 1985.

Kennedy, V.C., G.W. Zellweger, and B.F. Jones, Filter pore size effects on the analysis of Al, Fe, Mn, and Ti in water, *Water Resources Research*, 10(4), pp. 785–790, 1974.

Kent, R.T. and K.E. Payne, Sampling ground-water monitoring wells: special quality assurance and quality control considerations, in *Principles of Environmental Sampling*, American Chemical Society, Washington, DC, pp. 231–246, 1988.

Kim, J.I., G. Bachau, F. Baumgartner, H.C. Moon, and D. Lux, Colloid generation and the actinide migration in Gorbelen ground waters, in *Scientific Basis for Nuclear Waste Management*, Vol. 7, G.L. McVay, Ed., Elsevier Publishers, New York, NY, pp. 31–40, 1984.

Kinniburgh, D.G., M.L. Jackson, and J.K. Syers, Adsorption of alkaline earth, transition and heavy metal cations by hydrous gels of iron and aluminum, *Soil Science Society of America Journal* 40, pp. 796–799, 1976.

Laxen, D.P.H. and I.M. Chandler, Comparison of filtration techniques for size distribution in fresh waters, *Analytical Chemistry*, 54(8), 1350, pp, 1982.

McCarthy, J. and L. Shevenell, Obtaining representative ground-water samples in a fractured and karstic formation, *Ground Water*, 36(2), pp. 251–260, 1998.

NCASI, A Guide to Ground Water Sampling, Technical Bulletin 362 National Council of the Paper Industry for Air and Stream Improvement, 53 pp., 1982.

Nielsen, G.L., Ground-Water Sampling: Field Practices for Regulatory Compliance, Workshop Notebook, Tenth Annual Outdoor Action Conference, National Ground Water Association, Columbus, OH, pp. 89–97, 1996.

Pohlmann, K.F., G.A. Icopini, R.D. McArthur, and C.G. Rosal, Evaluation of Sampling and Field-Filtration Methods for Analysis of Trace Metals in Ground Water, EPA/600/R-94/119, U.S. Environmental Protection Agency, Office of Research and Development, Las Vegas, NV, 79 pp., 1994.

Posselt, H.S., F.J. Anderson, and W.J. Weber, Jr., Cation sorption on colloidal hydrous manganese dioxide, *Environmental Science and Technology*, 2, 1087, pp, 1968.

Puls, R.W., Adsorption of Heavy Metals on Soil Clays, Ph.D. dissertation, University of Arizona, Tucson, AZ, 144 pp., 1986.

Puls, R.W. and R.M. Powell, Acquisition of representative ground-water quality samples for metals, *Ground-Water Monitoring Review*, 12(3), pp. 167–176, 1992.

Puls, R.W. and J.H. Eychaner, Sampling Ground Water for Inorganics — Pumping Rate, Filtration and Oxidation Effects, Proceedings of the Fourth National Outdoor Action Conference on Aquifer Restoration, Ground Water Monitoring and Geophysical Methods, National Water Well Association, Dublin, OH, pp. 313–327, 1990.

Puls, R.W. and M.J. Barcelona, Ground-Water Sampling for Metals Analysis, EPA/540/4-89/001, Superfund Ground Water Issue, U.S. Environmental Protection Agency, Office of Solid Waste and Emergency Response, Washington, DC, 6 pp., 1989.

Puls, R.W. and M.J. Barcelona, Low-Flow (Minimal-Drawdown) Ground-Water Sampling Procedures, EPA/540/5-95/504, Ground Water Issue, U.S. Environmental Protection Agency, Office of Solid Waste and Emergency Response, Washington, DC, 12 pp., 1996.

Puls, R.W., D.A. Clark, B. Bledsoe, R.M. Powell, and C.J. Paul, Metals in Ground Water: Sampling Artifacts and Reproducibility, *Hazardous Waste and Hazardous Materials*, 9(9), pp. 149–162, 1992.

Puls, R.W., R.M. Powell, D.A. Clark, and C.J. Paul, Facilitated Transport of Inorganic Contaminants in Ground Water: Part II, Colloidal Transport, EPA/600/M-91/040, U.S. Environmental Protection Agency, Robert S. Kerr Environmental Research Laboratory, Ada, OK, 1991.

Reynolds, M.D., Colloids in Ground Water, Masters Thesis, Department of Civil Engineering, Massachusetts Institute of Technology, Cambridge, MA, 1985.

Ryan, J.N. and P.M. Gschwend, Colloidal mobilization in two atlantic coastal plain aquifers: field studies, Water Resources Research, 26, pp. 307–322, 1990.

Sawhney, B.L., Kinetics of cesium sorption by clay minerals, *Soil Science Society of America Proceedings*, 30, 565, pp., 1966.

Sheldon, R.W., Size separation of marine seston by membrane and glass-fiber filters, *Limnology and Oceanography*, 17 pp. 494–498, 1965.

Sheldon, R.W. and W.H. Sutcliffe, Jr., Retention of marine particles by screens and filters, *Limnology and Oceanography*, 14(3), pp. 441–444, 1969.

Smith, R.W. and J.D. Hem, Effects of Aging on Aluminum Hydroxide Complexes in Dilute Aqueous Solutions Water-Supply Paper 1827-D, U.S. Geological Survey, Reston, VA, 51 pp., 1972.

Sparks, D.L., L.W. Zelany, and D.C. Martens, Kinetics of potassium exchange in a paleodult from the coastal plain of virginia, *Soil Science Society of America Journal*, 44(1), pp. 37–40, 1980.

Stolzenburg, T.R. and D.G. Nichols, Preliminary Results on Chemical Changes in Ground-Water Samples Due to Sampling Devices, EPRI EA-4118, Project 2485–7, Electric Power Research Institute, Palo Alto, CA, 53 pp., 1985.

Stolzenburg, T.R. and D.G. Nichols, Effects of Filtration Method and Sampling on Inorganic Chemistry of Sampled Well Water, Proceedings of the Sixth National Symposium and Exposition on Aquifer Restoration and Ground Water Monitoring, National Water Well Association, Dublin, OH, pp. 216–234, 1986.

U.S. EPA, Compendium of ERT Ground Water Sampling Procedures, EPA/540/P-91/007, U.S. Environmental Protection Agency, Washington, DC, 63 pp., 1991.

U.S. EPA, Ground Water Sampling, A Workshop Summary, EPA/600/R-94/205, U.S. Environmental Protection Agency, Office of Research and Development, Washington, DC, 98 pp., 1995.

U.S. EPA, Test Methods for Evaluating Solid Waste, Physical/Chemical Methods, EPA SW846, U.S. Environmental Protection Agency, Office of Solid Waste and Emergency Response, Washington, DC, 1996.

Wagemann, R. and G.J. Brunskill, The effect of filter pore size on analytical concentrations of some trace elements in filtrates of natural water, *International Journal of Environmental Analytical Chemistry*, 4, pp. 75–84, 1975.

Yao, K., M.T. Habibian, and C.R. O'Melia, Water and wastewater filtration: concepts and applications, *Environmental Science and Technology*, 5(11), pp. 1105–1112, 1971.

6

Conducting a Ground-Water Sampling Event

David M. Nielsen and Gillian L. Nielsen

CONTENTS

Introduction

After the SAP has been prepared, and has received regulatory and client approval for implementation, the sampling team will travel to the field site to collect ground-water samples using the protocols and procedures contained in the SAP. Each sampling team member should have a personally assigned copy of the SAP on hand in the field to refer to if there is ever a need for clarification of a field procedure. Table 6.1 summarizes the field activities that will be implemented by a sampling team during a typical ground-water sampling event for a long-term ground-water monitoring program. A similar schedule of activities will be followed for most other ground-water sampling events. These activities are presented in the chronological order in which they would occur in the field. To be effective, sampling team members need to work together and develop a system to ensure efficient use of time and resources. This will come with time and experience as sampling team members work together at each field site. Sampling teams should create a system of checks and balances to ensure that errors are not made as a result of omission, use of improper protocols, unfamiliarity with equipment, or failure to read and understand the SAP. ASTM International has published a guide (ASTM Standard D 5903 [ASTM, 2006a]) on planning and preparing for a ground-water sampling event that provides an excellent "to do" list for sampling team members to follow to

TABLE 6.1

Typical Field Components of a Ground-Water Sampling Event for a Long-Term Monitoring Program

Review of site map for facility orientation
Prepare field notebook for daily entries
Check all sample containers supplied by the laboratory for breakage and to be sure all required containers (and a few extra containers) are available for each well to be sampled
Check all QC sample containers supplied by the laboratory
Calibrate all field equipment (e.g., PID and water-quality instrumentation)
Initial trip to inspect each well for structural integrity, measure headspace levels of volatile or combustible gases or vapors, take water-level or product-thickness measurements, and prepare the wellhead for sample collection by anchoring plastic sheeting on the ground surface (in the order from upgradient to downgradient locations)
Ensure that containment systems are in place to manage any purge water generated (e.g., 55 gal drums at each well) if required
Prepare purging and sampling equipment
Travel to first hydraulically upgradient well to be sampled with all equipment, sample containers, preservatives, filtration equipment, decontamination supplies, field notebook, sample labels, sample security tags and seals, chain-of-custody forms, shipping manifests, PPE, and garbage bags for solid waste
Take a second water-level measurement to provide information necessary for purging method (depth to static water level, height of the water column, and volume of the well)
Purge the well using the prescribed protocol, containing any purge water as necessary
Collect field water-quality data (i.e., indicator parameters)
Collect ground-water samples in the prescribed order of container filling, using required sample collection protocols
Filter and physically and chemically preserve any parameter specified in the SAP
Collect required field QC samples
Complete required labels, seals, and tags, affix to the appropriate sample bottles, and place the sample bottles in shipping containers for storage or transport during the rest of the day's field activities
Complete the chain-of-custody form
Dispose of any solid wastes generated during the sampling event (e.g., disposable gloves, plastic sheeting, disposable filters, disposable tubing, or suspension cord)
Clean any portable equipment prior to transporting it to the next sampling location
At the end of the day, complete the shipping manifest and ship samples to the laboratory

ensure that the sampling event is well organized and that delays associated with poor preparation are avoided. The guide also provides good check lists of typical sampling equipment and supplies needed in the field.

Field Preparation for a Ground-Water Sampling Event

Site Orientation and Sampling Event Preparation

Upon arrival at a field site, sampling team members should meet with facility personnel to learn of any changes related to facility operations, such as safety procedures, personnel changes, traffic pattern changes, locations of support systems such as fresh water or electricity for the sampling team to use, on-site construction, or other changes that have occurred at the site since the last sampling event. This is also a good time to be updated on field site conditions such as weather or any reports related to structural damage of any of the wells (e.g., if a well has been buried by waste or construction activities and if the protective casing of a well has been damaged by heavy equipment). It is helpful to have a current facility map available during this site orientation meeting to identify areas of concern on the map, especially if new sampling team members or facility personnel are present. This information can be of tremendous value to sampling teams as they plan their day and before they conduct an inspection of all wells scheduled for sampling during the event.

Following site orientation, the sampling team should relocate to either an established field office or to a predetermined area located upgradient and away from high traffic areas to prepare equipment and materials for the day's sampling event. This includes preparation of the field notebook; organization of sample containers, labels, and security tags and seals; organization of QC sample containers; and calibration of field instrumentation that will be used for headspace measurement in sampling points and for water-quality indicator parameter measurement. All instrumentation calibration should be done according to manufacturers' instructions under field conditions. The timing and frequency of calibration should be in accordance with the SAP, which will commonly require, at a minimum, daily calibration and periodic calibration checks of all equipment. After all equipment is checked and calibrated, the sampling team should visit each well to conduct the next phase of the sampling event.

Conducting the Sampling Event

Sampling Point Inspection

Prior to purging and sampling, the sampling team needs to physically inspect each well to ensure that it is structurally sound for sampling. As discussed earlier, reports on possible well damage from facility personnel are very helpful, especially if damage is severe enough to make location of the well difficult. Table 6.2 summarizes features that should be inspected by the sampling team prior to opening the protective casing and the well.

If the well inspection indicates that there is obvious or suspected damage, detailed field notes should be made and photographs taken (if permitted) to describe the damage, including mention of any report made by facility personnel. The SAP should be consulted to determine the most appropriate course of action. In general, if there is any possibility

TABLE 6.2

Well Inspection Checklist

Check identification markings on the well
Check the surface seal to ensure that it is intact with no cracks
Check the above-grade protective casing and the well cap to be sure neither has been damaged (or the cap removed)
Check to be sure that the locking mechanisms are in place and undamaged
Check to be sure that the protective bumper guards are in place and undamaged
Check to be sure that the protective painted surfaces are not weathered or altered or require repainting or etching
Check to be sure that the valve-box covers or vault lids are present and in good condition for flush-to-grade installations
Check to be sure that the vault lid security mechanisms are intact and rust-free for flush-to-grade completions
Check to be sure that the gasket seals in flush-to-grade completions are present and water tight
Check to be sure that there is no standing water inside the flush-to-grade vault; if water is present, note the depth, color, and appearance of any visible contamination of standing water (note especially if the water level is level with the top of the inner well casing cap)
Note if there is any flow of water into the vault from either ground surface or below ground surface (around the vault seal)

that the observed well damage could result in a detrimental impact on sample chemistry, the location should not be sampled because samples may not be representative of true formation water chemistry. An evaluation should be made as to whether the well can be repaired or whether damage is significant enough to warrant decommissioning of the existing well and replacing it with another. This determination may be made by project managers rather than sampling team personnel, but it should be made in a timely fashion to avoid the possibility of the damaged well acting as a conduit for downward movement of surface contaminants into the formation.

Well Headspace Screening

If the well appears to be structurally sound, instrumentation for well headspace screening should be used to take ambient or background readings and the readings recorded. After background has been established, the well cap should be removed and the probe quickly inserted into the well headspace. After the headspace reading has stabilized or peaked, the probe should be removed from the sampling point and allowed to cycle fresh ambient air through the instrument to purge any vapors that may be present in the instrumentation. The highest reading recorded and the type of instrument response (e.g., a rapid rise and drop or a gradual increase to stabilization) should be documented in the field notes. If readings indicate that volatile or combustible vapors or gases are present, the sampling team must refer to the SAP and the site health and safety plan to determine whether personnel should upgrade their PPE.

In some wells, samplers may hear the sound of air rushing out of the well or a whistling sound when the well cap is removed. This is usually indicative of a well not being properly vented and an air-pressure buildup occurring as a result of a hydraulic pressure increase in the formation since the last sampling event. The sound is the pressurized air being released from the well as it attempts to equilibrate with atmospheric pressure. Equilibration of air pressure should occur within the time taken to screen the wellhead, but the water level in the well may continue to recover for anywhere from several seconds or minutes (in high hydraulic conductivity formations) to several hours or even days (in low hydraulic conductivity formations).

Water-Level Measurement

After the water level in the well has stabilized, water-level measurements should be taken following the protocol documented in the SAP. To take a water-level measurement, samplers need to know the location of the surveyed reference measuring point on each well — this is the point to which all water-level measurements should be made. The reference measuring point should be physically marked on the well casing or outer protective casing but, in some cases, it may not be, so its description should be documented in the SAP. To ensure precision, the sampler should note the units of measure on the gauge tape and test the water-level gauge prior to lowering it into the well. After testing, the sampler should measure the depth to static water level and should take a minimum of three independent water-level readings (for precision) at least 5 to 10 sec apart. This is especially critical in unvented wells in which the water level may still be recovering; multiple measurements with different results indicate that the water level is not stable enough to record representative measurements. Results of water-level measurements should be recorded to an accuracy of ± 0.01 ft in the field notebook. Water levels should be measured in all wells at the site in as short a time interval as possible, before purging and sampling of any of the wells is attempted. These water-level data will be used for determining the direction, gradient, and rate of ground-water flow across the site. On large sites with many wells, a full day or more could be spent implementing this first phase of the ground-water sampling event. Between wells, the water-level gauge must be cleaned following protocols documented in the SAP to prevent potential cross-contamination of sampling locations.

Well Purging and Field Parameter Measurement

After well inspections are complete and the first set of water-level data is collected, the sampling team should return to the first upgradient sampling location to begin well purging and sampling. Prior to purging, a purge-water containment system must be in place. Commonly, this involves placing a 55 gal drum at each well or towing a 500 gal tank on a trailer behind the field truck to contain purge water.

The device selected for purging the well should be lowered into the well (if portable) or, in the case where dedicated pumps are installed, the accessory equipment required for pump operation should be brought to the well and set up on the plastic sheeting placed around it. For portable pumps, an effort should be made to closely match the length of the tubing used for the pump with the depth at which the pump will be set in the well. Excess tubing can affect the temperature of the water sampled and reduce the flow rate. Increases in temperature can affect dissolved gases and trace metals in samples (Parker, 1994; Stumm and Morgan, 1996).

Prior to purging the well, instrumentation to be used for water-quality analysis should be calibrated (Figure 6.1). If a flow-through cell is to be used, it should be assembled with the water-quality instrumentation, typically a multiparameter sonde, installed. The unit should be placed out of direct sunlight to avoid overheating of the cell and sensors. An effort should be made to keep tubing lengths that connect the flow-through cell to the pump discharge as short as possible. As mentioned earlier, excessive lengths of tubing may result in increases in temperature, which can have a detrimental effect on sample chemistry.

All sampling supplies, such as disposable gloves and paper towels, as well as sample containers, QC sample containers and supplies, preservatives, filtration equipment, labels, security tags and seals, the field notebook, and chain-of-custody forms, should be

FIGURE 6.1
Calibration of a multiparameter sonde to be used with a flow-through cell to measure indicator parameters during low-flow purging and sampling.

organized and ready for use at the well following purging. After all equipment is in place, purging should be conducted in accordance with the SAP.

Sample Collection Procedures

After a well has been purged and is deemed ready for sampling, the sampling team must disconnect any flow-through cell equipment from the pump discharge tube and prepare to collect samples for laboratory analysis, and field QC samples. Sampling team members must be consistent in the manner in which they collect and pretreat samples on a parameter-specific basis to ensure both accuracy and precision between sampling events. Table 6.3 provides a checklist of sample collection elements that sampling team members should verify at each well during a sampling event. Each of these items must be addressed in the site-specific SAP.

Order of Container Filling

The sampling team should assemble sample containers provided by the laboratory for each parameter or suite of parameters to be analyzed at that particular well and containers required for any field QC samples that will be collected at that location. The sampling team should verify against the SAP that containers provided are correct for the analytes of interest, and they should inspect each container to ensure that it is in good physical condition (clean, not damaged, good fitting caps and seals, etc.) and ensure that there are sufficient numbers of containers to meet the needs of the sampling program. The containers should then be arranged in the correct order for filling, particularly in situations where there may be an insufficient volume of water in the wells to fill all sample containers. U.S. EPA guidance recommends that ground-water samples be collected in a particular order, with those parameters that are most sensitive to handling being collected first, followed by those less sensitive to handling (U.S. EPA, 1991). Figure 6.2 illustrates the recommended order for sample container filling.

In addition to the sensitivity of a parameter to handling, it is critical that the relative importance or significance of each parameter be evaluated on a site-by-site basis when establishing the order of sample collection. For example, at a mining facility, there may be

TABLE 6.3

Checklist of Critical QA/QC Sample Collection Elements

All required water-quality measurements have been made, recorded, and checked for accuracy prior to disconnecting the flow-through cell
Laboratory analyses to be performed on samples from each well are confirmed
The correct sample containers and required sample volumes are checked and confirmed
The order of and methods for sample collection (bottle filling) are clearly documented
Field quality control samples to be collected (which kind, when, where, and how) are documented for each sampling location
The correct types of filtration equipment, including filters of the correct filter pore size, are present at the well head
Filter preconditioning procedures have been followed
The correct type and volumes of chemical preservatives (if required) are present
Procedures have been established to verify arrival temperature and end pH of samples requiring chemical preservation
Sample container labels and security tags and seals (if required) are ready to be completed
The appropriate chain-of-custody forms are available for completion immediately following sampling
Sample shipping containers, compliant with applicable DOT and IATA shipping regulations, are ready for delivering samples to the lab by hand, laboratory courier, or commercial carrier

no interest in VOCs at all, and total metals may be the major concern. In this situation, it is wise to collect this parameter first to ensure that a representative sample for the most important constituents can be submitted for analysis rather than leave it toward the end as is suggested by Figure 6.2. This is particularly critical when sampling low-yield wells, which may not have a sufficient volume of water available to fill all sample containers.

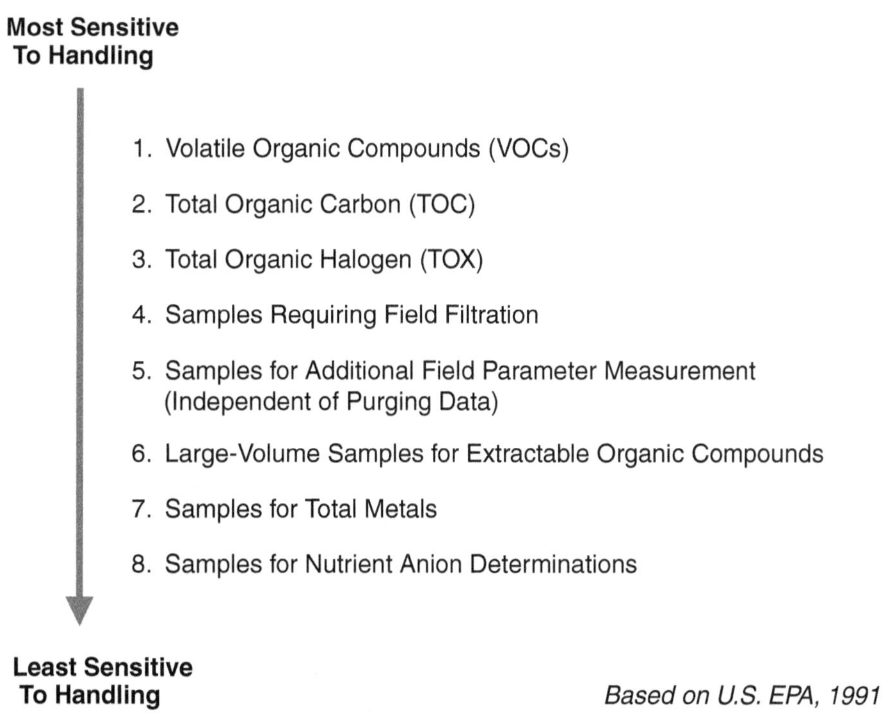

FIGURE 6.2
U.S. EPA guidelines on the order of sample container filling.

Sample Collection Protocols

In addition to establishing the order of sample container filling, sampling teams must follow correct procedures for collecting the ground-water samples. Without exception, ground-water samples should always be collected directly from the discharge tubing from the sampling device and at no time during container filling should anything but sample (e.g., discharge tubing, sampler's gloves, or filtration equipment) be allowed to enter the sample container or contact the mouth of the sample container. Use of funnels and transfer vessels should be avoided during sample collection because, as secondary forms of sample handling, they introduce potential sources of error and bias. Turbulent flow, aeration, and sample cross-contamination can result from the use of funnels and transfer vessels during sample decanting.

Caps should be kept on sample containers until the moment they are ready to be filled, and containers should be resealed immediately upon filling. At no time should the inside of the caps be allowed to come in contact with the ground surface, sampling equipment, or sampler's fingers. This can result in the transfer of contaminants to the inner cap surface and can introduce contaminants into the sample. If a cap is accidentally dropped onto the ground surface, it should be replaced with a new, clean cap.

Once delivered to ground surface, ground-water samples come into contact with atmospheric conditions. Sampling team members should make every attempt to minimize the time during which samples are exposed to atmospheric conditions, as a number of significant changes to the sample, affecting a wide range of analytes, may otherwise occur. Exposure of a sample to atmospheric conditions results in changes in the pressure and temperature of the sample. Additional changes include increases in the levels of DO and other gases and resultant changes in the redox state of ground-water samples. Most ground water is depleted in oxygen content due to chemical and biological reactions that occur during the infiltration process. When a ground-water sample is exposed to atmospheric conditions, the following processes may take place: oxidation of organics; oxidation of sulfide to sulfate; oxidation of ammonium to nitrate; and oxidation of dissolved metals to insoluble precipitates (Stumm and Morgan, 1996). The latter process is very important in terms of sample stability. Multivalent aqueous-phase species, such as Fe, Mn, and As, may be oxidized from a reduced state (Fe^{2+} to Fe^{3+}; Mn^{3+} to Mn^{4+}; and As^{3+} to As^{5+}), causing colloid-sized metal oxides and hydroxides to precipitate (Gillham et al., 1983; Puls et al., 1990; Ryan and Gschwend, 1990; Backhus et al., 1993). Because the kinetics of Mn oxidation are considerably slower than those for Fe, it is possible to collect a sample that is representative for one constituent (Mn) and not for another (Fe), depending on how rapidly the sample is preserved after collection (Gibb et al., 1981).

Because the oxidation of iron is particularly critical to maintaining sample integrity, it is worth discussing further. Under anoxic to suboxic conditions, ground water often contains high concentrations of dissolved iron (Fe^{2+}). Upon exposure to atmospheric conditions, Fe^{2+} oxidizes to Fe^{3+}, which precipitates as iron oxide or iron hydroxide, causes an increase in solution pH, and produces a rust-colored residue of colloid-sized particles in sample bottles. Iron hydroxide is known to adsorb or co-precipitate a number of other metals, including Cu, Zn, Co, Cd, As, Hg, Ag, Pb, V and even some organic species (Gibb et al., 1981; Gillham et al., 1983; Barcelona et al., 1984; Stoltzenburg and Nichols, 1986; Stumm and Morgan, 1996). The kinetics of oxidation-induced precipitation and subsequent sorption processes is such that they can occur within seconds or minutes (Reynolds, 1985; Puls et al., 1992). The end result is that a number of previously dissolved species are removed from solution and may be removed from the sample if it is filtered, resulting in significant negative bias for a number of analytes. Sample preservation methods (e.g., acidification) are meant to prevent such sample alteration, but are only effective if the sample is preserved prior to the occurrence of these reactions. The only commonly analyzed parameters that generally remain unaffected by exposure to atmospheric conditions are major ions.

A special sampling procedure must be used for samples collected for VOC analysis to prevent the loss of volatile constituents from the sample. VOC samples must be collected in specially designed 40 ml vials using a technique that is referred to as zero-headspace sampling, in which sample vials are filled at a relatively slow rate (Figure 6.3). Some U.S. EPA documents (i.e., U.S. EPA, 1986) recommend sampling at a rate of 100 ml/min, which is too slow for many sites, especially those where atmospheric contributions of contaminants are of concern. A sample collection rate of 200 to 250 ml/min is more reasonable for volatile constituents. This rate is fast enough to minimize contact with ambient air, but is not so fast that sample aeration, agitation, or turbulence occur during sample collection. To collect a zero-headspace sample, ground water is collected directly from the pump discharge tubing or the grab sampling device in a controlled manner. To fill the vial, the container is held on an angle and water is allowed to gently flow down the inside wall of the container. As the container fills, it is slowly straightened to vertical. Once vertical, the vial is filled until a positive meniscus forms on top of the water, taking care not to overfill the vial and wash out any chemical preservatives that may be in the vial (Figure 6.4). The cap is then carefully placed on top of the vial without disturbing the meniscus and tightened to the manufacturer-recommended degree of tightness. The samplers should then invert the vial and carefully tap it against the heel of their hand to check for the presence of any bubbles that may have been trapped in the vial during filling. If bubbles are detected, the sampler must refer to the SAP for procedures on how to handle the vials. If the bubbles are the size of a pinhead or smaller, one trick to implement in the field is to hold the vial vertically, carefully turn the cap one fourth to half turn back (as if taking the cap off) to release the pressure in the vial and then retighten the cap. Using this technique, most small bubbles will be lost and the sample will be zero headspace without having to open the vial and expose the sample to atmospheric conditions again. If, however, the bubbles are larger than pinhead size, it may be necessary to discard the vial and resample. Samplers should refrain from opening

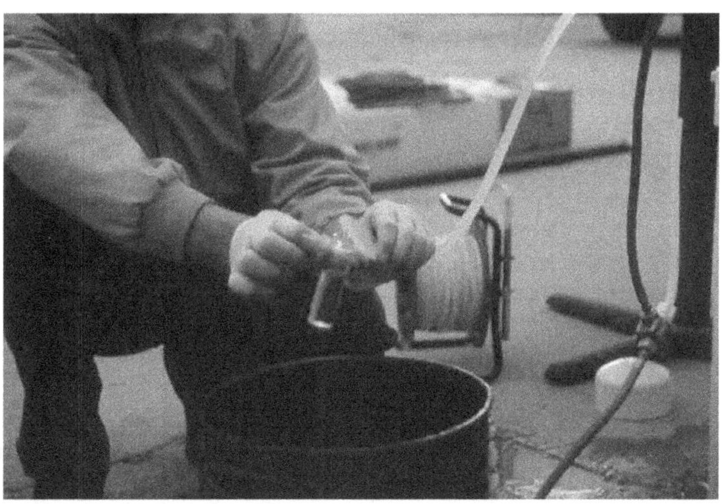

FIGURE 6.3
Collecting samples for VOC analysis requires use of a slow, controlled discharge rate. A rate of about 200 to 250 ml/min is a good compromise which allows collection of samples without agitation and turbulence, while minimizing the time with which the sample is in contact with atmospheric conditions.

FIGURE 6.4
To fill a VOC vial with zero-headspace, it is necessary to form a positive meniscus on the water surface prior to affixing the cap on the vial. For vials containing a chemical preservative, it is important not to overfill the vial to ensure that the preservative is not washed out.

a vial with headspace and "topping off" the sample until a zero-headspace sample is collected. This may result in either loss of constituents through volatilization into ambient air or contamination of the sample in some cases where atmospheric levels of volatile constituents are high. In addition, there is a risk that chemical preservatives will be washed out of the vial during this topping-off process, resulting in an improperly preserved sample.

For most parameters other than VOCs, sample collection rates of less than 500 ml/min are appropriate (Puls and Barcelona, 1996).

Protocols for Collecting Field QC Samples

Trip Blanks

A trip blank is prepared and provided by the laboratory as a standard QC sample. The laboratory ships a set of containers prepared for the required list of analytes that are filled with laboratory-prepared water (usually deionized water) of known and documented quality. The containers are labeled by the laboratory as being trip blanks and are shipped to the field along with the empty sample containers. Trip blanks should be documented by the laboratory on the accompanying chain-of-custody forms. After receipt in the field by the sampling team, the trip blanks should be inspected to ensure that all are present, that all containers are in good physical condition, and that there is no headspace in the VOC vials used for trip blanks. Any problems with trip blanks should be documented on the chain-of-custody forms and the laboratory should be notified immediately. The trip blanks should always be kept with the sample containers throughout the sample collection event and should be treated just like samples with regard to temperature control and packaging for shipment to the laboratory. At no time should any of the trip blank containers be opened and exposed to atmospheric conditions. If samples are sent to more than one laboratory, separate trip blanks should be submitted to each laboratory involved in sample analysis.

Temperature Blanks

To confirm that samples are appropriately preserved with respect to temperature ($4 \pm 2°C$), actual sample temperatures are measured upon arrival at the laboratory. Sample temperatures can be checked in one of three ways: (1) using a certified thermometer inserted into one or more sample containers; (2) using a calibrated infrared gun to determine the surface temperature of individual containers; or (3) using a specially prepared 40 ml sentry vial (see Figure 2.3) that contains a permanently affixed certified thermometer. Sentry vials, also referred to as temperature blanks, are usually supplied by the laboratory along with trip blanks, and they accompany samples throughout the sampling event. Temperature blanks should be packaged along with the ground-water samples and the rest of the QC samples and shipped to the laboratory. If the temperature measured upon arrival at the laboratory is outside of the allowable range, the sampling team will be contacted to discuss an appropriate course of action.

Field Blanks

To properly collect a field blank, the sampling team must order from the laboratory a set of containers prepared and filled in the same fashion as the trip blank described earlier, but these containers are labeled as field blanks. The deionized water-filled containers are accompanied by an identical but empty set of sample containers. To collect a field blank for parameters other than VOC analysis, the water-filled and empty containers are taken to the point of ground-water sample collection and the volume of water is transferred from the filled containers to the equivalent empty containers. The newly filled containers are labeled and sealed and the original containers are managed as part of the solid waste program for the sampling event. The purpose of this procedure is to expose the entire contents of the sample container to the same atmospheric conditions to which ground-water samples are exposed during sample collection. For VOCs, which require that samples be collected in 40 ml vials, it is strongly recommended that the sampling team make arrangements with the laboratory to have the deionized water sent in 60 ml vials rather than 40 ml vials, to ensure that the volume of water can be transferred with zero headspace. It is impossible to collect a zero-headspace sample when water is directly transferred from one 40 ml vial into another 40 ml vial. It is important that the chemistry of the 60 and 40 ml vials be identical to avoid introduction of a potential source of error.

Equipment Blanks

Two types of equipment blanks may be collected — a rinseate blank or a wipe or swipe blank. To collect a rinseate equipment blank, the field equipment is cleaned following the documented cleaning protocol. At the conclusion of cleaning, an aliquot of the final control water rinse is passed over and through the equipment just cleaned. The rinse water is collected directly into a sample container and is submitted to the laboratory for analysis. To collect a wipe or swipe equipment blank, the sampling team will make arrangements with the laboratory to provide a contaminant-specific wipe kit that contains some form of sterilized gauze or pad that is saturated in a contaminant-specific solvent (e.g., hexane for PCB determination). The sampling team will clean the equipment following the prescribed cleaning protocol. After cleaning, the saturated absorbent material is removed from its container (typically a threaded test tube, can, or vial) using forceps or tweezers and is wiped across the equipment covering a defined surface area (typically 100 cm^2), which may be delineated using a template. The absorbent material is then returned to its container and submitted to the laboratory for analysis. The laboratory analyzes the extract fluid and provides a count per surface area to indicate if a residue is present on the surface of the equipment following cleaning. These types of equipment blanks are appropriate to use on equipment that has a large surface area or has a surface area that is subject to trapping contaminants such as the exterior of pump tubing bundles.

Blind Duplicate Samples

There are two procedures for collecting blind duplicates, one for nonvolatile parameters and the other for volatile parameters requiring the use of 40 ml vials. To collect a blind duplicate sample for nonvolatile parameters, two identical sample containers are alternately filled until both are full. The number of times the samplers go back and forth between the two containers is largely a function of the type of sampling device used. If a pump is used, it is easy to go back and forth with the discharge tubing frequently until both containers are filled. If a grab-sampling device such as a bailer is used, samplers must first determine whether a bottom-emptying device is to be used to decant the sample from the device. If so, then samplers must alternate between containers frequently to avoid filling one container with the bottom portion of the water column in the bailer and the other with water from the top of the water column in the bailer. Alternating between containers will ensure that both containers receive an equal mix of upper and lower portions of the bailer water column. If a top-emptying procedure is used, the water column within the device is mixed during decanting, so the sampler will decant the sample into the two containers by going back and forth between them until they both are filled.

In cases where VOCs are sampled using 40 ml vials, a second strategy for sample collection must be implemented. As already discussed, the key to the successful collection of VOC samples is to ensure that the sample is collected with zero headspace and minimal exposure of the sample to ambient conditions. A typical sample for VOCs will consist of from two to four 40 ml vials from a single sampling point. To collect a duplicate sample for VOCs, the sampling team must assemble all of the sample containers for both the primary set and the duplicate set of samples. They must then alternately fill one vial from the primary set of containers, then the other vial from the duplicate set of containers. This process continues until all of the vials are filled in succession. All samples should be collected using the zero-headspace sampling technique described earlier.

Duplicate samples are referred to as "blind" samples because the sampling team should not indicate on the sample container label or chain-of-custody forms that one sample is a duplicate of another. This is done to prevent possibly biasing the laboratory's handling and analysis of the duplicate sample. It is recommended that a code of some sort be used by samplers to indicate which sample is a duplicate of another. The code selected should be consistent between sampling events and must be documented in field notes. Samplers should resist the temptation to change the time of sample collection for the duplicate sample to ensure that the laboratory will not determine which sample is a duplicate of another. This may be interpreted in a legal context as sample tampering or falsification of data, which can lead to serious consequences for the sampling team if the data are obtained through discovery as evidence in a court case.

When duplicate samples are analyzed, results reported should be within acceptable ranges for the analytical method used. If results do not meet this requirement, the challenge is then to determine why the duplicates were not close. Was there an error in how samples were collected that introduced sampling imprecision? Was there a problem in how the laboratory analyzed the samples? This determination can be difficult, but it is critical to make to assign responsibility and to implement corrective action for future sampling events.

Field Spiked Samples

To prepare a spiked sample, the sampling team collects a second set of duplicate samples as described earlier. To one of the duplicate samples, a spiking solution of one or more known compounds of known concentration is added. Typically, this spiking solution is a commercially prepared and certified matrix of compounds that are close relatives of compounds (e.g., different isomers) analyzed in the ground-water samples by the laboratory. This permits detection of the spiking compounds using the same analytical method used for the samples and avoids masking concentrations of related constituents in the samples. In theory, if there is no microbial activity and the sample has been appropriately preserved, the concentrations of the spiking compounds should be detected during the analysis of the spiked sample and there should be little difference between spiked concentrations and detected concentrations. If there is a large difference between the two values, it is interpreted that microbial activity has occurred and that sample preservation methods need to be improved or altered.

Field Split Samples

Field split samples are collected for the purpose of verifying the performance of one laboratory against a laboratory of known performance. Typically, field split samples are collected when a regulatory agency wishes to evaluate the performance of a new or unknown laboratory against the regulatory agency's approved or internal laboratory to ensure the accuracy of sample analysis. To collect field split samples, a team of samplers from the visiting (i.e., regulatory agency) sampling team will collect duplicate samples with the facility sampling team during a routine sampling event. The facility sampling team collects and handles its samples according to its SAP and sends samples to its laboratory, and the visiting sampling team submits their samples to their "known" laboratory.

Difficulties can arise in cases where sampling points have very low yield or insufficient water volume to permit collection of two full sets of samples. Preliminary planning for the sampling event must include a strategy on how to handle this situation. In most cases, the facility sampling team will have priority over the visiting sampling team to obtain a full set of samples in low-yield or low-volume wells and the visiting sampling team will not collect samples from those locations. This ensures that samples are submitted to the laboratory for analysis so data are reported to the regulatory agency as required for compliance. The alternate approach is to have the facility sampling team to collect all of their samples first and then allow the visiting sampling team to collect their samples from whatever water remains in the wells. Unfortunately, this approach introduces an uncontrollable source of variability, as this water may be of different chemistry and thus may make a meaningful comparison of laboratory results difficult or impossible.

If everything is as it should be, data generated by the two labs should be similar (at least within the performance standards of the analytical method used). If results are not close, then the challenge is to determine why there is a difference. Is it a case where the new laboratory's performance is substandard? Were there differences in laboratory methods used for sample preparation, extraction, or analysis? Where there differences in handling and field pretreatment of samples? It is important to make this determination to implement corrective action prior to the next sampling event.

Ground-Water Sample Pretreatment Procedures

Sample Filtration

As discussed in Chapter 5, positive-pressure filtration methods are preferred for ground-water sample filtration (U.S. EPA, 1991). There are two general categories of positive-pressure filtration equipment: (1) in-line filtration equipment used with pumping devices; and (2) remote pressurized filtration equipment that is not in line with a pumping device.

FIGURE 6.5
Positive-pressure filtration using an in-line cartridge filter is the preferred method for sample filtration. Here, the filter is being preconditioned prior to being used for sample filtration.

In-Line Filtration

In-line filtration (Figure 6.5) is recommended for ground-water sample filtration because it provides better consistency through less sample handling and minimized sample exposure to the atmosphere (Stolzenburg and Nichols, 1986; Puls and Barcelona, 1989, 1996). In-line filters eliminate the effects of turbulent discharge and can reveal the amount of aeration caused solely by the sampling mechanism (Stolzenburg and Nichols, 1986). These filters are available in both disposable (barrel filters) and nondisposable (in-line filter holder using flat membrane filters) formats and in various filter pore sizes (see Table 5.3).

Ground-water samples filtered using in-line filtration systems should be collected according to the following sequence:

1a. If using disk filters, assemble the disk filter holder and filtration equipment, so it is leak tight and the filter (handled with forceps) is centered on the holding device. Connect the filtration equipment to the discharge tubing of the pumping device.

1b. If using capsule filters, attach the filter directly to the discharge tubing of the pumping device.

2. Precondition the filter as previously described.
3. Initiate and gradually increase the flow of water through the filter to reach the appropriate rate and pressure, not to exceed the maximum recommended by the filtration equipment manufacturer (e.g., <65 psi) pressure for many capsule filters).
4. Collect the filtered ground-water sample directly into a prepared sample bottle.
5. Preserve the filtered ground-water sample as required on a parameter-specific basis.
6. Release the pressure from the filtration equipment and disconnect it from the sampling device discharge tubing.

7. Discard any disposable materials (e.g., filter media) in accordance with the site-specific waste-management provisions of the SAP.
8. Clean any equipment used for filtration of the next sample following decontamination procedures outlined in the SAP.

Positive-Pressure Filtration

Ground-water samples can also be filtered using positive-pressure equipment that is not operated in line with a pumping device. To operate this equipment, the following procedures should be implemented:

1. Assemble the filter holder and support equipment, checking to make sure the system is leak tight.
2. Precondition the filter medium as previously described.
3. Remove a sample of ground water from the monitoring well.
4. Carefully decant the sample into the filtration vessel (if not using the sampling device itself as the vessel) to minimize aeration, agitation, and turbulence and to prevent introduction of airborne contaminants into the sample during transfer.
5. Pressurize the filtration vessel using oil-free inert gas (e.g., nitrogen) or some type of oil-free air pump (e.g., hand-operated positive-pressure pump). Pressure should not exceed manufacturers' guidelines for the equipment in use.
6. Collect the sample directly into a prepared sample container.
7. Preserve the filtered ground-water sample as required on a parameter-specific basis.
8. Release the pressure from the filtration equipment.
9. Discard any disposable materials (e.g., filter media) in accordance with the site-specific waste-management provisions of the SAP.
10. Clean any equipment used for filtration of the next sample, following decontamination procedures outlined in the SAP.

Negative-Pressure Filtration

Negative-pressure filtration systems that require applying a vacuum or suction to draw samples through a filter medium are available for ground-water samples. However, because of the detrimental effects on sample chemistry caused by applying negative pressure (discussed earlier in this chapter), vacuum filtration of ground-water samples is not recommended (U.S. EPA, 1991). If, for some reason, negative-pressure equipment must be used, the following procedures should be implemented:

1. Assemble the filter holder and support equipment making sure it is leak tight.
2. Precondition the filter medium using the methods described previously.
3. Remove a sample of ground water from the monitoring well.
4. Decant the sample into the filtration vessel, taking care not to agitate the sample, increase turbulence, or introduce airborne contaminants into the sample during transfer.
5. Apply a negative pressure to the filtration vessel using a vacuum pump. The negative pressure applied should not exceed manufacturers guidelines for the equipment in use.

6. Collect the filtrate into a flask or other transfer vessel.
7. Release the negative pressure at the vacuum pump connected to the filtration equipment.
8. Transfer the filtrate into a prepared sample container, taking care not to agitate the sample, increase turbulence, or introduce airborne contaminants into the sample.
9. Preserve the filtered ground-water sample as required on a parameter-specific basis.
10. Discard any disposable materials in accordance with the site-specific waste-management provisions of the SAP.
11. Clean any equipment used for filtration of the next sample following decontamination procedures outlined in the SAP.

Sample Preservation

After samples are collected and, if required, filtered, most must be chemically preserved to protect the physical and chemical integrity of the sample. As mentioned earlier in this chapter, sample chemistry can change between the time of sample container filling and sample extraction or analysis in the laboratory. Samplers must be sure to add the required parameter-specific chemical preservative as documented in the SAP in accordance with the prescribed method (i.e., titration, addition of preservatives using ampoules or vials, or using prepreserved containers). To ensure adequate mixing of the preservative within the sample container, samplers should gently invert (not vigorously shake) the sealed sample container to mix the sample and preservative.

Following mixing the preservative, the end pH of the sample must be verified in the field to ensure that sufficient preservative has been added to those containers requiring pH adjustment. Details on how this should be performed are provided in Chapter 5.

Preparation of Sample Containers for Shipment

Following field verification of sample pH, the sampling team must prepare all sample containers for shipment or hand delivery to the laboratory. After a full set of samples has been collected from a well, all sample containers must be recorded in the field notebook and on the chain-of-custody form. The exterior surface of the sample containers must be wiped clean to ensure that adhesive labels (that are completed with pertinent sample-related information) will adhere to the container. This can be problematic when condensation forms on the exterior of sample containers. When this occurs, samplers should dry the container surface and then quickly place the label on the dried surface. After the label is in place, it can be covered with a single layer of clear packing tape around the entire container to form a water-proof seal over the label and the tape will help secure the label to the container. Packing tape should never be placed over the top of a container lid, especially two-part septum caps used for VOC vials. This can result in contamination of the sample during sample analysis.

In cases in which there is a need to verify that samples have not been tampered with following collection, it may be necessary to affix a security tag or seal around the lower edge of the container cap. Security seals must be affixed properly. For most sample containers, the seal may be draped over the container cap and attached to the neck of the container. However, for VOC vials and other containers that have a septum in the cap, the seal must be affixed around the cap and must not cover the septum (Figure 6.6). An improperly affixed seal (Figure 6.7) may compromise sample integrity through inclusion of the adhesives on the back of the seal during analysis. Security seals and tags are used

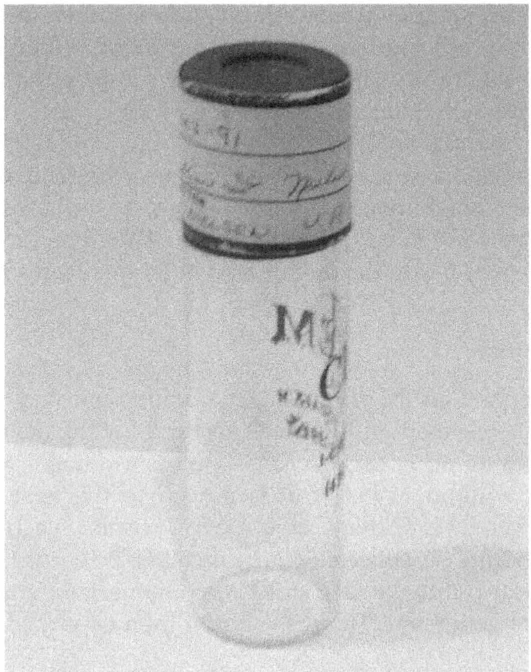

FIGURE 6.6
Sample seals on VOC vials must be affixed around the lid as shown here.

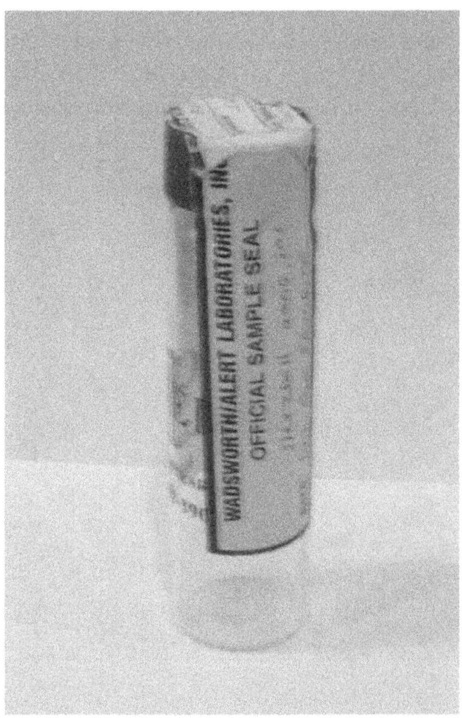

FIGURE 6.7
If the sample seal on a VOC vial covers the septum, the constituents of the adhesive from the seal may be detected in the analysis as the needle from the analytical instrument penetrates the septum to collect the sample.

as physical deterrents to sample tampering. When containers arrive at the laboratory, seals will be inspected in the sample reception department. If they are broken or missing, that is interpreted by laboratory staff to mean that samples may have been tampered with. At that point, a decision will need to be made by the project manager regarding how (or if) to proceed with sample analysis.

After samples are labeled and sealed, the containers will generally be placed on ice in some form of shipping container, such as a cooler. As indicated in Table 5.5, many parameters are required to be stored and shipped at 4°C. Temperature blanks should be placed in each sample shipper to document arrival temperatures of samples.

Cleanup of the Work Area

Following sample preparation for shipment, the equipment and work area around the sampling point must be cleaned and the well secured for the next sampling event. This typically involves cleaning all water-level measurement devices, purging and sampling devices, and support equipment before it is moved to the next sampling location. In addition, disposable items (e.g., disposable gloves, disposable filters, plastic sheeting, paper towels, etc.) should be collected and managed in accordance with the SAP. If required by the SAP, all purge water should be containerized and records made of the volume of purge water generated. The well should then be locked or otherwise secured.

Delivery or Shipment of Samples to the Laboratory

At the end of the day, samples are either hand delivered (preferred to avoid loss or damage of samples) or shipped to the laboratory for analysis. Samplers must be prepared to have to drive samples to the local station of a commercial carrier, which may be an hour or more from the field site, so time must be well managed in the field. ASTM Standard D 6911 (ASTM, 2006b) provides guidance on appropriate packaging and shipping of both regulated and unregulated ground-water samples.

FIGURE 6.8
All sampling teams should relax after a day of hard work and have a cold beverage to quench their thirst once the samples have been safely delivered to the commercial carrier or lab. Creative application of sampling equipment in this example has uncovered yet another viable use for the equipment. Any equipment used in this manner should be thoroughly cleaned prior to the next day's work.

After samplers have delivered the samples either to the laboratory or to the commercial carrier, they should hit the showers and grab a well-deserved beverage (Figure 6.8) after a day of hard work.

References

ASTM, Standard Guide for Planning and Preparing for a Ground-Water Sampling Event, ASTM Standard D 5903, ASTM International, West Conshohocken, PA, 4 pp., 2006a.

ASTM, Standard Guide for Packaging and Shipping Environmental Samples for Laboratory Analysis, ASTM Standard D 6911, ASTM International, West Conshohocken, PA, 6 pp., 2006b.

Backhus, D.A., J.N. Ryan, D.M. Groher, J.K. MacFarlane, and P.M. Gschwend, Sampling colloids and colloid-associated contaminants in ground water, *Ground Water*, 31(3), pp. 466–479, 1993.

Barcelona, M.J., J.A. Helfrich, E.E. Garske, and J.P. Gibb, A laboratory evaluation of ground-water sampling mechanisms, *Ground-Water Monitoring Review*, 4(2), pp. 32–41, 1984.

Gibb, J.P., R.M. Schuller, and R.A. Griffin, Procedures for the Collection of Representative Water Quality Data from Monitoring Wells, Cooperative Ground Water Report 7, Illinois State Water and Geological Surveys, Champaign, IL, 61 pp., 1981.

Gillham, R.W., M.J.L. Robin, J.F. Barker, and J.A. Cherry, Ground-Water Monitoring and Sample Bias, American Petroleum Institute Publication, Environmental Affairs Department, Washington, DC, 206 pp., 1983.

Parker, L., The effects of ground-water sampling devices on water quality: a literature review, *Ground-Water Monitoring and Remediation*, 14(2), pp. 130–141, 1994.

Puls, R.W., J.H. Eychaner, and R.M. Powell, Facilitated Transport of Inorganic Contaminants in Ground Water: Part I. Sampling Considerations, EPA/600/M-90/023, U.S. Environmental Protection Agency, Robert S. Kerr Laboratory, Ada, Oklahoma, 12 pp., 1990.

Puls, R.W. and M.J. Barcelona, Ground-Water Sampling for Metals Analysis, EPA/540/4-89/001, Superfund Ground Water Issue, U.S. Environmental Protection Agency, Office of Solid Waste and Emergency Response, Washington, DC, 6 pp., 1989.

Puls, R.W. and M.J. Barcelona, Low-Flow (Minimal-Drawdown) Ground-Water Sampling Procedures, EPA/540/5-95/504, Ground Water Issue, U.S. Environmental Protection Agency, Office of Solid Waste and Emergency Response, Washington, DC, 12 pp., 1996.

Ryan, J.N. and P.M. Gschwend, Colloidal mobilization in two atlantic coastal plain aquifers: field studies, *Water Resources Research*, 26, pp. 307–322, 1990.

Stolzenburg, T.R. and D.G. Nichols, Effects of Filtration Method and Sampling on Inorganic Chemistry of Sampled Well Water, Proceedings of the Sixth National Symposium and Exposition on Aquifer Restoration and Ground Water Monitoring, National Water Well Association, Dublin, OH, pp. 216–234, 1986.

Stumm, W. and J.J. Morgan, *Aquatic Chemistry: Chemical Equilibria and Rates in Natural Waters*, John Wiley and Sons, Inc, New York, NY, 1996.

U.S. EPA, RCRA Technical Enforcement Guidance Document, OSWER-9950, U.S. Environmental Protection Agency, Office of Waste Programs Enforcement, Office of Solid Waste and Emergency Response, Washington, DC, 208 pp., 1986.

U.S. EPA, Compendium of ERT Ground Water Sampling Procedures, EPA/540/P-91/007, U.S. Environmental Protection Agency, Washington, DC, 63 pp., 1991.

7

Acquisition and Interpretation of Water-Level Data

Matthew G. Dalton, Brent E. Huntsman, and Ken Bradbury

CONTENTS

Introduction

Importance of Water-Level Data

The acquisition and interpretation of water-level data are essential parts of any environmental site characterization or ground-water monitoring program. When translated into values of hydraulic head, water-level measurements are used to determine the distribution of hydraulic head in one or more formations. This information is used, in turn, to assess ground-water flow velocities and directions within a three-dimensional framework. When referenced to changes in time, water-level measurements can reveal changes in ground-water flow regimes brought about by natural or human influences. When measured as part of an *in situ* well or aquifer pumping test, water levels provide information needed to evaluate the hydraulic properties of ground-water systems.

Water-Level and Hydraulic-Head Relationships

Hydraulic head is the driving force for ground-water movement and varies both spatially and temporally. A piezometer is a monitoring device specifically designed to measure hydraulic head at a discrete point in a ground-water system. Figure 7.1 shows water-level and hydraulic-head relationships at a simple vertical standpipe piezometer (A). The piezometer consists of a hollow vertical casing with a short screen open at point P. The piezometer measures total hydraulic head at point P. *Total hydraulic head* (h_t) has two components — *elevation head* (h_e) and *pressure head* (h_p).

$$h_t = h_e + h_p$$

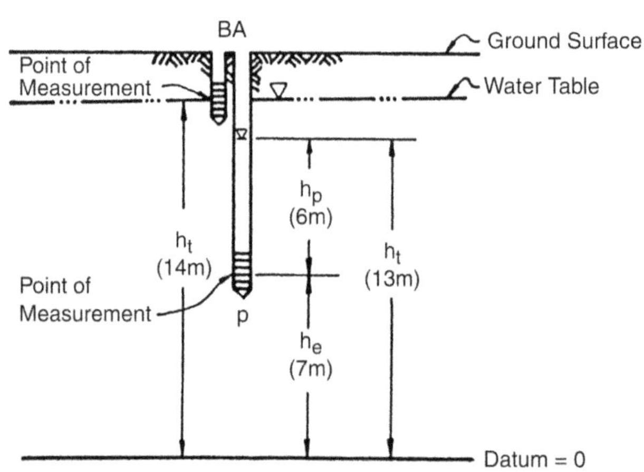

FIGURE 7.1
Hydraulic-head relationship at a field piezometer. (Adapted from Freeze and Cherry (1979). With permission.)

Elevation head (h_e) refers to the potential energy that ground water possesses by virtue of its elevation above a reference datum. Elevation head is caused by the gravitational attraction between water and earth. In Figure 7.1, the elevation head (h_e) at point P is 7 m.

Pressure head (h_e) refers to the force exerted on water at the measuring point by the height of the static fluid column above it (in this discussion, atmospheric pressure is neglected). In Figure 7.1, the pressure head (h_p) at point P is 6 m. Note that h_p is measured inside the piezometer and corresponds to the distance between point P and the water level in the piezometer.

Total hydraulic head (h_t) is the sum of elevation head (h_e) and pressure head (h_p). The total hydraulic head at point P in Figure 7.1 is $7 + 6 = 13$ m relative to the datum.

The water level in piezometer A is lower than the water level (at the water table) measured in piezometer B. The difference in elevation between the water-table piezometer (B) and the water level in the deeper piezometer (A) corresponds to the hydraulic gradient between the two piezometers. In this case, there is a downward vertical gradient because total hydraulic head decreases from top to bottom.

Hydraulic Media and Aquifer Systems

The "classic" definition of an *aquifer* as "a water-bearing layer of geologic material, which will yield water in a usable quantity to a well or spring" (Heath, 1983) was developed to address water-supply issues, but it is less useful for describing materials in terms of modern ground-water monitoring. Today, ground-water monitoring (including well installation, water-level measurement, and water-quality assessment) occurs in hydrogeologic media ranging from very low hydraulic conductivity shales, clays, and granites to very high hydraulic conductivity sands and gravels. The term aquifer (in ground-water monitoring) is used as a relative term to describe any and all of these materials in various settings.

Aquifers are also generally classified based on where a water level lies with respect to the top of the geologic unit. Figure 7.2 shows an example of layered hydrogeologic media forming both *confined* and *unconfined* aquifers. The confined aquifer is a relatively high hydraulic conductivity unit, bounded on its upper surface by a relatively lower hydraulic conductivity layer. Hydraulic head in the confined aquifer is described by a potentiometric surface, which is an imaginary surface representing the distribution of total hydraulic head (h_t) in the aquifer and which is higher in elevation than the physical top of the aquifer.

The sand layer in the upper part of Figure 7.2 contains an unconfined aquifer, which has the water table as its upper boundary. The water table is a surface corresponding to

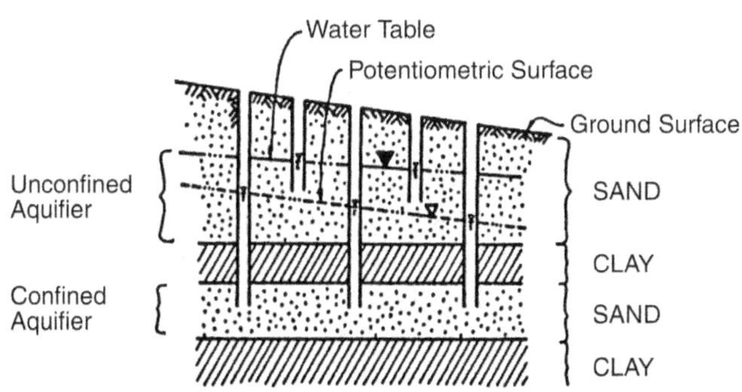

FIGURE 7.2
Unconfined aquifer and its water table; confined aquifer and its potentiometric surface. (Adapted from Freeze and Cherry [1979]. With permission.)

the top of the unconfined aquifer where total hydraulic head is zero relative to atmospheric pressure or the hydrostatic pressure is equal to the atmospheric pressure.

Notice that water levels in the piezometers in Figure 7.2 vary with the depth and position of the piezometer. This variation corresponds to the variation of total hydraulic head throughout the saturated system. Hydraulic head often varies greatly in three dimensions over small areas. Thus, the design and placement of water-level monitoring equipment is critical for a proper understanding of the ground-water system.

Design Features for Water-Level Monitoring Systems

An important use of ground-water level (hydraulic head) data from wells or piezometers is assessment of ground-water flow directions and hydraulic gradients. The design of ground-water monitoring systems must usually consider requirements for both water-level monitoring and ground-water sampling. In many cases, both needs can be accommodated with one set of wells and without installing separate systems. However, to collect acceptable water-level data, certain requirements need to be met, which may not always be consistent with the requirements for collecting ground-water samples. For example, additional wells may be required to fully assess the configuration of a water table or potentiometric surface over and above the wells that might be required to collect ground-water samples. Conversely, the design of wells to collect ground-water samples may differ from wells that are used solely to collect ground-water level data.

Water-level monitoring data are generally collected during two phases of a monitoring program. The initial phase is when the site to be monitored is being characterized to provide data to design a monitoring system. The second phase is when water-level data are being collected as part of the actual monitoring program to assess whether changes in ground-water flow directions are occurring and to confirm that wells used to provide ground-water samples are properly located (i.e., hydraulically upgradient and downgradient of a facility that requires monitoring). The latter data also provide a basis to determine the cause of flow-direction changes and to assess whether the monitoring system needs to be reconfigured to account for these changes.

To design a water-level monitoring system, a detailed understanding of the site geology is necessary. The site geology is the physical structure in which ground-water flows and, as such, has a profound influence on water-level data. It is very important that reliable geologic data be collected so that the water-level monitoring system can be properly designed and the water-level data can be accurately interpreted.

Sites at which there is a high degree of geologic variation require more extensive (and costly) water-level monitoring systems than sites that are comparatively more homogeneous in nature. The degree of geologic complexity is often not known or appreciated during the early phases of a site-characterization program, and it may require several stages of drilling, well installation, water-level measurement, and analysis of hydrogeologic data before the required level of understanding is achieved.

Piezometers or Wells?

Ground-water level measurements are typically made in piezometers or wells. Most ground-water monitoring systems associated with assessing ground-water quality are composed of wells rather than piezometers.

Piezometers are specialized monitoring installations; the primary purpose of which is the measurement of hydraulic head. Generally, these installations are relatively small in diameter (less than 1 in. in diameter if a well casing is used), or in some applications, it may not include a well casing and just consist of tubes or electrical wires connected to pressure or electrical transducers. Piezometers are not typically designed to obtain ground-water samples for chemical analysis, although the term piezometer has been applied to pressure measuring devices which have been modified to collect ground-water samples (Maslansky et al., 1987). Piezometers have traditionally had the greatest application in geotechnical engineering for measuring hydraulic heads in dams and embankments.

Wells are normally the primary devices in which water levels are measured as part of a monitoring system. They differ from piezometers in that they are typically designed so ground-water samples can be collected. To accommodate this objective, wells are larger in diameter than piezometers (usually larger than 1.5 in. in diameter), although sampling devices have been developed, which allow ground-water samples to be obtained from small-diameter wells (see Chapter 3).

Approach to System Design

Design of a water-level monitoring system should begin with a thorough review of available existing data. This review should be directed toward developing a conceptual model of the site geologic and hydrologic conditions. The conceptual model of the hydrogeologic system is used to determine the locations of an initial array of wells. Tentative decisions regarding drilling depths and the zone or zones to be screened should also be made using existing data. Existing wells may be incorporated into the array if suitable information regarding well construction details is available. Boring and well construction logs, surficial geologic and topographic maps, drainage features, cultural features (e.g., well fields, irrigation, and buried water pipes), and rainfall and recharge patterns (both natural and man-induced) are several of the major factors that need to be assessed as completely as possible.

The available data should be reviewed to identify:

- The depth and characteristics of relatively high hydraulic conductivity geologic materials (aquifers) and low hydraulic conductivity confining beds that may be present beneath a site
- Depth to the water table and the likelihood of encountering perched or intermittently saturated zones above the water table
- Probable ground-water flow directions
- Presence of vertical hydraulic gradients
- Features that might cause ground-water levels to fluctuate, such as well-field pumping, fluctuating river stages, unlined ditches or impoundments, or tides
- Probable frequency of fluctuation
- Existing wells that may be incorporated into the water-level monitoring program

The practical limitations of where wells can be located on a site should not be overlooked during this phase of the system design. Wells can be located almost anywhere on some sites; however, on other sites, buildings, buried utilities, and other site features can impose limitations on siting wells.

Number and Placement of Wells

The number of wells required to assess ground-water flow directions beneath a site is dependent on the size and complexity of the site conditions. Simple and smaller sites require fewer wells than larger or more hydrogeologically complex sites.

Many sites have more than one saturated zone of interest in which ground-water flow directions need to be assessed. High hydraulic conductivity zones may be separated by lower hydraulic conductivity zones. In these cases, several wells screened at different depths may be required at several locations to adequately assess flow directions in, and between, each of the saturated zones of interest.

The minimum number of wells required to estimate a ground-water flow direction within a zone is three (Todd, 1980; Driscoll, 1986). However, the use of just three wells is only appropriate for relatively small sites with very simple geology, where the configuration of the water table or potentiometric surface is essentially planar in nature, as shown in Figure 7.3.

Generally, conditions beneath most sites require more than three wells. Lateral variations in the hydraulic conductivity of subsurface materials, localized recharge patterns, drainage channels, and other factors can cause the potentiometric or water-table surface to be nonplanar.

On large or more geologically complex sites, an initial grid of six to nine wells is usually sufficient to provide a preliminary indication of ground-water flow directions within a target ground-water zone. Such a configuration will generally allow the complexities in the water table or potentiometric surface to be identified. After an initial set of data is collected and analyzed, the need for and placement of additional monitoring installations can be assessed to fill in data gaps or to further refine the assessment of the potentiometric or water-table surface.

Figure 7.4 shows a site at which leakage from a buried pipe has caused a ground-water mound to form. In this situation, a three-well array would not provide sufficient data to detect the presence of the mound and could result in a faulty assessment of the groundwater flow direction beneath the site.

Screen Depth and Length

After well locations are established, well screen depths and lengths should be chosen. Screen depths are generally determined during the drilling operation after a geologic log

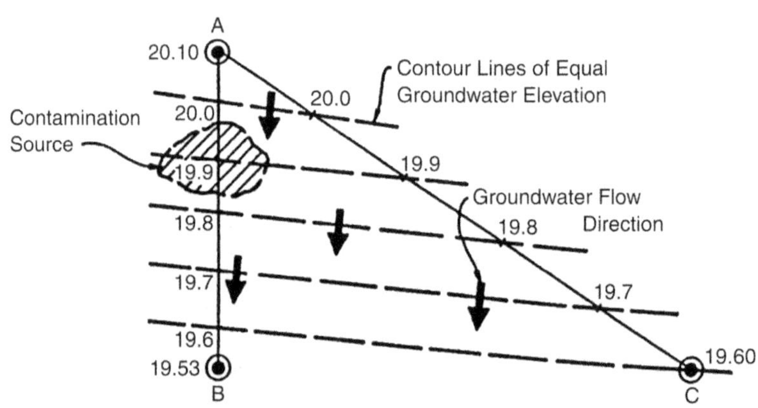

FIGURE 7.3
Assessing ground-water flow directions at a small site with a planar water-table surface.

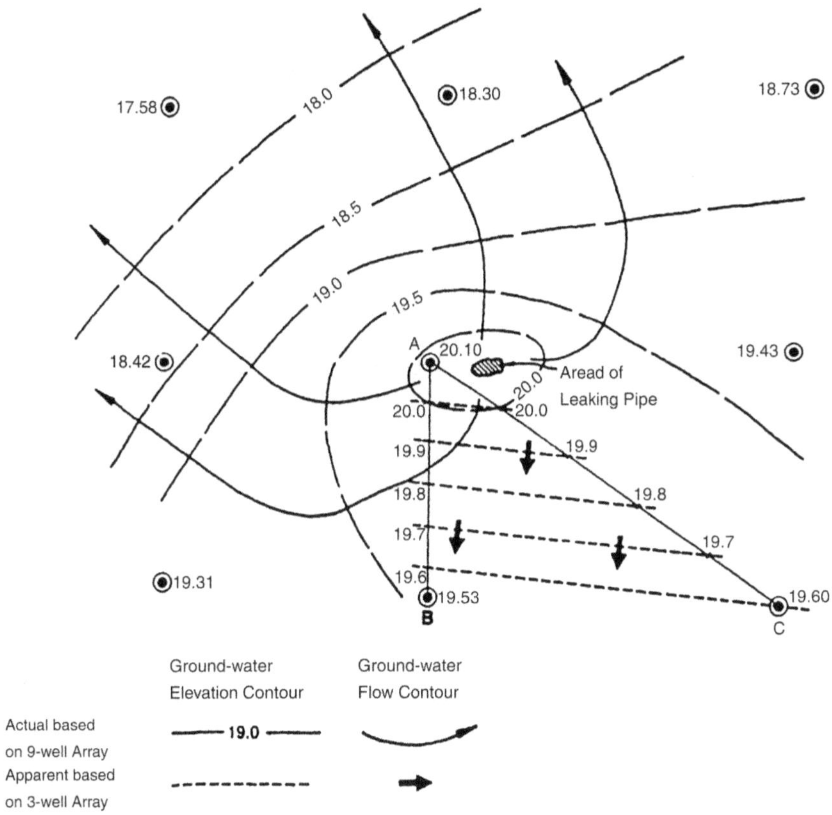

FIGURE 7.4
Estimation of ground-water flow directions with a three-well and a nine-well array.

has been prepared, depending on the amount and quality of the data available prior to drilling.

Wells used to assess flow directions within a zone are usually screened within that zone at similar elevations. Highly layered units may require screens in each depth zone that is isolated by lower hydraulic conductivity layers (Figure 7.5a). Where the units are dipping, it is generally more important to place the screens in the same zone even if the screens are not placed at similar elevations (Figure 7.5b).

Similar well-screen lengths should be used and the screen (and filter pack) should be placed entirely within the zone to be monitored. This will allow field personnel to obtain a water level that is representative of the zone being monitored and will minimize the possibility of allowing contaminants, if present, to migrate between zones screened by the well. If the well screen is open to several zones, then a composite or average water level will be measured, which will not be representative of any single zone, and will add to the difficulty in interpreting the water-level data. Typical commercially available well screens are 5 or 10 ft long, although it is possible to construct wells with longer or shorter screens, to meet specific project objectives.

If multiple saturated zones are present beneath a site, it is generally necessary to install either several wells screened at different depths at a single location or a multilevel monitoring system. Such installations allow the assessment of both horizontal and vertical hydraulic gradients. If few reliable data are available for a site, it is desirable that the initial hydrologic characterization starts with the uppermost zone of interest. During this initial work, a limited number of deeper installations can be installed to provide data

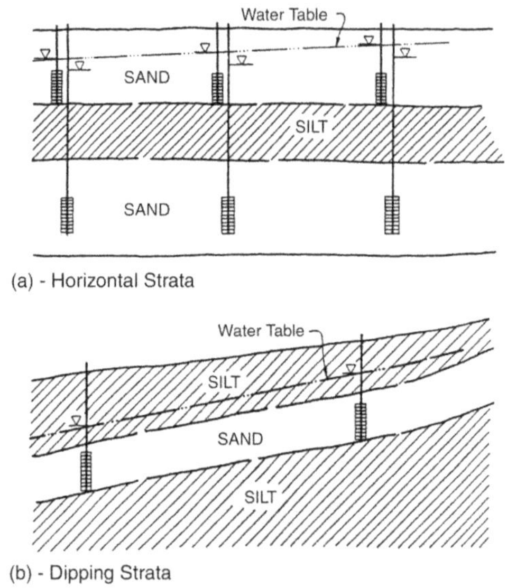

FIGURE 7.5
Well screen placement in horizontal and dipping strata.

to assess the need for additional deeper installations. In situations in which contamination is present in a shallow aquifer, extreme care must be exercised with regard to installing deeper wells, to prevent the possible downward movement of contamination into deeper zones.

Construction Features

Water-level monitoring points can be installed using a variety of methods and configurations (Figure 7.6). Typically, the installations are constructed in drilled boreholes, although

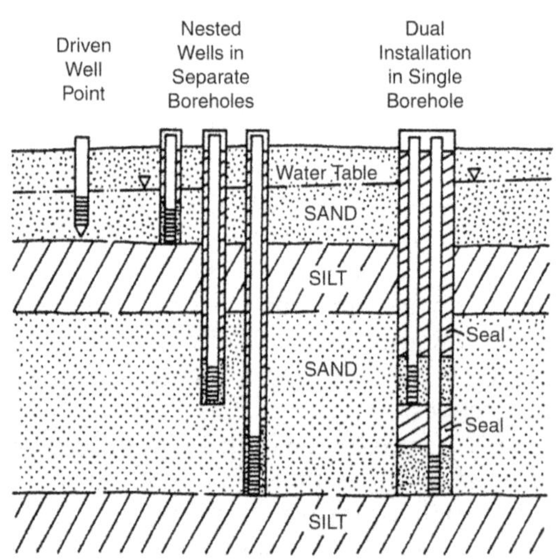

FIGURE 7.6
Typical monitoring well installation configurations.

driven well points can be used to provide water-level data in shallow, unconfined saturated zones.

At locations where multiple zones are to be monitored, single or multiple installations in the same borehole or multilevel systems can be used. If a single well is installed in a borehole, several boreholes will be necessary to monitor multiple zones.

A single installation in a single borehole is often preferred because it is easier to install a reliable annular seal above the well screen when only one well is completed in a borehole. An annular seal is necessary to ensure that the water-level data are representative of the zone being monitored and to ensure that contaminants do not move between zones within the borehole. In many situations, especially if a hollow-stem auger is being used to install the well, the cost of installing single installations is only marginally higher than multiple installations in a single borehole.

Multiple installations in a single borehole have been installed successfully as long as an adequate borehole or drill casing diameter is used and care is taken in installing the wells. Installing two 2 in. diameter wells per borehole should be feasible within 6 to 8 in. diameter boreholes or drill casings. While multiple installations in the same borehole may be technically feasible, some local well-drilling regulations may preclude or restrict such installations.

Water-Level Measurement Precision and Intervals

Wells should be accurately located horizontally and vertically, although horizontal surveying is not always required, depending on the size of the site and available base maps. The precision of the horizontal locations is generally not as important as the precision of the elevation survey and water-level measurements.

The top of the well casing (or other convenient water-level measuring point) should be surveyed to a common datum (usually National Geodetic Vertical Datum or NGVD) so that water-level measurements can be converted to water-level elevations. The reference point for water-level measurements should be clearly marked at a convenient location on each well casing. This will facilitate reducing measurement error.

The precision of the elevation survey and water-level measurements depends on the slope of the potentiometric or water-table surface and the distance between wells. Greater precision is required at sites where the surface is gradual or the wells are close together. Generally, reference point elevations should be surveyed and water levels measured with a precision ranging between ± 0.1 and ± 0.01 ft.

For example, if water-level fluctuations are occurring over a short period of time, it may be more important to obtain a set of less precise measurements in a short period of time rather than a very precise set of measurements over a longer period of time. In such cases, measurements made to 0.1 ft may be appropriate. In contrast, if the slope of the potentiometric surface or water-table surface is very gradual, more precise elevation control and water-level measurements may be required.

Current environmental regulations generally require that water levels be monitored and reported on a quarterly basis. A quarterly monitoring schedule may be appropriate for sites at which water levels fluctuate only in response to seasonal conditions, such as precipitation or irrigation recharge. However, water levels at many sites respond not only to seasonal factors but also to factors of shorter duration or greater frequency. These factors may include fluctuations caused by tides in coastal areas, changes in river stage, and daily well pumping, among others. Separate zones may also respond differently to the cause of the fluctuations.

During site-characterization activities, factors that may cause water levels to fluctuate need to be assessed and their importance evaluated with respect to two issues:

- The time in which a set of water-level measurements needs to be obtained
- How the flow directions may change as the water levels fluctuate

With the advent of computer technology, our ability to analyze complex systems at a reasonable cost has increased dramatically. Microprocessors connected to transducers allow the collection and analysis of water-level data over extended periods of time. To determine a site-specific monitoring interval, continuous monitoring can be economically accomplished in selected wells screened at different depths and at varying distances from the cause of the fluctuation. These data can then be used to determine the time frame and intervals in which to obtain water-level measurements and to determine how the various zones beneath the site respond to the cause of the fluctuation.

The period in which the continuous monitoring should be conducted depends on the frequency and duration of the fluctuations. If possible, monitoring should be conducted at times of representative fluctuation. For example, on sites affected by tides, monitoring over several tidal cycles during relatively high and low tides may be warranted.

Reporting of Data

Interpretation of water-level data requires that information be available about the monitoring installations and the conditions in which the water-level measurements were made. This information includes:

- Monitoring installations
  a. Geologic sequence
  b. Well construction features, especially screen and sand pack length, and geologic strata in which the screen is situated
  c. Depth and elevation of the top and bottom of the screen and sand pack
  d. Measuring point location and elevation
  e. Casing stickup above ground surface
- Water-level data
  a. Date and time of measurement
  b. Method used to obtain the measurement
  c. Other conditions in the area that might be affecting the water-level data, such as tidal or river stage, well pumping, storm events, etc.

Water-Level Data Acquisition

For many purposes in ground-water investigations, the accurate determination of water levels in wells or piezometers is paramount. Without accurate measurements, it is not possible to interpret the data to assess conditions such as ground-water flow directions, ground-water flow velocities, seasonal variations in water levels, aquifer hydraulic conductivity, and other important features.

Depending upon the ultimate use of the water-level data, the methods and instruments used to collect and record changes in ground-water levels may vary substantially. Water-level data acquisition techniques are divided into two major categories for discussion purposes: manual measurements or typically nonrecording methods and continuous measurements using instruments that provide a record. Although not exhaustive, the following discussion describes techniques most frequently used by the practicing hydrogeologist. These methods are summarized in Table 7.1.

Manual Measurements in Nonflowing Wells

Accurate manual measurements of water levels in wells and piezometers should be a core skill for any practicing hydrologist, hydrogeologist, ground-water scientist, or technician. Regardless of the method used, repeated measurements of water levels in wells made within a few minutes and within 200 ft of the top of casing should agree within 0.01 or 0.02 ft. As a standard of good practice, Thornhill (1989) suggests that anyone obtaining a water-level measurement in a well should take at least two readings. If they differ by more than 0.02 ft, then continue to measure until the reason for the lack of agreement is determined or until the results are shown to be stable.

TABLE 7.1

Summary of Methods for Manual Measurement of Well-Water Levels in Nonflowing and Flowing Wells

Measurement Method	Measurement Accuracy (ft)	Major Interference or Disadvantage
Nonflowing Wells		
Wetted chalked tape	0.01	Cascading water or casing wall water and chalk in water
Air line	0.01–0.25	Air line or fitting leaks; gage inaccuracies and air source
Electrical probes	0.02–0.1	Cable wear; hydrocarbons on water surface and turbulence
Transducer	0.01–0.1	Temperature changes; electronic drift; blocked capillary
Float	0.02–0.5	Float or cable drag; float size and lag
Popper	0.1	Well noise; well pipes and pumps; well depth
Acoustic probe	0.02	Cascading water; hydrocarbon on water surface
Ultrasonic	0.02–2.4	Temperature changes; well pipes and pumps; casing joints
Radar	0.01–0.02	Temperature; humidity; well pipes and pumps, small wells
Laser	0.01	Nonstraight wells; beam penetration through water
Flowing Wells		
Casing extensions	0.1	Limited range; awkward to implement
Manometer and pressure gage	0.1	Gage inaccuracies; calibration required
Transducers	0.02–0.1	Temperature changes; electronic drift

Wetted Chalked Tape Method

Although less commonly utilized in today's hydrologic assessments, one of the most accurate techniques used to manually measure ground-water levels is the wetted chalked tape method (ASTM D 4750; ASTM, 2006a). The equipment needed to make a measurement using this method consists of a standard steel surveyor's tape, a block of carpenter's chalk, and a slender lead or stainless-steel weight. Steel tapes and hand reels are commercially available in lengths up to 1000 ft. It is recommended, however, that shorter standard lengths (100, 200, 300, and 500 ft) be used because of weight and cost. Steel-tape markings are usually divided only into tenths of feet or inches and fractions of an inch. Interpolation to the nearest 0.01 ft is possible.

The weight is attached to the steel tape end clip with sufficient wire for support, but not enough to be stronger than the tape. This allows the tape to be pulled free if the weight become snagged on something in the well. The bottom 2 or 3 ft of the tape is coated with carpenter's chalk. A water-level measurement is made by lowering the tape slowly into the well, about 1 or 2 ft into the water. It is convenient to lower the tape into the water a sufficient distance to allow the tape to read an even foot mark at the top of the well casing or the reference measuring point at the surface. The water-level measurement is calculated by subtracting the submerged distance, as indicated by the absence of change in chalk color, from the reference point at the top of the well.

The practical limit of measurement precision for this method is ± 0.01 ft (U.S. Geological Survey, 1980). Coefficients of stretch and temperature expansion of the steel tape become a concern when water-level measurements are made in wells that have higher temperatures or at depths greater than 1000 ft (Garber and Koopman, 1968). For most ground-water investigations, corrections for these errors are not necessary.

A disadvantage of using the wetted chalked tape method is that if the approximate depth to water is unknown, too short or too long a length of chalked tape may be lowered into the well, thereby necessitating a number of attempts. In addition, water condensed on the side of the casing, or cascading water, may wet the tape above the actual water level and result in errors in measurement (Everett, 1980). When compared with other manual measurement techniques, this method is more time consuming. Proficiency in obtaining water levels with a wetted chalked tape requires practice. In addition, the introduction of chalk into a well that is used to obtain water-quality samples is discouraged.

Air-Line Submergence Method

The air-line submergence method, although less precise than other manual water-level measurement methods, continues to be a preferred technique in wells that are being pumped. To make an air-line measurement of water level in a well, a straight, small-diameter tube of accurately known length is installed in the well. This tube, usually 0.375 in. or less in diameter, can be made of plastic, copper, or steel. The air line and all connections must be air tight, without bends or kinks, and installed to several feet below the lowest anticipated water level. A pressure gage (preferably calibrated in feet of water), along with a fitting for an air source, is attached to this line. In deep wells or where multiple water-level measurements are needed, a small air compressor is useful. In shallow wells, a hand-operated air pump is typically used.

A water-level measurement is made when air is pumped into the small tube and the pressure is monitored. Air pressure will continue to increase until it expels all water from the line. Air pressure, which is determined when the pressure gage stabilizes, is used to calculate the height of the water in the tube. If the pressure gage is calibrated in pounds per square inch (psi), a conversion is made to feet by multiplying the psi reading by 2.31.

The actual water level in the well is determined by subtracting the calculated distance from the air line's length. According to Driscoll (1986), the dependability of measurements made by the air-line method varies with the accuracy of the pressure gage and the care used in determining the initial pressure reading. Depth to water can usually be determined to within 0.25 ft of the true water level. Garber and Koopman (1968) have also shown that the precision of the measurement is mainly dependent upon the accuracy of the pressure gage. They state that even with gages having gradations as small as 0.1 psi, the maximum possible resolution would be 0.23 ft. Digital quartz pressure transducers and specialized data loggers have been tested as replacements for standard pressure gages. Water levels from 0 to 50 ft have been measured with an accuracy of better than 0.01 ft (Paroscientific, Inc., 2002). However, these precision pressure measurement systems are designed for more permanent installations, not for portable applications. Unless the air-line method is used in wells of substantial depth, corrections for thermal expansion, hysteresis, fluid density, and barometric pressure are not necessary.

Electrical Methods

Currently, the most favored technique for manual water-level measurement is the use of an electrical probe. The most widely used instrument of this type is one that operates on the principle that a circuit is completed when two electrodes come in contact with the water surface in the well, which is conductive. Other instruments rely on physical characteristics such as resistance, capacitance, or self-potential to produce a signal. Many of these instruments employ a two-wire conductor that is marked every foot, with minor interval markings of 0.01 ft. Some instruments use vinyl-, epoxy-, or Teflon-clad steel tapes as an insulated electrode and the well casing or grounding wire as the other electrode. Because of weight and the amount of potential cable stretch, most commercial electrical probes are designed for water-level measurements within several hundred feet of the top of casing.

Water-level probes that use self-potential typically have one electrode made of magnesium and the other made of brass or steel. When the probe comes into contact with water, a potential between the two dissimilar metals is measured at the surface on a voltmeter (generally in millivolts).

If a battery is added to the circuit, the two electrodes may be of the same material, usually brass, lead, or ferrous alloy. When the electrodes come into contact with the water surface, the water conducts the current and a meter, light, or buzzer is activated at the ground surface.

The principles of capacitance and inductance have been used by the U.S. Geological Survey to detect water surfaces (Garber and Koopman, 1968). These are basically specialty instruments and few are available commercially for common water-level measurements. However, some units that employ capacitance or inductance are used for detection of water levels and hydrocarbons in wells. These units have the same apparent accuracy and precision as other electrical probes because the sensing elements are suspended in the well via multiwire conductors.

Errors in water-level measurements using electrical probes result from changes in the cable length and diameter as a function of use, depth, and temperature. After repeated use, the markings on the drop line often have a tendency to become loose and slide (if banded) or become illegible from wear (if embossed). Shallow measurements made with well-maintained electrical probes are typically reproducible to within ± 0.02 ft. Because of kinks in the cable and less than vertical suspension in a well, Barcelona et al. (1985) stated that the accuracy of electrical probes is about 0.1 ft. Plazak (1994) showed that even the same make and model of electrical probes may vary ±0.01 to ±0.11 ft in precision and accuracy, depending on the depth of the water-level measurement.

A disadvantage inherent in most electrical probe instruments is that if substantial amounts of oil or other nonconductive materials (i.e., oils) are floating upon the water surface, contact cannot be reliably made. This is a major concern in ground-water investigations involving petroleum hydrocarbon releases. Special sensing probes utilizing an optical or infrared sensor in conjunction with electrical conductivity are commercially available to measure the hydrocarbon-water interface. Because this type of sensing probe is also suspended from multiconductor wire, the same errors as previously discussed for electrical probes apply.

Pressure Transducer Methods

With the advent of reliable silicon-based strain gage pressure sensors and vibrating-wire transducers, a unique type of instrument is being commercially marketed for measuring changes in water levels. These transducers contain a 4–20 mA current transmitter and a strain gage sensor or a vibrating wire in an electromagnetic coil with frequency measurement circuitry. The current transmitter circuitry in both types of transducers prevents measurement sensitivity from being affected by cable length. Because all sensitive electronics are in the transducer and submerged in a constant temperature environment (the well water), errors due to temperature fluctuations are negligible (In Situ, Inc., 1983; Zarriello, 1995). The simultaneous measurement of temperature and water level is becoming a standard feature for most of the transducers used in hydrogeologic studies.

Many transducers used for measuring ground-water levels have a small capillary tube shielded in the support cable leading from one side of a differential pressure sensor. This tube is vented to the atmosphere, which provides automatic compensation of barometric pressure. Care must be taken when working with transducer cables that contain capillaries to avoid kinking, crushing, or allowing condensate to form in the vent tube. Blocked vent tubes may result in erroneous water-level measurements. A signal conditioning unit and a power source are required ancillary equipment to make a water-level measurement.

To avoid the need for electrical cables to transmit signals from the transducer to the data storage unit, some manufacturers have totally sealed the data logger, battery, pressure transducer, and temperature sensor in a small stainless-steel case for total submersion in a well. Communication with the data logger is established via an infrared optical port, either with a cradle component or through extension cables connected to a host computer. These units use an absolute pressure sensor to avoid the need for a vent tube to the surface. However, all water-level readings obtained by this type of monitor will require subsequent corrections for barometric pressure changes (Solinst Ltd., 2001). For a discrete water-level measurement, the transducer is lowered a known distance into the well and allowed to equilibrate to the fluid temperature. The distance of submergence of the transducer is read on the signal conditioning unit and is subtracted from the known cable length referenced at the top of the well.

This technique is easily adaptable to continuous monitoring. It also offers several advantages in ease of accurate measurement in both pumping wells and wells with cascading water. Sources of error in this type of instrument include the electronics (linearity, accuracy, temperature coefficient, etc.), temperature changes, and inappropriate application (i.e., range and material of construction) of a transducer in a given medium (Sheingold, 1980; Zarriello, 1995). Because of the sensitive electronics, rough handling of the transducers in the field or in storage should be avoided.

The accuracy of water-level transducers is dependent upon the type and range (sensitivity) of the device used. Most transducers are rated in terms of a percent of their full-scale capability. For example, a 0 to 5 psi tranducer rated at 0.01% will provide measurements to the nearest 0.01 ft. In contrast, a 0 to 25 psi transducer rated at 0.01% will provide measurements to the nearest 0.05 ft (Barcelona et al., 1985). Standard practices for the static calibration of electronic transducers used for obtaining field pressure measurements have been developed and should be used to document the accuracy of the instrumentation system (ASTM D 5720; ASTM, 2006b). These calibration procedures are typically included in the standard operating procedures prepared for any large-scale hydrogeologic investigation in which electronic pressure transducers are used.

Float Method

As the name implies, a float is attached to a length of steel tape and suspended over a pulley into the well. At the opposite end of the steel tape, a counterweight is attached. The depth to water is read directly from the steel tape at a known reference point at the top of the casing.

To obtain an accurate measurement using this technique, the absolute length of the float assembly must be measured and subtracted from the steel-tape measurement. For greater accuracy, the total amount of float submergence should be calculated and a correction factor applied. This becomes more critical with smaller diameter floats (Leupold and Stevens, Inc., 1978). This method is used principally to obtain continuous water-level measurements. The accuracy and errors in float-operated devices will be discussed in greater detail in the following section.

Sonic or Audible Methods

Virtually every practicing hydrogeologist has (but should not have) dropped a rock down a well, at one time or another, to determine whether water is present and to estimate the depth to water. Stewart (1970) investigated and developed a technique to determine the depth to water by timing the fall of a BB (air rifle shot) or a glass marble and by recording the time of the return sound of impact. This sonic technique will not be discussed here in detail because of the rather large range of error in measurement (± 5 ft), but interested readers are referred to Stewart (1970). Other sonic methods are described subsequently.

Popper

The most simplistic device used to audibly determine the depth to water in a well casing is a popper (also called a plopper). This is a metal cylinder from 1 to 1.5 in. in diameter and generally 2 to 6 in. in length, with a concave bottom. The popper is attached to a steel tape and lowered to within a few inches of the water surface in the well. By repeatedly dropping the popper onto the water surface and noting the tape reading at which a distinctive "pop" is heard, the depth to water is determined (U.S. Bureau of Reclamation, 2001).

Because of noise and the lack of clearance, the use of poppers in pumping wells is limited. The accuracy of water-level measurements made by this technique is highly dependent upon the skill of the measurer and the depth of the well. Determination of the water level to within 0.02 ft is usually the detection limit of this procedure.

Acoustic Probe

A unique adaptation of the popper principle was developed by Schrale and Brandwyk (1979), with the construction of an acoustic probe. This electronic device is attached to a steel tape and lowered into the well until an audible sound is emitted from a battery-powered transducer contained in the probe. The electric circuit is completed when the two electrodes placed in the bottom of the probe come in contact with the water level in the well. As with the previously discussed electrical methods, problems with measurements can occur when hydrocarbons are present or if the well has cascading water. According to the developers of this instrument, a water-level determination is possible to within ±0.02 ft.

Ultrasonic Methods

Instruments that measure the arrival time of a reflected transmitted sonic or ultrasonic wave pulse are becoming more common in the measurement of water levels. These instruments electronically determine the amount of time it takes for a sound wave to travel down the well casing, reflect off the water surface, and return to the surface. Because the electronic circuitry typically uses microprocessors, this signal is transmitted, received, and averaged many times a second. The microprocessor also calculates the depth to water and displays it in various units. Several of the commercially available instruments simply rest on top of the well casing with nothing being lowered into the well. Rapid determination of water depths in deep wells is a distinct advantage of this technique.

The presence of hydrocarbons on the water surface usually has no effect on the measurement. Accuracy can be limited by change of temperature in the path of the sound wave and other reflective surfaces in the well (i.e., pipes, casing burrs, pumps, samplers, crooked casing, etc.). Large variations in humidity will also effect readings. Most commercially available hand-held units can measure the depth to water within 0.1 ft if the well's temperature gradient is uniform. Usually, the greater the depth to water, the less accurate the measurement. One manufacturer reports a ±0.2% accuracy over a range of 25 to 1200 ft. Specialized installations, however, have repeatedly provided water-level measurements accurate to within ±0.02 ft (Alderman, 1986).

Radar Methods

Similar to the ultrasonic measurement instrumentation, radar-based portable units use a pulsed or continuous high-frequency wave to reflect off the water surface in a well. Depth to water is calculated by determining the travel time of the pulse or wave and electronically converting the signal to a depth measurement. Range of measurement to water is typically limited to larger wells and water levels about 100 ft or less from the top of casing. These limitations are the result of a need to maintain a focused beam width.

Accuracy of commercial units is reportedly good, from ±0.01 to ±0.02 ft over the range of measurement. As with other acoustic methods, temperature, humidity, and obstacles in the beam pathway all will have an effect on the quality of the water-level measurement (Ross, 2001).

Laser Methods

Lasers have been used in the food, chemical, and energy industries for over a decade as a method of noncontact level monitoring of liquids and solids in tanks. Advances in laser technology have allowed the manufacturing of battery-powered units potentially capable of obtaining water-level measurements in wells and piezometers. Tests of prototype instrumentation show promise for use in well-monitoring applications, but further development is needed to bring this technology into common use by the ground-water professional.

One of the significant advantages of laser technology for obtaining water level measurements is an unparalleled accuracy to depth range. Ross (2001) reported an accuracy of ± 0.01 ft for distances greater than 1000 ft. Because of the very high frequency of the laser pulse, humidity, and temperature variations in a typical well would not significantly effect the signal. However, the use of the laser requires a clear beam pathway. If a well is not plumb or if obstacles in the well prevent a clean line of sight down the well, a measurement cannot be made. Other issues include scattering of the reflected laser beam from the water surface due to turbulence or the beam penetrating through the target water surface without reflection (Ross, 2001).

Manual Measurements in Flowing Wells

Casing Extension

When the pressure of a flowing well is sufficiently low, a simple extension of the well casing allows the water level to stabilize so that a water-level measurement can be made. The direct measurement of the piezometric level by casing extension is practical when the additional height requirement is several feet or less. A water-level measurement using this technique should be accurate to within ± 0.1 ft because flowing well water levels tend to fluctuate.

Manometers and Pressure Gages

If the pressure of the flowing well is sufficiently high, the use of a casing extension is usually not practical. To measure the piezometric level in such circumstances, the well is sealed or "shut-in" and the resulting pressure of the water in the well casing is measured. Two commonly used instruments to monitor the well pressure are manometers and pressure gages.

A mercury manometer, when properly installed and maintained, has a sensitivity of ± 0.005 ft of water, and these devices have been constructed to measure ranges in water levels in excess of 120 ft (Rantz, 1982). When used to monitor shut-in pressure of wells, an accuracy of ± 0.1 ft is typical (U.S. Geological Survey, 1980).

Pressure gages are typically less sensitive to head pressure changes than mercury manometers and, therefore, have only a routine accuracy of ± 0.2 ft under ideal conditions when calibrated to the nearest tenth of a foot of water. According to the U.S. Geological Survey (1980), probable accuracy of measuring the pressure of a shut-in well with pressure gages is about 0.5 ft with these older style units. Many of these less sensitive gages are still in use today. Design advances during the last decade in both mechanical and electronic gages used as replacements for mercury manometers have increased the measurement accuracy to better than ± 0.01 ft of the gage range (Paroscientific, Inc., 2002). However, because well shut-in pressures typically fluctuate, a practical accuracy still remains at about ± 0.1 ft for this technique.

When using either of these instruments to measure well pressure, care should be taken to avoid rapid pressure change caused by opening or closing the valves used in sealing the well. This could create a water-hammer effect and cause subsequent damage to the manometer or pressure gage. In addition, field instruments used to monitor pressure should be checked periodically against master gages and standards.

Pressure Transducers

As previously described, pressure transducers can accurately monitor changes in pressure over a wide range. Transducers have been installed in place of pressure gages to determine the potentiometric level. If the pressure transducer range is carefully matched with the shut-in well pressure, measurements to ± 0.02 ft can be obtained. One source of error in these measurements results from changes in temperature in the transducer. Either a transducer unit that has some form of electronic temperature compensation or a unit that is totally submerged in the well should be used. Again, due to fluctuations in well shut-in pressures, the apparent measurement accuracy of this method will be about ± 0.1 ft.

Applications and Limitations of Manual Methods

No single method for determining water levels in wells is applicable to all monitoring situations, nor do all monitoring situations require the accuracy and precision of the most sensitive manual measurement technique. The practicing hydrogeologist should become familiar with the various techniques using two or more of these methods to obtain water levels on the same well. By doing so, the strengths and weaknesses of the monitoring methods will quickly become evident.

Table 7.1 is a summary of the manual measurement techniques discussed earlier, with their reported accuracies. Also presented in this summary are several of the principal sources of error or interference relevant to each technique. This table should be used only as a guide because each monitoring application and the skill of the measurer can result in greater or lesser measurement accuracy than stated.

Continuous Measurements of Ground-Water Levels

The collection of long-term water-level data is a necessary component of many hydrogeologic investigations. A commonly employed technique is the use of mechanical float recording systems. These devices typically produce a continuous analog record, usually on a strip chart, which is directly proportional to the water-level change.

Electromechanical instruments that use a conductance probe with a feedback circuit to drive a strip chart or a punched tape can successfully monitor rapid changes in water levels. These are used where float-operated systems fail to follow water-level fluctuations as expected.

With the development of field-operable solid-state data loggers and portable computers, long-term monitoring systems using pressure transducers are favored among those conducting hydrogeologic investigations. As with manual water-level measurements, the type of long-term monitoring system employed is dependent upon the investigator's data needs.

Methods of Continuous Measurement

Mechanical: Float Recorder Systems

Instruments that use a float to operate a chart recorder (a drum or wheel covered with chart paper and containing a time-driven marking pen) have been used to measure water levels since the early 1900s. These devices produce a continuous analog record of water-level change, usually as a graph. Depending upon the gage scale and time-scale gearing, a single chart may record many months of water-level fluctuations. To augment or even replace the analog record of float recorder systems, digital encoders and data loggers have been added to many of these systems. If properly installed and maintained, float recorder systems are very reliable, as is evidenced by their continued use in many municipal well-field monitoring programs. Mechanical systems are also useful when interfering electromagnetic currents or other harsh environmental conditions preclude the use of electronic-based units.

Float-operated devices are subject to several sources of error, which include float lag, line shift, submergence of counterweight, temperature, and humidity. Leupold and Stevens (1978) detail these errors and suggest methods to correct them. The reader should consult this reference for additional details. For purposes of this discussion, it is noted that when smaller floats are used, the magnitude of error is greatest. For example, float lag, or the lag of the indicated water level behind the true water level due to the mechanical work required by the float to move the instrument gears, can be as much as 0.5 ft for a 1.5 in. float if the force to move the instrument is 3 oz. This is contrasted to a 0.07 ft error for a 4 in. float and 0.03 ft error for a 6 in. float on an instrument requiring the same 3 oz of force (Leupold and Stevens, 1978). This error is magnified if the float or float cable is allowed to drag against the well casing. Shuter and Johnson (1961) discuss these problems in measuring water levels in small-diameter wells and offer several devices to improve recorder performance. Because many of the wells constructed in today's ground-water monitoring programs are 2 in. in diameter, caution should be used if a float recording system is installed to obtain continuous water-level measurements.

According to Rantz (1982), if a mechanical float recording system is properly installed and operated, long-term water-level measurements in wells are obtainable to an accuracy of about ± 0.01 ft. This accuracy is based on measurements made in stilling wells used for long-term monitoring of stage height of rivers. Because the piezometers and wells typically utilized in monitoring well networks are smaller in diameter, the accuracy for float recording systems used to measure ground-water fluctuations will usually be greater than ± 0.01 ft.

Electromechanical: Iterative Conductance Probes (Dippers)

Iterative conductance probes, commonly referred to as dipping probes or dippers, are electromechanical devices that use an electronic feedback circuit to measure the water level in a well. A probe is lowered on a wire by a stepping motor until a sensor in the probe makes electrical contact with the water. This generates a signal that causes the motor to reverse and retract the probe slightly. After a set time period, the probe is lowered again until it makes contact with the surface, retracts, etc., thus repeating the iterative cycle. The wire cable is connected to either a drum used for chart recording or a potentiometer whose output signal is proportional to the water level (Grant, 1978).

Dipping probes have several advantages over float recording systems. The well can be of smaller diameter and the system can accommodate some tortuosity in the well casing. Because the sensing probe is electromechanical, greater depths to water can be monitored without the mechanical losses associated with float systems. When water-level fluctuations are cyclic or change moderately rapidly, the dipping probe better reflects the oscillations in the water levels of smaller diameter wells.

Data Loggers

Data loggers consist of microprocessors connected to transducers that are installed in the well. The microprocessors consist of hardware and software that allow the automated collection of water-level data over various time periods. Data can be easily manipulated after transfer to a computer database. The use of this equipment is common, and a variety of equipment systems are commercially available.

Variations of data-logger based systems have been installed to better access and process water-level data. From the transducer at the wellhead, data is transferred to a data logger or signal processor to a central computer via hardwire, line-of-sight radio, satellite radio, or phone lines. At some of these installations, the central computer can query each remote well unit at any desired frequency including a continuous data scan mode (U.S. Bureau of Reclamation, 2001).

Analysis, Interpretation, and Presentation of Water-Level Data

The primary use of ground-water level data is to assess in which direction ground-water is flowing beneath a site. The usual procedure is to plot the location of wells on a base map, convert the depth-to-water measurements to elevations, plot the water-level elevations on the base map, and then construct a ground-water elevation contour map. The direction of ground-water flow is estimated by drawing ground-water flow lines perpendicular to the ground-water elevation contours (Figure 7.4).

The relatively simple approach to estimating ground-water flow directions described earlier is suitable where geologic media are assumed to be isotropic, wells are screened in the same zone, and the flow of ground-water is predominantly horizontal. However, with the increased emphasis on detecting the subsurface positions of contaminant plumes or in predicting possible contaminant migration pathways, it is evident that the assumptions of isotropy and horizontal flow beneath a site are not always valid. Increasingly, flow lines shown on vertical sections are required to complement the planar maps showing horizontal flow directions (Figure 7.7) to illustrate how ground water is flowing either upward or downward beneath a site (Figure 7.8).

Ground water flows in three dimensions and as such can have both horizontal and vertical (either upward or downward) flow components. The magnitude of either the horizontal or the vertical flow component and the direction of ground-water flow is dependent on several factors.

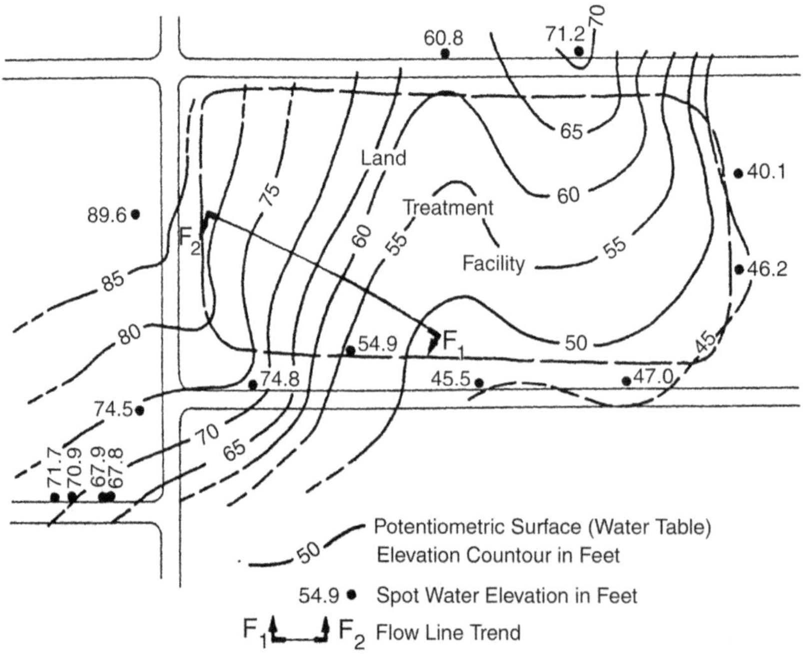

FIGURE 7.7
Potentiometric surface elevation contour map. (Adapted from Rathnayake et al. [1987]. With permission.)

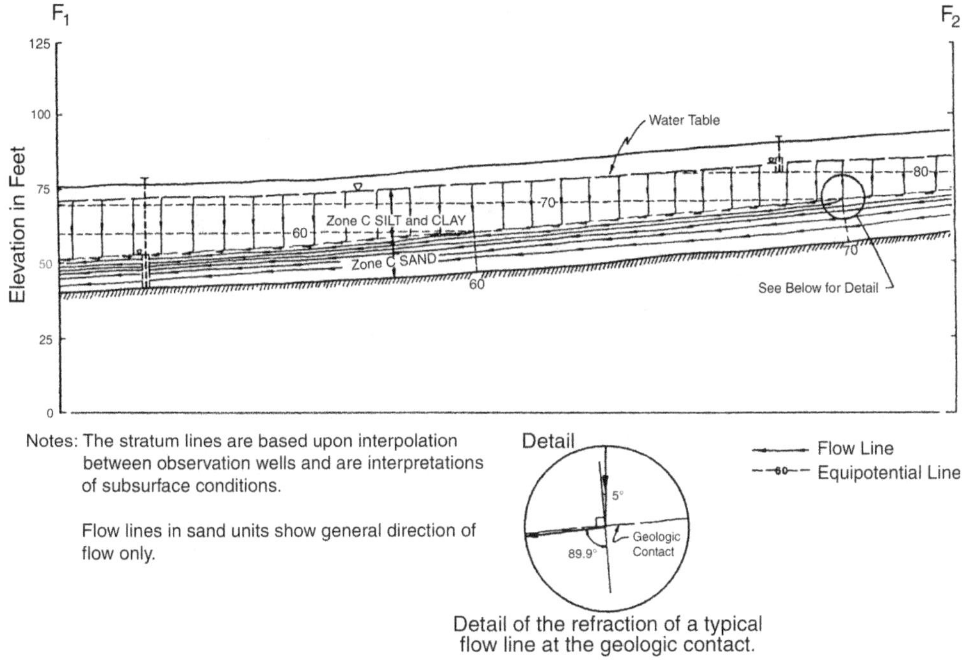

FIGURE 7.8
Cross-section showing vertical flow directions.

Recharge and Discharge Conditions

In recharge areas, ground water flows downward (or away from the water table), while in discharge areas, ground water flows upward (or toward the water table). Ground water migrates nearly horizontally in areas between where recharge or discharge conditions prevail. For example, in Figure 7.9 well cluster A is located in a recharge area, well cluster B is located in an area where flow is predominantly horizontal, and well cluster C is located in a discharge area. Note that in Figure 7.9, wells located adjacent to one another, and at different depths, display different water-level elevations.

Aquifer heterogeneity refers to an aquifer condition in which aquifer properties are dependent on *position* within a geologic formation (Freeze and Cherry, 1979), which is an important consideration when evaluating water-level data. While recharge or discharge may cause vertical gradients to be present within a discrete geologic zone, vertical gradients may be caused by the contrast in hydraulic conductivity between aquifer zones. This is especially evident where a deposit of low hydraulic conductivity material overlies a deposit of relatively higher hydraulic conductivity material, as shown in Figure 7.8.

Aquifer anisotropy refers to an aquifer condition in which aquifer properties vary with direction at a point within a geologic formation (Freeze and Cherry, 1979). For example, many aquifer materials were deposited in more or less horizontal layers, causing the horizontal hydraulic conductivity to be greater than the vertical hydraulic conductivity. This condition tends to create more pronounced vertical gradients (Fetter, 1980) that are not indicative of the actual direction of ground-water flow. In anisotropic zones, flow lines do not cross potential lines at right angles and flow will be restricted to higher elevations than that in isotropic zones showing the same water-level conditions.

Detailed discussions of each of these factors are beyond the scope of this section. The reader is referred to Fetter (1980) and Freeze and Cherry (1979) for more detailed discussions of the effects of these aquifer conditions on ground-water flow.

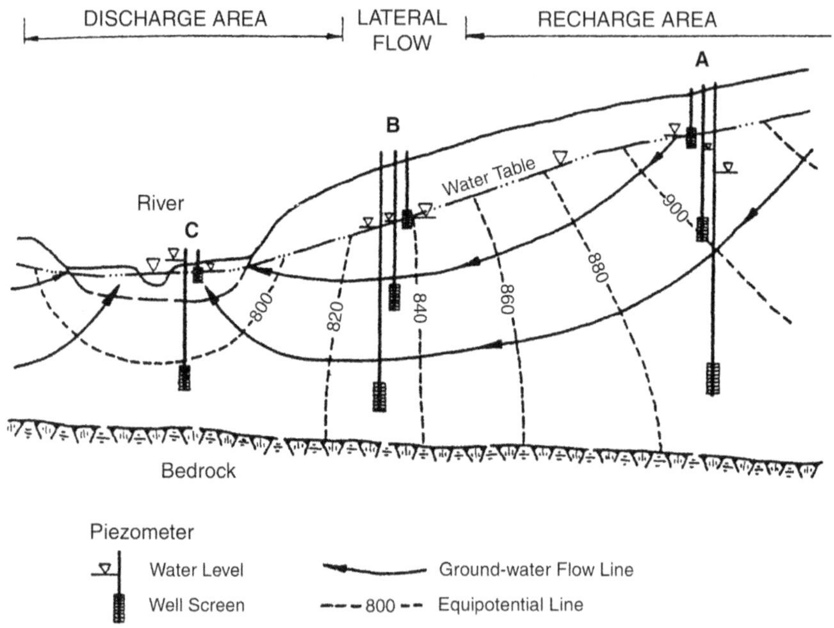

FIGURE 7.9
Ideal flow system showing recharge and discharge relationships. (Adapted from Saines [1981]. With permission.)

The practical significance of the three factors discussed earlier is that ground-water levels can be a function of either well-screen depth or well position along a ground-water flow line or, more commonly, a combination of both. For these reasons, considerable care needs to be taken in evaluating water-level data.

Approach to Interpreting Water-Level Data

The first step in interpreting ground-water-level data is to conduct a thorough assessment of the site geology. The vertical and horizontal extent and relative positions of aquifer zones and the hydrologic properties of each zone should be determined to the extent possible. It is difficult to overemphasize how important it is to have as detailed an understanding of the site geology as possible. Detailed surficial geologic maps and geologic sections should be constructed to provide the framework to interpret ground-water-level data. Man-made features that could influence ground-water levels should also be identified at this stage.

The next step in interpreting ground-water level data is to review monitoring well installation features with respect to screen elevations and the various zones in which the screens are situated. The objective of this review is to identify whether vertical hydraulic gradients are present beneath the site and to determine the probable cause of the gradients.

One method that can be used to assess the distribution of hydraulic head beneath a site is to plot water-level elevations versus screen midpoint elevations. An example of such a plot is shown in Figure 7.10 for wells completed within a layered geologic sequence. Figure 7.10 indicates that a steep downward hydraulic gradient, on the order of 0.85, exists within the sandy silt to silty clay layer. However, in the lower layers, the vertical component of flow is substantially less both within and between the layers.

Once the presence and magnitude of vertical gradients and the distribution of data with respect to each zone are established, the direction of ground-water flow can be

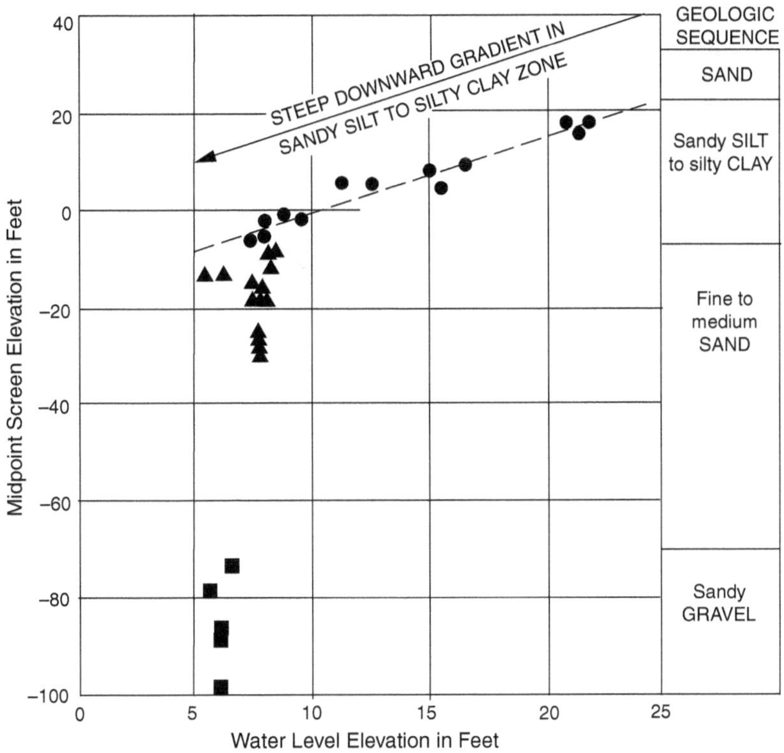

FIGURE 7.10
Water-level elevation versus midpoint screen elevation for a well screened in a stratified geologic sequence.

assessed. If the geologic system is relatively simple and if substantial vertical gradients are not present, a planar ground-water elevation contour map can be prepared to show the direction of ground-water flow. However, if multiple zones of differing hydraulic conductivity are present beneath the site, several planar maps may be required to show the horizontal component of flow within each zone (typically the zones of relatively higher hydraulic conductivity). Vertical cross-sections are required to illustrate how ground water flows between each zone.

For the example presented in Figure 7.10, the data indicate that flow is predominantly downward within the upper silt or clay zone. Flow within the lower zone appears to be largely horizontal, although a vertical component of flow is indicated between the sand and the underlying gravel layer.

The examples presented earlier show downward vertical gradients that are indicative of recharge areas. Sites can also be situated within discharge areas where the vertical components of flow are in an upward direction.

The presence of vertical gradients can be anticipated in areas where sites are:

- Underlain by a layered (heterogeneous) geologic sequence, especially where deposits of lower hydraulic conductivity overlie deposits of substantially higher hydraulic conductivity
- Located within recharge or discharge areas

It should be noted that site activities often locally modify site conditions to such an extent that ground water flows in directions contrary to what would be expected for

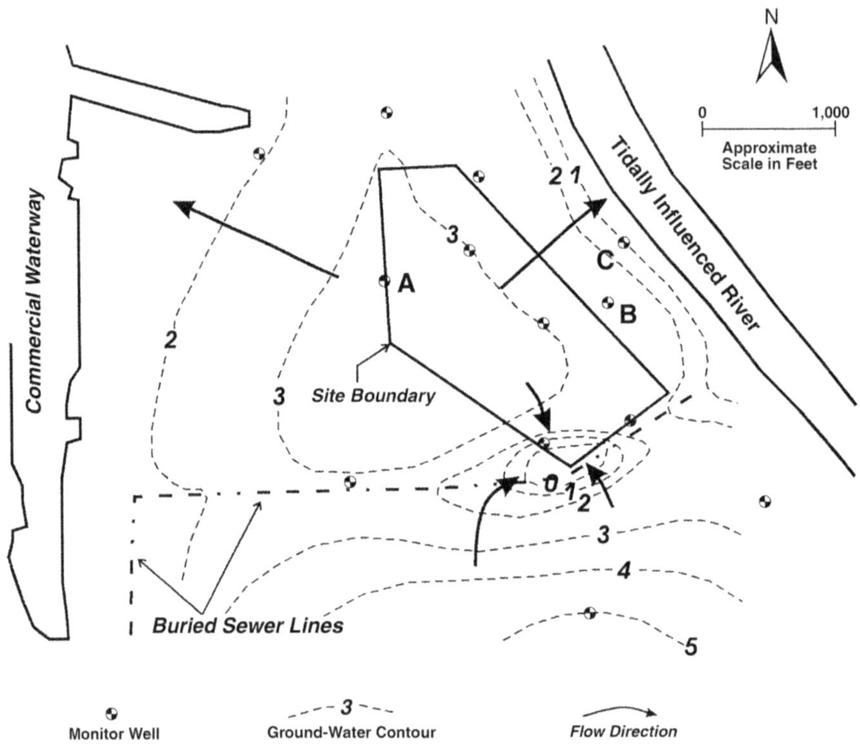

FIGURE 7.11
Average ground-water elevation contours — deeper aquifer.

"natural" conditions. Drainage ditches, buried pipelines, and other features can modify flow within near-surface deposits, and facility-induced recharge (e.g., from unlined ponds) can create local downward gradients in regional discharge areas among others.

Figure 7.11 shows the average ground-water elevation contours in a relatively complex hydrogeologic setting. The site lies between two water bodies that are tidally influenced and deep sewer lines are located near the southeast corner of the site. The aquifer of interest lies below a shallow water-table aquifer. A discontinuous aquitard separates the aquifers. The position of the site with respect to the water bodies would suggest that a ground-water divide is present near the site. On the west side of the site, ground water would flow toward the commercial waterway, and on the east side of the divide, ground water would flow toward the river.

Water levels were measured using pressure transducers and data loggers over several days because the site location suggested that tidal fluctuations could affect ground-water levels. Well locations in which transducers were installed are illustrated in Figure 7.11 and some of the transducer data are shown in Figure 7.12. Average water levels and elevations were calculated for each well (see Transient Effects) and were used to construct the ground-water elevation contour map.

Water levels in nested wells screened in the shallow and deeper aquifers indicated the presence of downward vertical gradients (i.e., water-level elevations in the shallower aquifer wells were higher than elevations in wells screened in the deeper aquifer). Analysis of the ground-water contours (for the deeper aquifer) in Figure 7.11 shows that a portion of the site (near well A) lies near the center of a ground-water mound generally defined by the 3 ft elevation contour. Evaluation of boring logs indicated that the mound lies in an area where the aquitard appears to be absent. Interpretation of the available data

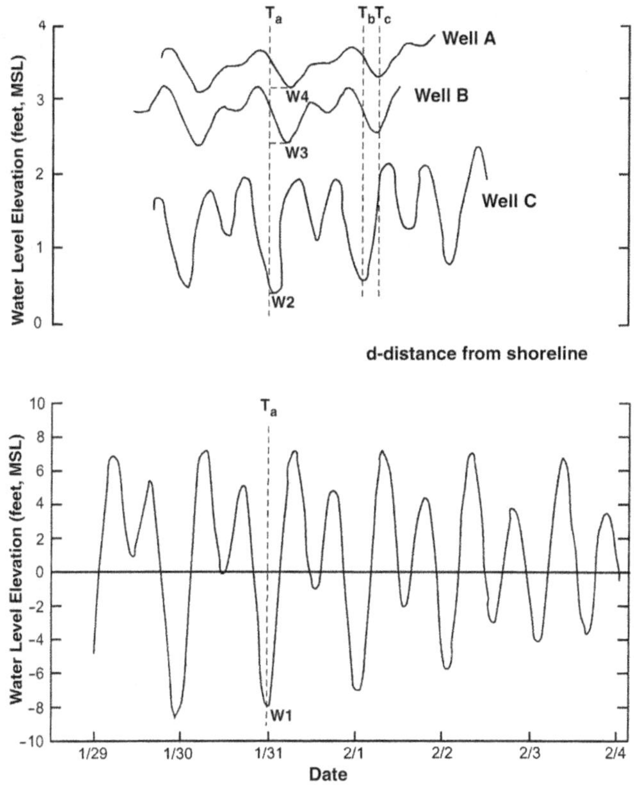

FIGURE 7.12
Influence observed in wells due to tidal fluctuations.

indicates that a partial cause of the ground-water mound was water flowing downward from the shallow aquifer into the deeper aquifer where the aquitard is absent.

As expected, some ground water in the vicinity of the site flows to the east and to the west. However, ground-water contours in the southeastern portion of the site indicated the presence of a low ground-water elevation, where ground water flows in a southerly direction. Two deep buried sewer lines are present near the southeastern site boundary. Review of construction drawings shows that excavation for the sewers penetrated into the deeper aquifer. Interpretation of the water-level elevation data strongly suggests that the sewer lines are acting as drains (i.e., are intercepting ground water). These man-made features appear to have substantially modified the ground-water flow patterns compared to what would be expected under natural conditions.

Transient Effects

Ground-water flow directions and water levels are not static and can change in response to a variety of factors such as seasonal precipitation, irrigation, well pumping, changing river stages, and tidal fluctuations. Fluctuations caused by these factors can modify, or even reverse, horizontal and vertical gradients and thus alter ground-water flow directions. For example, in areas influenced by tides, the net flow of ground water will typically be toward the tidally affected water body. However, during certain portions of the tidal cycle (i.e., during higher tidal levels), there may be a temporary reversal in flow along and some distance inland from the shoreline. Even if significant flow reversals do not occur, hydraulic gradients can change as tidal levels change. Gradients will typically be steeper during lower tides and flatter during periods of higher tides.

Time series water-level data are required to assess how ground-water flow directions change in response to these factors. Figure 7.13 shows data for several wells finished at different depths in an area influenced by changing river stage. The data indicate that river stage affects water levels but that the direction of flow and the horizontal and vertical gradients do not substantially change with river fluctuation. However, the fluctuations do affect the length of time over which each set of ground-water level measurements should be made. In this case, measurements were made in less than 1 h to minimize the effects of the fluctuations on the interpretation of ground-water flow directions.

Figure 7.12 shows hydrographs of water levels in three wells located at varying distances from the shorelines influenced by tides. Well locations are shown in Figure 7.11. Water levels fluctuate in a regular manner but the fluctuations in the wells lag behind the fluctuating tide. In the case illustrated in Figure 7.12, at time T_a, low water levels in the wells occur approximately 2 h (point W2) to 6 h (point W4) after the tidal low (point W1). Several other conclusions can be made using data illustrated in Figure 7.12:

- The mean tidal fluctuation (difference between the mean higher high tide and mean lower low tide) in the area where the data were collected is approximately 11.8 ft. Tidal fluctuations during the measurement period were greater than 15 ft. This means that the water-level measurements are representative of a period of the year when tidal fluctuations are somewhat greater than the mean or average range.
- Water-level elevations in the wells indicate that ground-water flow reversals in the area of interest do not occur. Elevations in well A are always higher than that in well B. Similarly, elevations in well B are always higher than that in well C. This does not mean that flow reversals do not occur nearer to the shoreline, rather it does not occur in the area where the wells are installed.
- Assuming that a sufficient number of wells were instrumented, transducer data can be used to calculate an average water level for each well, and, using the averages, ground-water contour maps can be prepared, which show the average flow direction for the time period in which the data were collected (as shown in Figure 7.11). "Spot" measurements can also be extracted from the hydrographs to construct contour maps representative of tidal highs/lows or ground water highs/lows.
- The time interval in which water-level measurements are taken may affect analysis of flow directions and will affect analysis of hydraulic gradients. For example, if the water level in well A is measured at time T_b, and the water level in well C is measured at time T_c (approximately 5 h later), the water-level elevation will have risen more than 1 ft during the intervening period, which will introduce some error in the analysis.
- Ideally, water levels would be measured in all wells at the same instant (such as at time T_b) to assess flow directions and gradients. As noted earlier, this is a relatively easy matter to resolve if water-level fluctuation data similar to those shown in Figure 7.12 are available for all wells during the same time interval. However, this type of data is seldom available on a routine basis at most sites, especially on a large site with numerous wells. In these cases, if some hydrographic data are available, the data can be used to develop a strategy to

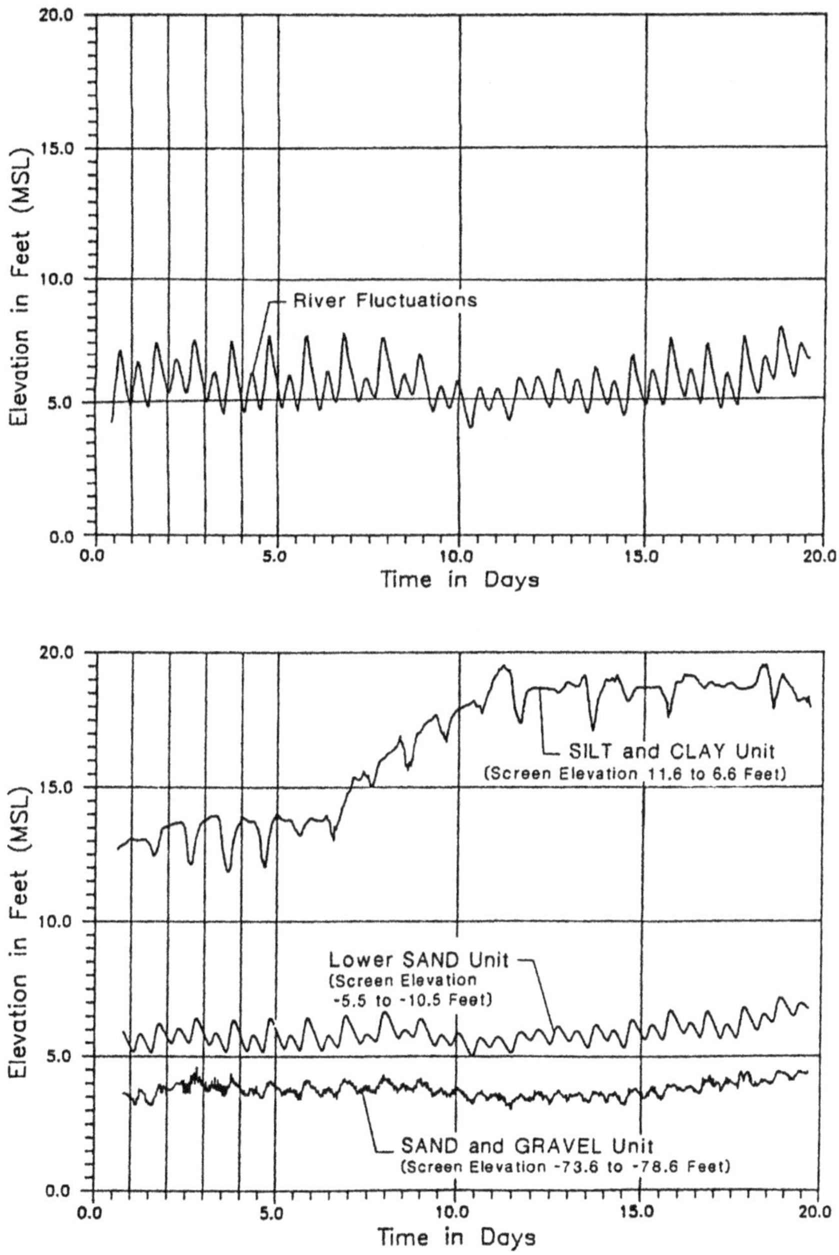

FIGURE 7.13
Influence observed in wells due to river level fluctuations.

minimize the error caused by the regular fluctuations. At the site illustrated in Figure 7.11 and Figure 7.12, these strategies might include:

a. Using several persons to measure water levels in as short a time as possible (i.e., get all of the water-level measurements done before ground-water quality samples are obtained)
b. Selecting a measurement period during a time when the least amount of fluctuation is expected to occur. This might be near tidal high or tidal low periods

c. Initially measuring water levels in wells with the greatest expected fluctuation and moving toward wells where the fluctuations are expected to be less.

Contouring of Water-Level Elevation Data

Typically, ground-water flow directions are assessed after preparing ground-water elevation contour maps. Water-level elevations are plotted on base maps and linear interpolations of data between measuring points are made to construct contours of equal elevation (Figure 7.7). These maps should be prepared using data from measuring points screened in the same zone where the horizontal component of the ground-water flow gradient is greater than the vertical gradient. The greatest amount of interpretation is typically required at the periphery of the data set. A reliable interpretation requires that at least a conceptual analysis of the hydrogeologic system has been conducted. The probable effects of aquifer boundaries, such as valley walls or drainage features, need to be considered.

In areas where substantial vertical gradients are present, the areal ground-water flow maps need to be supplemented with vertical cross-sections that show how ground water flows vertically within and between zones (Figure 7.8). These cross-sections should be oriented parallel to the general direction of ground-water flow and should account for the effects of anisotropy.

Computer contouring and statistical analysis (such as kriging) of water-level elevation data have become more popular (McKown et al., 1987). These tools offer several advantages, especially with large data sets. However, the approach and assumptions that underlie these methods should be thoroughly understood before they are applied and the output from the computer should be critically reviewed. The most desirable approach would be to interpret the water-level data using both manual and computer techniques. If different interpretations result, then the discrepancy between the interpretations should be resolved by further analysis of the geologic and water-level data.

The final evaluation of water-level data should encompass a review of geologic and water-quality data to confirm that a consistent interpretation is being made. For example, at a site where contamination has occurred, wells that are contaminated should be downgradient of the site (based on the water-level data). If this is the case, then a consistent interpretation is indicated. However, if wells that are contaminated are not downgradient of the site, based on water-level data, then further evaluation is required.

References

Alderman, J.W., FM radiotelemetry coupled with sonic transducers for remote monitoring of water levels in deep aquifers, *Ground-Water Monitoring Review*, 6(2), pp. 114–116, 1986.

ASTM, Standard Test Method for Determining Subsurface Liquid Levels in a Borehole or Monitoring Well (Observation Well), ASTM Standard D 4750, ASTM International, West Conshohocken, PA, 2006a.

ASTM, Standard Practice for Static Calibration of Electronic Transducer-Based Pressure Measurement Systems for Geothechnical Purposes, ASTM Standard D 5720, ASTM International, West Conshohocken, PA, 2006b.

Barcelona, M.J., J.P. Gibb, J.A. Helfrich, and E.E. Garske, Practical Guide for Ground-Water Sampling, EPA-600/2-85–104, U.S., Environmental Protection Agency, Robert S. Kerr Environmental Research Laboratory, Ada, OK, pp. 78–80, 1985.

Driscoll, F.G., *Ground Water and Wells*, Johnson Division, St. Paul, MN, 1089 pp, 1986.

Everett, L.G., *Ground Water Monitoring*, General Electric Company, Schenectady, NY, pp. 196–198, 1980.

Fetter, C.W., *Applied Hydrogeology*, C. E. Merrill Publishing Co, Columbus, OH, 1980.

Freeze, R.A. and J.A. Cherry, *Groundwater*, Prentice Hall, Englewood Cliffs, NJ, 1979.

Garber, M.S. and F.C. Koopman, Methods of measuring water levels in deep wells, *Techniques of Water Resources Investigations*, Book 8, U.S. Geological Survey, Reston, VA, chap. A-1, 1968.

Grant, D.M., *Open Channel Flow Measurement Handbook*, Instrumentation Specialties Company (ISCO, Inc.), Lincoln, NE, pp. 6–7, 1978.

Heath, R.C., Basic Ground-Water Hydrology, U.S. Geological Survey, Water Supply Paper 2220, 1983.

In-Situ, Inc, Owner's Manual: Hydrologic Analysis System, Model SE200, In-Situ, Inc, Laramie, WY, pp. 7–11, 1983.

Leupold and Stevens, Inc, *Stevens Water Resources Data Book*, Leupold & Stevens, Inc, Beaverton, OR, 1978.

Maslansky, S.P., C.A. Kraemer, and J.C. Henningson, An Evaluation of Nested Monitoring Well Systems, Ground-Water Monitoring Seminar Series Technical Papers, U.S. Environmental Protection Agency, EPA-CERI-87-7, 1987.

McKown, G.L., G.W. Dawson, and C.J. English, Critical Elements in Site Characterization, Ground-Water Monitoring Seminar Series Technical Papers, U.S. Environmental Protection Agency, EPA-CERI-87-7, 1987.

Paroscientific, Inc., Digital Quartz Pressure Transmitters for Accurate Water Level Measurements, Paroscientific, Inc., pp. 2–7 (Available at www.paroscientific.com/waterlevel.htm), 2002.

Plazak, D., Differences between water-level probes, *Ground-Water Monitoring and Remediation*, 14(1), 84 pp, 1994.

Rantz, S.E., Measurement and Computation of Streamflow: Volume 1. Measurement of Stage and Discharge, U.S. Geological Survey, Water Supply Paper 2175, U.S. Govt. Printing Office, Washington, DC, pp. 63–64, 1982.

Rathnayake, D., C.D. Stanley, and D.H. Fujita, Ground water flow and contaminant transport analysis in glacially deposited fine grained soils: a case study, *Proceedings of the FOCUS Conference on Northwestern Ground Water Issues*, National Water Well Association, Dublin, OH, pp. 125–151, 1987.

Ross, J.H., Evaluation of Non-Contact Measurement Instrumentation for Ground Water Wells, *Presentation made at the AGWSE Annual Meeting and Conference*, Nashville, TN, December, 2001.

Saines, M., Errors in interpretation of ground-water level data, *Ground-Water Monitoring Review*, 1(1), pp. 56–61, 1981.

Schrale, G. and J.F. Brandwyk, An acoustic probe for precise determination of deep water levels in boreholes, *Ground-Water*, 17(1), pp. 110–111, 1979.

Sheingold, D.H., *Transducer Interfacing Handbook*, Analog Devices, Inc, Norwood, MA, 1980.

Shuter, E. and A.I. Johnson, Evaluation of Equipment for Measurement of Water Levels in Wells of Small Diameter, U.S. Geological Survey Circular 453, 1961.

Solinst, Ltd, Levelogger Series Model 3001 Data Sheet, Solinst Canada Ltd, Georgetown, ON, Canada, 4 pp, 2001.

Stewart, D.M., The rock and bong techniques of measuring water levels in wells, *Ground-Water*, 8(6), pp. 14–18, 1970.

Thornhill, J.T., Accuracy of depth to water measurements, U.S. EPA Superfund Ground-Water Issue, EPA/540/4-89-002, Robert S. Kerr Environmental Research Laboratory, Ada, OK, 3 pp, 1989.

Todd, D.K., Groundwater Hydrology (2nd ed.). John Wiley and Sons, New York, NY, 1980.

U.S. Bureau of Reclamation, Engineering Geology Field Manual, U.S. Department of Interior, Bureau of Reclamation, U.S. Govt. Printing Office, Denver, CO, Chap. 9, pp. 227–247, 2001.

U.S. Geological Survey, National Handbook of Recommended Methods for Water-Data Acquisition: Chap. 2—Ground-Water, U.S. Department of Interior, Geological Survey, Reston, VA, 1980.

Zarriello, P.J., Accuracy, precision, and stability of a vibrating-wire transducer measurement system to measure hydraulic head, *Ground-Water Monitoring and Remediation*, 15(2), pp. 157–168, 1995.

8

Decontamination of Field Equipment Used in Ground-Water Sampling Programs

Gillian L. Nielsen

CONTENTS

Introduction

As discussed throughout this text, many different types of equipment are used during a ground-water sampling event to collect samples and generate field data. Often, this equipment is used at more than one site and in more than one well, necessitating the development and implementation of an effective program for equipment cleaning, or decontamination, on a site-by-site (and parameter-specific) basis. Equipment cleaning is a critical component of any ground-water sampling program to ensure that samples obtained for field or laboratory analysis be representative, accurate and precise, and not be influenced by bias or error associated with sample collection. The economic and technical consequences associated with making decisions based on field and laboratory analyses of samples that are not representative can be substantial.

One major source of bias or error that has the potential to influence the quality and representative nature of samples collected for chemical and physical analyses, is the presence of contamination on field equipment. If the equipment used to collect samples or generate field analytical data is not appropriately cleaned to remove potential contaminants, the data collected with that equipment could be erroneous. Failure to adequately clean equipment used to collect environmental samples between sampling points, such as split-spoon samplers or ground-water sampling pumps, could result in the cross-contamination of individual samples or sampling locations. This would, in turn, make any information obtained from these samples unrepresentative of actual *in situ* physical and chemical properties of the material being sampled. Data derived from the analysis of these samples would not accurately reflect actual site conditions and would, therefore, be virtually meaningless to interpret and use for the purpose of making important decisions for a site under investigation.

This chapter will focus on the objectives and methods available for decontamination of field equipment. Personnel decontamination practices are not addressed here.

Objectives of Equipment Decontamination

An effective equipment decontamination protocol must be designed to meet the following objectives:

- Prevent introduction of contaminants from one site to another site.
- Prevent contamination of areas on a site designated as being "clean" work or equipment storage areas.
- Prevent cross-contamination of individual sampling locations at a single site.
- Prevent cross-contamination of individual samples from a single sampling location as a result of using common or portable sampling equipment at more than one sampling location or to collect more than one sample at a single location.
- Ensure proper operation of equipment.
- Prevent accidental exposure of workers to contaminants that may be distributed on equipment through unprotected handling of equipment.

If these objectives are successfully met, samples should not be impacted by either negative or positive bias associated with poor equipment cleaning practices, provided the chosen cleaning protocols are effective and implemented correctly. Consequently, field or lab analysis of samples should accurately reflect *in situ* chemistry.

Current Status of Equipment Decontamination Protocols

Environmental scientists have many sources of protocols for decontamination of field equipment. Protocols are available from several federal as well as most state regulatory and nonregulatory agencies, equipment manufacturers, corporate standard operating procedures, and manufacturers of cleaning solutions and equipment. Unfortunately, there is a lack of continuity between these protocols, making it difficult to establish a single standard protocol to follow (Mickam et al., 1989; Parker, 1995). Extensive surveys of federal and state agencies across the U.S. have been conducted to evaluate the status of current field equipment decontamination procedures. The need for some form of standardization of decontamination methods became readily apparent during these surveys. Of the federal agencies interviewed (including the Department of the Army, Office of the Chief of Engineers, the Nuclear Regulatory Commission, and the National Science Foundation), none had any specific guidance addressing field equipment decontamination protocol (Parker, 1995). This survey also found that the U.S. Environmental Protection Agency (U.S. EPA, 1996) had no document used nationally to furnish guidance on field equipment decontamination protocols. While this survey found there were numerous decontamination methods in the published literature, there was a significant disparity between the protocols and there was no systematic study on the relative effectiveness of the various procedures (Parker, 1995).

In response to the lack of a standard equipment decontamination protocol, ASTM International developed two standards on field equipment decontamination in the early 1990s. The primary objective of these standards is to provide a basis for standardized protocols for effective equipment decontamination that could be used at a wide variety of facilities for a wide variety of equipment. The ASTM standards on equipment decontamination are:

- ASTM Standard D 5088 — Standard Guide for Decontamination of Field Equipment Used at Non-Radioactive Waste Sites (ASTM, 2006a).
- ASTM Standard D 5608 — Standard Guide for Decontamination of Field Equipment Used at Low-Level Radioactive Waste Sites (ASTM, 2006b).

These standards are now widely referred to by regulatory agencies as standard methods for equipment decontamination for use in a wide variety of environmental projects. Their use requires an understanding of which procedures are most appropriate for specific applications at any given site.

Preparing an Effective Decontamination Protocol for Field Equipment

The field equipment decontamination protocol is an important component of any site-specific sampling and analysis plan, regardless of the simplicity or intricacy of the ground-water investigation. The decontamination protocol must provide easily understood and implemented procedures for all aspects of equipment cleaning, including: (1) what equipment should be cleaned; (2) whether disposable equipment can be used in lieu of cleaning between uses; (3) when and where equipment cleaning should take place; (4) what cleaning protocols (equipment, solutions, other materials, techniques) should be used on a parameter- and equipment-specific basis; and (5) what should be done with any waste materials generated by equipment-cleaning activities. The decontamination protocol must be written in sufficient detail to ensure that the selected protocol will be effectively and consistently implemented. As part of a field quality assurance/quality control (QA/QC) program, it is essential to include the collection of equipment blanks to verify the effectiveness of the decontamination protocol.

What Equipment Requires Field Decontamination?

A typical environmental investigation involves several different phases of activity in the field. In each phase of investigation, a wide variety of equipment may be used, and most of this equipment requires decontamination at one point or another in the investigation. During the development of a field equipment decontamination program, it is important to develop an itemized list of equipment that will require field decontamination. Equipment should be evaluated with regard to its role in the investigation from the perspective of actual or potential contact with a sample analyzed either in a lab or in the field. Three general categories can be created for the evaluation of how field equipment is used: (1) equipment that directly contacts a sample being collected for physical or chemical analysis; (2) equipment that facilitates sample collection but does not contact the sample directly; and (3) equipment that is used for measurement or analysis of some type of parameter. Table 8.1 provides examples of equipment that would require field decontamination for each of these three equipment-use categories.

TABLE 8.1

Examples of Field Equipment That May Require Decontamination

	Examples of equipment that contacts samples collected for physical or chemical analysis
Soil sampling	Split-spoon samplers
	Thin-wall (Shelby) tube samplers
	Direct-push soil samplers
	Hand auger barrels or bits
	Continuous tube samplers
	Sample inspection tools (e.g., knives, metal spatulas)
Monitoring well installation	Well screen
	Well casing
	Well screen centralizers
	Field sieves for determining grain-size distribution
Ground-water sample collection	Well purging and sampling devices
	Pump tubing
	Sample filtration apparatus
	Equipment that facilitates sample collection but does not contact the sample
Soil sampling	Drilling rig and drill rod
	Hand auger rods and handles
	Direct-push rig and rod
Monitoring well installation	Drilling rig and associated tools
	Auger flights
	Well development equipment
Ground-water sample collection	Reels for pump tubing
	Support vehicle
	Suspension cable
	Rope, cord or line attached to grab sampling devices
	Flow-through cell and associated discharge tubing
	Equipment used for field parameter measurement or analysis
Soil sample collection	Tape measure
	X-ray fluorescence devices
	Field-portable analytical balance
Monitoring well installation	Tape measure
	Borehole TV camera
	Borehole geophysical equipment
	Pump discharge flow gauges
Ground-water sample collection	Flow-through cells
	Multi-parameter sondes
	Single-parameter meters
	Beakers or open containers
	Water-level gauges
	Oil– water interface probes

Whether or not equipment contacts a sample directly is one criterion used to determine the most appropriate method for equipment decontamination. This is discussed in greater detail later in this chapter.

Using Disposable Equipment to Avoid Equipment Decontamination Issues

In some programs where the level of contamination is high, it is often desirable to identify ways to minimize or eliminate the need for field equipment decontamination to prevent cross-contamination of samples being collected or measurements being taken. There are two primary options available to meet this objective. One option is to use disposable equipment that is brought to the site in a sealed package or container as shipped by the manufacturer, and then is used to collect one sample only, after which the equipment is discarded. Two common examples of disposable equipment used for ground-water sample collection are disposable bailers and disposable filtration media (see Chapters 3 and 5 for more information on ground-water sampling equipment and sample pretreatment methods). Examples of disposable equipment used in soil sample collection are polyvinyl chloride (PVC) or Teflon™ core barrel liners and sample retainer baskets used in devices such as split-spoon samplers. Disposable equipment can offer a number of advantages as summarized in Table 8.2.

One of the major limitations of attempting to incorporate disposable equipment in ground-water investigations is that disposable equipment options are somewhat limited. In general, support equipment such as sampling vehicles, sample collection devices involving pumping mechanisms, and field parameter measurement instrumentation are not disposable and must therefore be cleaned. Other limitations of disposable equipment are presented in Table 8.3.

The second alternative to field equipment decontamination is to use equipment that is "dedicated" or "designated" for use at a single location at a single site. Dedicated equipment is equipment that is permanently installed within a single monitoring or sampling location and is never exposed to atmospheric conditions during operation or use of that equipment (ASTM, 2006c). Using dedicated equipment can virtually eliminate the potential for cross-contamination of sampling locations and samples associated with contact with the sampling device itself. Examples of dedicated equipment used in ground-water investigations include dedicated bladder pumps installed in a ground-water monitoring well or a bubbler system permanently installed in a monitoring well for long-term water-level measurement.

Designated equipment is defined by ASTM (ASTM, 2006c) as equipment that is restricted to use at a single location. Designated equipment is differentiated from dedicated

TABLE 8.2

Advantages of Using Disposable Field Equipment

Saves time associated with field equipment cleaning
Reduces the number of field quality control samples required to verify the effectiveness of field equipment cleaning
Minimizes the potential for cross-contamination of samples and sampling locations
Reduces the volume of liquid waste generated by field decontamination activities
Equipment is generally simple to operate
Precleaned equipment options may be available
Some equipment is available in a variety of different materials (e.g., polytetrafluoroethylene [PTFE], PVC, high-density polyethylene), making it possible to select equipment with chemical compatibility in mind
Convenience

TABLE 8.3

Limitations of Disposable Field Equipment

Increased volume of solid waste generated in the field, which may require handling and disposal as a hazardous waste, and may increase overall costs of the ground-water investigation
May become very expensive when a large number of samples must be collected
Potential for residual contamination as a result of manufacture of the equipment (i.e., extrusion agents or mold-release compounds for plastic equipment) if not precleaned prior to use
Some cleaning protocols may require the collection of a rinseate blank on a per-lot or per-manufacturer basis to quantify the presence of any surface residues (if any) on equipment prior to use
Limited selection of types of disposable equipment available
Very rarely can all samples or field data measurements be collected using disposable equipment entirely; therefore, there will still be a need to implement some level of field equipment decontamination for every project
There is a temptation to try to clean and reuse disposable equipment to save money

equipment in that designated equipment typically comes in contact with atmospheric conditions during use or storage between sampling events. The advantage of using designated equipment is that it removes the potential for introduction of contaminants from a remote site to the site under investigation and helps to ensure precision in field measurements and sample collection. In some long-term environmental investigations, it is common to assign specific pieces of equipment to be used at that single site exclusively. This can include equipment such as photoionization detectors for sample screening, pH meters for sample analysis, or water-level gauges. On a sample-collection location basis, it is common to designate lengths of tubing to individual wells when a portable pump is used for purging and sample collection at a number of wells. This eliminates the need to try to clean lengths of pump tubing between wells, which can be very difficult to do successfully. Bailers are also commonly designated, although it is a common error to refer to these devices as dedicated equipment. Under the ASTM definition, bailers cannot be dedicated because they must come into contact with atmospheric conditions during use.

When equipment is designated for use at a specific location, it is critical that control is maintained over the equipment when it is in storage to ensure that it does not become contaminated as a result of contact with atmospheric or surface contaminants, or use at a location other than the one for which it is intended. At sites where atmospheric contributions of contaminants are of concern, it may be necessary to clean at least the exterior surfaces of designated equipment prior to use and prior to putting it into the storage container after use. It is necessary to address these issues on a site-specific basis within the field QA/QC program.

When and Where Should Equipment Be Decontaminated?

Remote Equipment Cleaning

Equipment decontamination can be performed in a remote location such as a laboratory. When equipment is cleaned in a remote location, it is precleaned prior to shipment to or use in the field. In theory, this approach ensures the highest level of equipment cleaning possible because the equipment is cleaned in a controlled indoor environment with ideal facilities for both chemical and physical cleaning procedures (McLaughlin and Levin, 1995). During cleaning at a remote location, a piece of equipment can also be inspected and repaired as necessary prior to shipment to the field. This should ensure optimal operation and performance of any piece of field equipment. From the field perspective, remote cleaning is perhaps the most convenient option for field equipment cleaning. Precleaning equipment can save time in the field for investigators because they do not need to create a formal decontamination area at a site, haul and store cleaning and rinsing solutions into the field, or deal with the generation of investigation-derived waste (IDW) that may need to be containerized and managed as a hazardous waste.

There are several drawbacks to this approach to cleaning. Typically, during environmental investigations, multiple samples are collected daily, therefore, if equipment is cleaned prior to use in the field exclusively, more equipment will need to be cleaned and shipped to a site during the investigation because the field team will not have cleaning solutions in the field. Without having cleaning solutions and rinse water available in the field, it would be impossible to clean equipment that has been accidentally contaminated. An example of accidental contamination would be if a precleaned recovery auger was accidentally dropped onto the ground in an area that is not going to be sampled or if a sampling device was exposed to high concentrations of atmospheric contamination prior to use. While the field team may not generate IDW as a result of equipment cleaning, the remote cleaning location would, and it may not be as well equipped to manage this waste. Equipment that has been used in the field would be returned to the remote location for cleaning after a single use. This would increase the potential of worker exposure to contaminants because the equipment will not have been cleaned prior to transport to and storage at the remote location. It also represents a real potential for contamination of a previously uncontaminated work area. Cost evaluations of remote versus in-field cleaning need to be conducted to determine which option is least expensive.

Field Equipment Cleaning

More commonly, field equipment is cleaned in the field, either at a designated decontamination area located at some point on site that is determined to be free of contamination (normally a central location), or at the point of equipment use. The decision regarding which approach to take is largely dependent on the nature of the equipment to be cleaned and the characteristics of the contaminants of concern. For example, if heavy equipment such as a drilling rig is to be cleaned, it may be necessary to construct a centrally located decontamination pad with access to power and water supplies to support a portable power washer unit. Many of these decontamination pads are designed to facilitate complete containment and collection of any IDW generated, so it may be subjected to on-site or off-site treatment or disposal (see Figure 8.1).

In cases where sample-collection equipment is being cleaned, a smaller-scale decontamination area is created at each sampling location to facilitate cleaning of all equipment immediately prior to use or movement to the next sampling location. As illustrated in Figure 8.2, these decontamination areas typically consist of a series of buckets or pails placed on heavy-gage plastic sheeting. All cleaning supplies such as detergent solutions and disposable supplies are also placed on this plastic sheeting. One common error observed in the field when using this type of set up is that sampling team members sometimes mistakenly run equipment through the decontamination line backwards (i.e., they begin equipment cleaning at the point of the final control water rinse and end in the bucket containing the detergent solutions and most-contaminated control water). Such an error can result in use of improperly decontaminated equipment and cross-contamination of samples. One solution to this problem is to use color-coded pails where, in the example illustrated in Figure 8.3, equipment cleaning begins in the red pail containing a detergent solution, progressing to the yellow pail with a rinse solution, and finally the green pail where the final equipment rinse water is contained. If this is not an option, an alternative is to create a directional arrow using duct tape on the plastic sheeting to direct personnel through the decontamination line correctly.

FIGURE 8.1
This decontamination area was designed to contain all waste water generated from cleaning heavy equipment used during site remediation. A french drain system was built beneath the gravel pad in which a sump pump was used to transfer all waste water into the 500-gallon poly tank seen to the right of the decontamination pad.

Advantages of field-cleaning equipment rather than cleaning it in a remote location include: (1) ability to clean equipment that may have been accidentally contaminated in the field; (2) ability to adjust cleaning protocols if in-field QC samples indicate that the equipment cleaning protocol being used proves to be ineffective; (3) sampling team members are assured that equipment is correctly cleaned immediately prior to use; (4) it is commonly less expensive to clean equipment in the field because it is not necessary to have as many pieces of equipment as is necessary when all equipment is cleaned

FIGURE 8.2
A series of buckets or pails is commonly used in the field for containing various cleaning solutions when equipment is cleaned at the point of use rather than remotely. Often the buckets are similarly colored, which can lead to confusion and error, resulting in sampling team members going through the decontamination line backwards.

FIGURE 8.3
A series of color-coded buckets is a good solution to prevent errors in equipment decontamination. The red bucket in the foreground indicates that that is the pail with the detergent solution and is the place to begin equipment cleaning. The middle pail is yellow, indicating that it contains control water rinse liquids; the green pail at the opposite end of the decontamination line indicates that that is the location of the final control water rinse. In this example, chemical desorbing agents were not required.

remotely; and (5) sampling team members are not required to move amounts of increasingly contaminated equipment in field vehicles to a remote location for cleaning, thereby reducing the potential for accidental personnel exposure to contaminants, contamination of the field vehicle and support materials and supplies that may be stored in the field support vehicle, and cross-contamination of samples that may be transported in the field vehicle.

Field decontamination of equipment does, however, require that sampling team members spend time cleaning equipment under field conditions, which are often not ideal. Problems associated with poor weather (e.g., wind, precipitation, extreme temperatures, high relative humidity) and less than ideal support facilities (e.g., unsuitable water supplies, lack of electricity) can result in less than optimal effectiveness of equipment cleaning. In some cases, for example, when collecting continuous split-spoon soil samples, it may be necessary to have extra personnel in the field to clean equipment, to prevent down time in the field, which can cause major cost overruns on some projects. Sampling team members will typically need to make arrangements for managing and testing waste water generated by equipment-cleaning activities.

Selecting an Appropriate Decontamination Protocol

Factors to Evaluate on a Task-Specific Basis

A number of different protocols exist for equipment decontamination. It is the responsibility of the project manager to determine which method or methods are most appropriate for all pieces of equipment that are to be used for sample collection, field measurements, or sampling point construction (e.g., ground-water monitoring wells). To make those decisions, a number of factors must be evaluated. These factors are summarized in Table 8.4.

Of utmost importance in developing an effective decontamination program is establishing the purpose of the environmental investigation. This is directly linked to the level of QA/QC demanded by the investigation. For example, during the installation of a groundwater monitoring system to act as a leak-detection system around a newly installed underground storage tank system, the required level of QA/QC might be low. No subsurface contamination would be expected at the site unless the tanks were being installed as replacements for old tanks. Therefore, decontamination of equipment used at the site may not be an issue. In contrast, however, installation of ground-water monitoring wells at hazardous waste (Comprehensive Environmental Response, Compensation and Liability Act [CERCLA], Resource Conservation and Recovery Act [RCRA]) sites would require a higher level of QA/QC throughout every aspect of the investigation to ensure the collection of representative data. The level of QA/QC can be further intensified when an investigation is conducted at a site under litigation. Under these circumstances, not only is sample integrity of concern, but also all data must prove to be legally defensible in terms of validity and reproducibility.

After the purpose of the investigation has been established, the suspected site contaminants must be identified and a representative list of parameters that will be analyzed must be established. This process requires an evaluation of a number of contaminant-specific physical and chemical properties to provide information that will be incorporated into the decontamination protocol design. These properties include:

TABLE 8.4

Criteria for Selection of Field Equipment Decontamination Protocol

Existence of federal, state, and/or regional regulatory guidelines that must be followed
Purpose of the investigation (e.g., initial site assessment, long-term monitoring, site remediation design, litigation-driven monitoring or sampling)
Media to be sampled (soil, soil gas, ground water, surface water, waste)
Does the equipment requiring cleaning actually contact the sample
Does the equipment requiring cleaning facilitate sample collection but does not contact the sample itself
Nature and anticipated concentrations of expected contaminants (chemical species, carrier chemicals such as petroleum hydrocarbons or solvents, chemical properties of contaminants)
How will the contaminants be physically distributed (i.e., in air, fill, soil, or ground water)
Physical features of the equipment to be cleaned and its associated support equipment including:
Materials of construction, including inner parts, seals, and external components
Ease of disassembly and reassembly for cleaning
Ability to withstand the rigors of cleaning
Size
Dedicated versus portable devices
Management of decontamination wastes generated
Site support equipment requirements (e.g., power, water, site security)
Site accessibility (political and physical, seasonal variability)
Cost
Health and safety concerns when using chemical desorbing agents such as acids or solvents

- Contaminant physical distribution (i.e., in air, fill, soil, ground water, or surface water)
- Chemical matrix (i.e., interfering constituents, or carrier chemicals such as petroleum hydrocarbons or solvents)
- Chemical species (i.e., volatile organics, nonvolatile organics, heavy metals, inorganic nonmetals, or others)
- Physical properties (density, volatility, flammability, corrosivity, viscosity, reactivity, natural decomposition, or transformation rates)

These are critical factors that direct the selection of decontamination procedures. Decontamination activities must be selected based upon chemical suitability and compatibility with the constituents to be removed during decontamination, and with the concentrations of constituents anticipated. For example, decontamination protocols for an investigation being conducted at a fuel distribution terminal would have to incorporate decontamination procedures such as solvent rinses or special degreasing detergents effective in removing oily substances from all equipment used in the project. However, for an investigation conducted at a metal sludge surface impoundment, metals would be of prime concern and would require different decontamination procedures such as dilute acid rinses. Method selection can become more complicated if more than one contaminant group is of concern at a site. For example, if petroleum hydrocarbons and solvents are of concern in an investigation, use of solvents such as acetone or hexane to degrease oily equipment may interfere with efforts to characterize solvent contamination at the site.

Relative concentrations of contaminants influence many components of an environmental investigation, from establishing health and safety guidelines, to sample collection techniques, to decontamination of personnel and equipment. Anticipated concentration levels (i.e., percent-range concentrations versus parts-per-million or parts-per-billion levels) must be considered along with project objectives and QA/QC controls to identify any physical limitations to decontamination.

Matteoli and Noonan (1987) determined through controlled field testing that the time required for effective decontamination was directly related to the construction materials of the equipment being decontaminated. They found that more than 3 hours of rinsing with clean water was required to lower trichloroethylene (TCE) levels below detection limits for a submersible pump equipped with a rubber discharge hose, compared to 90 min for the same pump equipped with a Teflon® hose. They also determined that individual parameters had unique responses to decontamination. For example, they found that decontamination times for Freon 113 were longer than those for TCE, while decontamination times for 1,1-dichloroethylene, selenium, and chromium were shorter.

Identifying physical limitations or logistical problems associated with decontamination for a project is essential if the decontamination protocol is to be workable. Many very elaborate decontamination procedures developed and approved for theoretical use may not actually be implemented on a project, due to factors such as time and budget constraints, incompatibility with equipment to be decontaminated, inconvenience, and inability to manage the wastes generated. To avoid this problem, a realistic protocol must be developed that recognizes and accounts for myriad project-specific logistical constraints listed in Table 8.4. It is neither logistically reasonable nor safe, for example, to specify that a drilling rig and all necessary support tools and equipment undergo decontamination by solvent or acid washing on a project. Conversely, it is also not reasonable to expect a drilling rig to be "dedicated" to a site unless the project is conducted under the most drastic conditions, such as a high-level radioactive waste site, where it may not be physically possible to decontaminate the rig sufficiently to permit demobilization to another site. Some element of compromise must be incorporated into the decontamination protocol to make the program workable and allow it to meet the objectives of the investigation.

As a case in point, Keely and Boateng (1987) found that procedures originally developed for field decontamination of an electric submersible pump were not workable due to the time and patience required on the part of the sampling team to completely disassemble the sampling pump, scrub each individual component, and then reassemble the pump for use in the next monitoring well. The original decontamination procedures were modified to permit circulation of decontamination solutions through the pump and tubing, thereby avoiding the need to disassemble the multi-component submersible pump. The compromise, however, as indicated by the authors, was the potential for carryover of the solvent (acetone) used in the decontamination procedure. The argument used to justify the change in protocol was that the amount of carryover was limited to a few milliliters of water wetting the surface of the pump, and that the device would be immersed in many gallons of water in the casing of the next monitoring well, resulting in potential residual concentrations of acetone in the sub-parts-per-trillion level. This type of compromise may jeopardize the ability to meet the objectives of QA/QC standards of a highly sensitive analytical program. However, it does illustrate the need to anticipate actual field conditions as opposed to ideal field conditions. Under actual field sampling conditions, performance factors such as weather, operator skill, and pressure to complete assigned tasks within a limited time frame can have a significant impact on the quality of actual decontamination techniques.

The most widely applied solution to this problem is to use dedicated equipment, thereby avoiding all but initial equipment decontamination prior to installation. The second alternative is to reevaluate and modify the program's original decontamination QA/QC specifications.

It is possible to avoid the need to decontaminate well construction materials, such as casing and screen, prior to installation of a well. Many manufacturers can supply precleaned casing and screen that is delivered to a job site in sealed containers. While the initial costs of these precleaned materials may be somewhat high, the higher cost associated with on-site cleaning of these materials is eliminated, as long as the sealed containers are not opened until use.

It is evident that a variety of factors must be considered when developing a decontamination program. It is also readily apparent that those factors are not only site-specific, but are work-task-specific as well. For this reason, it is often necessary to prepare a document that details decontamination procedures on a task-by-task basis. Some states, such as Florida, require that these work plans go one step further and address decontamination procedures for equipment before it is brought to a site for use, as well as on-site field procedures. In a case study of decontamination procedures implemented at a Superfund site in Indiana, Fetter and Griffin (1988) discussed no fewer than eight task-specific components that were incorporated into the field investigation, from well installation to ground-water sample collection.

Available Decontamination Procedures

Methods for Larger Support Equipment

Most larger equipment, such as or a support truck, is cleaned between sampling locations using a high-pressure, hot-water power wash system as described in ASTM Standard D5608 (ASTM, 2006a). This system frequently employs an initial soapy wash and scrubbing with brushes to remove larger soil particles and contaminants from surface areas. This is followed by a high-pressure, clean-water (potable water) wash to remove soap and contaminants (see Figure 8.4). Equipment is usually allowed to air dry before being placed onto elevated racks or plastic sheeting for storage before use at the next location.

This type of power wash system should not be confused with or replaced by the type of power wash system found at the local car wash. Most car washes recirculate water and use additives such as a glycerin to provide a shine to cleaned vehicles. Both these practices make it impossible to have good quality control on decontamination efforts in an environmental investigation.

Hot-water power wash systems are often incorrectly referred to as "steam cleaning," probably because the fine, mist-like hot-water spray that is generated by the high-pressure spray nozzle resembles steam. Steam cleaning technically refers to the application of high-pressure steam to remove contaminants and solid particles from larger pieces of field equipment such as drilling rigs or direct-push rods. The primary advantage of steam cleaning is that the volume of decontamination wastewater is minimized. There are several disadvantages associated with steam cleaning, however, which include: (1) the lack of adequate pressure to effectively dislodge large particles or sticky substances such as clayey soils or residues such as oils; (2) logistical difficulties in generating a sufficient source of steam under field conditions; and (3) temperatures that are so high as to cause degradation of flexible materials such as hydraulic lines or seals on equipment and failure of the equipment.

Most state and federal regulatory agencies will specify the use of "steam cleaning" or high-pressure hot-water power wash methods for drilling rigs, associated tools, and support vehicles. There are exceptions, however. Mickam et al. (1989) indicate that in one state's decontamination program, drilling tools, including augers and split-spoon samplers, must undergo the following decontamination procedures before steam-cleaning:

- Immersion and scrubbing in a mixture of detergent and water
- Rinse with clean water

FIGURE 8.4
Power washers such as the one illustrated here are commonly used to decontaminate larger support equipment like drilling rigs and field vehicles. Often confused with "steam cleaning," these power washers are effective at removing particulate contaminants as well as chemical contaminants from surfaces. This cleaning method is not appropriate for smaller, more delicate instrumentation such as pH meters or for personnel.

- Rinse with isopropyl alcohol, methanol, or acetone
- Multiple rinses with distilled water

This may be a manageable procedure for small equipment such as split-spoon samplers and wrenches. However, a significant exposure hazard can be associated with working with the large volumes of concentrated solvents necessary for decontaminating larger equipment such as augers, drill rod, and bits. In addition, large quantities of potentially hazardous waste would be generated by this cleaning procedure; these materials would be costly to manage and dispose of properly.

Methods for Sample Collection or Analysis Equipment

The variability in decontamination procedures employed for cleaning small equipment, such as sampling devices, is considerable. Some agencies have developed analyte-specific protocols, while others have developed procedures on the basis of equipment type and materials of construction of the equipment to be cleaned. Usually, one or a combination of several of five solutions are typically specified in field decontamination procedures: tap (potable) water, dilute acid, solvent, distilled or deionized water, and laboratory-grade phosphate-free detergent. The number and sequence of use of these solutions is usually the largest source of variation between decontamination protocols.

To generalize, for sampling devices used to sample for inorganics, such as metals, the most commonly used decontamination procedure is as follows:

- Initial wash with water (tap, distilled, or deionized) and laboratory-grade detergent
- Rinse with control water (i.e., water of a known and acceptable chemistry for a particular application)
- Rinse with dilute acid solution (10% nitric acid or hydrochloric acid)
- Final rinses with distilled water or deionized water

When sampling devices are used to sample for organics, the decontamination procedure typically includes:

- Initial wash with water (tap, distilled, or deionized) and laboratory-grade detergent
- Rinse with control water
- Solvent rinse (pesticide-grade acetone, hexane, isopropyl alcohol, methanol alone, or in some combination)
- Final rinses with distilled water or deionized water rinses

It is important to understand the purposes of each individual step in these protocols. Aqueous cleaning is used to initially remove gross contamination and particles from equipment. The control water used to make up the detergent solutions acts as a solvent for water-soluble contaminants and as a dispersal medium for insoluble substances that can be carried in suspension (Parker and Ranney, 1997a, 2000). Detergents are used to improve the "wetting" ability of the cleaning solution and to aid cleaning by separating the contaminant from the solid surface and keeping contaminants in suspension to prevent redeposition onto the equipment. The detergent should be selected with consideration of:

- Method of cleaning and detergent use (manual versus machine, water temperature)
- Will the detergent be able to remove contaminants of concern from equipment at the anticipated contaminant concentrations
- Does the detergent pose any health hazards associated with its use (e.g., high alkalinity, corrosivity, flammability, reactivity)
- Will the detergent be able to physically remove particulates from equipment (e.g., mild alkaline cleaners containing a blend of surfactants and sequestering agents are effective at removing a broad range of organic and inorganic soils)
- Will the detergent leave a residue that could potentially interfere with sample analyses (McLaughlin and Levin, 1995)

In some cases, it may be necessary to use a chemical desorbing agent to thoroughly remove a contaminant that a detergent is unable to fully remove from a piece of equipment. Acid rinses are typically used to desorb metal ions from non-metal surfaces such as polymers and glass. Commonly used acids include dilute (5–10%) solutions of nitric acid and hydrochloric acid. Organic solvent rinses are used to remove residual organic contaminants by dissolving them. The general rule of thumb when selecting an appropriate chemical desorbing agent is "like dissolves like." That is, polar solvents dissolve polar contaminants and non-polar solvents dissolve non-polar or less polar solvents (Parker and Ranney, 1997b, 2000). Because water is a very polar solvent, non-polar or less polar solvents are typically used to remove residual nonpolar organic contaminants such as oils and tars. Commonly used solvents include pesticide-grade isopropanol, methanol, acetone, and hexane.

Considering the number of subsurface investigations associated with underground petroleum product storage systems, pipelines, and terminals, it is unfortunate that decontamination procedures addressing petroleum hydrocarbons are commonly overlooked in decontamination protocols. Solvents such as acetone and hexane are commonly used to degrease equipment. However, this practice can cause interference in analyses for dissolved-phase hydrocarbons and product-specific additives. Other problems include generating hazardous wastes, the potential for the solvents to degrade flexible (i.e., rubber or plastic) parts on some field equipment, and the potential to damage sensitive hydrocarbon sensors on oil/water interface probes. Specially formulated laboratory-grade detergents, such as "Detergent 8," are more effective than most solvents at removing petroleum hydrocarbon residues from the surfaces of field equipment (ASTM, 2006a). Additionally, use of detergents eliminates personnel exposure hazards associated with solvents and greatly reduces the cost associated with wastewater disposal.

For practical purposes, decontamination programs can often be simplified to using an initial rinse with water (tap, distilled, or deionized), followed by a wash and scrubbing with a phosphate-free laboratory-grade detergent (e.g., Liquinox, or Detergent 8), followed by two to three rinses with distilled water. This method has been used in ground-water monitoring programs for inorganics, organics, pesticides, and petroleum hydrocarbons, and for concentrations of these compounds ranging from percent to low parts-per-billion levels. Equipment blanks should be used to determine the effectiveness of this and other decontamination practices (refer to Chapter 6 for additional detail on how to collect equipment blanks). Table 8.5 presents a summary of decontamination procedures that have been implemented during a variety of field investigations.

TABLE 8.5

Currently Available Decontamination Protocol Options

Physical decontamination
Air blasting
Wet blasting
Dry ice blasting
High-pressure Freon cleaning
Ultrasonic cleaning
Vacuum cleaning
Physical removal or scrubbing
Chemical decontamination
Water wash
Tap water followed by deionized and distilled water rinses
Water, laboratory-grade detergent wash, distilled and deionized water rinses
Pesticide-grade solvent rinses in combination with distilled and deionized water rinses
High-pressure steam cleaning
High-pressure hot-water power wash
Hydrolazer
Acid rinses in combination with distilled or deionized water rinses

Inherent Problems with Decontamination Techniques

A review of the various federal and state guidelines demonstrates that materials that are themselves hazardous by definition (i.e., acids and solvents) are commonly incorporated into many decontamination procedures. Other methods, such as steam cleaning or hot-water power washes, require the use of potable water supplies and a variety of support equipment. It is, therefore, apparent that the method of decontamination selected for a given project must be evaluated with respect to its potential to impact sample integrity by contributing potential contaminants to the equipment being cleaned. Table 8.6 presents a summary of some of the more common potential sources of contamination associated with decontamination procedures.

With any decontamination program, solid and/or liquid wastes will be generated. The exact nature and volume of the waste generated is dependent on the equipment decontaminated, the method of decontamination, QA/QC performance standards, the amount of decontamination required, and the contaminants of concern. Depending upon the decontamination methods used and the type of investigation being conducted, it is possible that decontamination wastes may be classified as hazardous by virtue of the

TABLE 8.6

Potential Sources of Contamination Associated with Decontamination Procedures

Use of "contaminated" potable water supplies (i.e., contaminated due to the presence of bacteria, organic compounds, metals, or other objectionable substances)
Use of contaminated supplies of commercially prepared distilled water (i.e., contaminated due to the presence of plasticizers and organic compounds)
Contamination of samples with residues of fluids used for decontamination such as dilute acids or solvents
Oil spray from unfiltered exhaust from generators used as power sources
Volatile organic contamination associated with equipment exhaust systems (steam cleaners, generators, support trucks, etc.)
Use of ethylene glycol (antifreeze) in decontamination equipment (e.g., hot-water power wash spray nozzles and hoses) to prevent freezing in extremely cold weather
Hydraulic fluids, oils, gasoline, or diesel fuel used to operate generators and other support equipment

decontamination fluids used (i.e., nitric acid, acetone, methanol, etc.) and contaminants encountered during the investigation. Generally, when working at hazardous waste sites, management of hazardous decontamination wastes is not a major issue, although it can result in substantial budget increases when specially designed decontamination facilities must be constructed and wastes must be containerized and sampled and the samples analyzed to provide for appropriate disposal. However, at nonhazardous sites where the monitoring program is designed to monitor mainly non-hazardous constituents (i.e., inorganic parameters such as chloride or nitrate), the potentially hazardous wastes generated by decontamination procedures can pose a major problem. Compliance with federal hazardous waste regulations becomes a major issue that must be resolved. This is typically very time-consuming and costly and, consequently, can have the undesired effect of encouraging improper management of hazardous materials and wastes. Because of these potentially significant problems, an attempt is usually made to avoid the use of solvents and acids in decontamination programs.

Quality Assurance/Quality Control Components of Decontamination Protocols

As with all other components of an environmental investigation, it is necessary to monitor the effectiveness of decontamination protocols. This is done to verify that the contaminants of concern are removed from all equipment being decontaminated so that any data generated from samples collected for chemical analysis during the investigation can be considered valid and uncompromised.

The QA/QC segment of the decontamination protocol must address several key issues including:

- Location and construction of a decontamination area.
- Movement of clean and contaminated equipment in and out of the decontamination area.
- Preliminary cleaning of all equipment to be used in a project at an off-site location prior to being permitted access to the site.
- Segregation of clean and contaminated equipment.
- Controls to ensure that cleaned equipment does not become contaminated prior to use (i.e., placing cleaned auger flights on elevated racks or plastic sheeting).
- Controls to ensure that the equipment used for decontamination will not in itself act as a source of contamination (i.e., installing exhaust collectors on generators).
- Chemical verification of suitability of the potable water supply.
- Use of rinse (equipment) blanks and "wipe" samples to verify the effectiveness of decontamination procedures, some of which may be analyzed in the field to provide real-time indications of the effectiveness of the method being implemented.

Should any of the approved decontamination procedures be modified during the course of the field investigation, it is critical to thoroughly document all changes, and provide justification for any changes. Under these circumstances, the use of rinse blanks or wipe samples can be even more critical.

Summary

It is not possible to write a single decontamination program that will be applicable to every field situation. Each environmental investigation will be governed by site-specific physical and chemical variables that will direct the process of selection of the most effective decontamination method on a task-specific and project-specific basis. Once these variables have been defined, however, it is possible to develop a workable decontamination program incorporating procedures that will ensure that all data generated by the investigation are representative of site conditions and that the results of the study are not compromised.

References

ASTM, Standard Practice for Decontamination of Field Equipment Used at Non-Radioactive Waste Sites, ASTM Standard D 5088, ASTM International, West Conshohocken, PA, 8 pp., 2006a.

ASTM, Standard Practice for Decontamination of Field Equipment Used at Low-Level Radioactive Waste Sites, ASTM Standard D 5608, ASTM International, West Conshohocken, PA, 8 pp., 2006b.

ASTM, Standard Guide for the Selection of Purging and Sampling Devices for Ground-Water Monitoring Wells, ASTM Standard D 6634, ASTM International, West Conshohocken, PA, 14 pp., 2006c.

Fetter, C.W. and R.A. Griffin, Field Verification of Noncontaminating Methodology for Installation of Monitoring Wells and Collection of Ground Water Samples, Proceedings of the Second National Outdoor Action Conference on Aquifer Restoration, Ground Water Monitoring and Geophysical Methods, 1, National Water Well Association, Dublin, OH, pp. 437–444, 1988.

Keely, J.F. and K. Boateng, Monitoring Well Installation, Purging and Sampling Techniques Part 1: Conceptualizations, Proceedings of the FOCUS Conference on Northwestern Ground Water Issues, National Water Well Association, Dublin, OH, pp. 443–472, 1987.

Matteoli, R.J. and J.M. Noonan, Decontamination of Rubber Hose and "Teflon" Tubing for Ground Water Sampling, Proceedings of the First National Outdoor Action Conference on Aquifer Restoration, Ground Water Monitoring and Geophysical Methods, National Water Well Association, Dublin, OH, pp. 159–183, 1987.

McLaughlin, M. and P. Levin, Reduce the risk of cross-contamination, *International Ground Water Technology*, pp. 13–15, March 1995.

Mickam, J.T., R. Bellandi, and E.C. Tifft, Jr., Equipment decontamination procedures for ground water and vadose zone monitoring programs: status and prospects, *Ground-Water Monitoring Review*, 9(2), pp. 100–121, 1989.

Parker, L.V., A Literature Review on Decontaminating Ground-Water Sampling Devices: Organic Pollutants, CRREL Report 95-14, U.S. Army Corps of Engineers, Cold Regions Research and Engineering Lab, Hanover, NH, 22 pp., 1995.

Parker, L.V. and T.A. Ranney, Decontaminating Ground-Water Sampling Devices, Special Report 97-25, U.S. Army Corps of Engineers, Cold Regions Research and Engineering Lab, Hanover, NH, 28 pp., 1997a.

Parker, L.V. and T.A. Ranney, Decontaminating Materials Used in Ground-Water Sampling Devices, Special Report 97-24, U.S. Army Corps of Engineers, Cold Regions Research and Engineering Lab, Hanover, NH, 35 pp., 1997b.

Parker, L.V. and T.A. Ranney, Decontaminating materials used in ground-water sampling devices: organic compounds, *Ground-Water Monitoring and Remediation*, 20(1), pp. 56–68, 2000.

U.S. EPA, Test Methods for Evaluating Solid Waste: Physical/Chemical Methods, SW 846, U.S. Environmental Protection Agency, Office of Solid Waste and Emergency Response, Washington, DC, 1996.

9

Ground-Water Sample Analysis

Rock J. Vitale and Olin C. Braids

CONTENTS

Introduction

With each passing year, advancements in technologies and the resultant analytical capabilities of laboratories have been realized for the handling, preparation, and analysis of water samples. During the early 1970s, while techniques and instrumentation were available for the analysis of common ions and trace metals, analytical techniques and instrumentation for determining specific organic species were extremely limited, both in sensitivity and scope. At that time, general methods (e.g., total organic carbon [TOC], chemical oxygen demand [COD], biochemical oxygen demand [BOD], etc.) were extensively used to approximate the gross amount of carbon in a water sample. By today's standards, these methods, although still used for certain legitimate general water-quality purposes, only provide a general noncompound-specific indication of the presence of organic materials in water samples.

A limited determination of specific organic compounds in water was possible in the early 1970s through the use of gas chromatographs. Earlier organic analytical protocols involved extracting the organic substances from the water using solvents, which would be concentrated and then injected into a gas chromatograph. For volatile organic compounds (VOCs), headspace analysis was the usual approach. Some analysts measured headspace at ambient temperature, while others placed the water sample in a temperature-controlled bath and measured headspace at an elevated temperature. This approach resulted in varying sensitivities with the VOCs because of their differences in aqueous solubility and volatility. Analyte identification depended on matching chromatographic retention times with known standards.

It was not until the research of Bellar and Lichtenberg (1974) resulted in a method for VOCs that released these compounds from water by purging the sample with air, followed by capturing the released compounds on an exchange resin, that the method achieved uniformity. This important development enabled the analysis of VOCs to be done rapidly and with significantly improved sensitivity. Surveys of public water supplies that were made following this analytical development resulted in the detection of VOCs (i.e., chloroform) in many public water supplies in the U.S. (Federal Register, 1985).

The discovery of VOC contaminants in public water supplies and in ground water that had been contaminated by chemicals associated with industrial processes, wastes, and other anthropogenic sources has resulted in continuing developmental challenges to qualitatively and quantitatively detect lower and lower amounts of pollutants, currently at the parts-per-trillion (ppt) and even the parts-per-quadrillion (ppq) level.

Selection of Analytical Parameters

The selection of analytical parameters for a ground-water investigation is primarily driven by the purpose and objectives of the investigation, which is often affected by the site's regulatory status, existing site conditions, knowledge of past site practices, and a number of other considerations. During the past decade, transfers of commercial properties have included due diligence investigations of existing environmental conditions as a condition of sale. Several states have made these investigations mandatory.

Ground-water investigations can be done to determine the natural quality of ground water for academic interest or to evaluate its potential as a potable water supply. Alternatively, ground-water investigations can be done to determine whether chemical contaminants are present and, if so, to what extent. Regardless of the category or reason, the list of analytes may not be appreciably different because anthropogenic sources of contaminants are so widespread that ground water completely unaffected by industrial, agricultural, or municipal practices is extraordinarily rare.

A detailed discussion of the common types of investigations and the typical lists of analytical parameters that are analyzed is presented below. In most cases, the various required parameters consist of a mixture of organic and inorganic constituents in addition to measures of esthetic water-quality parameters such as color, turbidity, and odor.

Ground-Water Investigations Governed by a Regulatory Agency

There are a number of federal regulations that have established lists of parameters for analysis of ground-water samples including: the Resource Conservation and Recovery Act (RCRA), the Safe Drinking Water Act (SDWA), the Clean Water Act (CWA), and the Comprehensive Environmental Response, Compensation, and Liability Act (CERCLA) and its Superfund Amendments and Reauthorization Act (SARA). Individual state regulatory agencies may also have variations on these lists, separate lists, and analytical method or sensitivity requirements.

Analytical Requirements Under RCRA

RCRA was enacted to regulate activities related to the transport, storage, and disposal of hazardous wastes. As part of the overall regulation, ground water is specifically addressed. Under RCRA, typically hazardous-waste disposal and storage facilities are required to have ground-water monitoring wells. Water-quality parameters required under RCRA are divided into several categories with different requirements for replication and frequency of analysis. The parameters that indicate if ground water is an acceptable drinking water source are included in the U.S. EPA Primary Drinking Water Standards, which were established under the SDWA of 1974. Parameters establishing ground-water quality include analytes such as chloride, iron, manganese, phenols, sodium, and sulfate. Parameters designated as general indicators of ground-water contamination include pH, specific conductance, total organic carbon, and total organic halogen.

Under certain conditions, analytical requirements under RCRA may include the analysis of a very extensive list of organic and inorganic parameters included in RCRA Appendix IX Constituents.

Many states have adopted the National Primary and Secondary Drinking Water Regulations or have modified them in part to become more stringent and applied them to ground-water investigations within the state. Although ground water may not meet drinking-water standards in all places, the objective of applying drinking-water standards is to provide a goal to which ground water should be treated in the event that it has become contaminated.

Under a Remedial Investigation/Feasibility Study (RI/FS) at a Superfund site, the standard analytical suite is presently referred to as the Toxic Compound List (TCL). Although the TCL includes many parameters, additional parameters could be added if there is information that indicates the possible presence of specific compounds at the site (i.e., waste products known to be present at the site). As part of the investigation, records of the potentially responsible parties (PRPs) and those of waste handlers are reviewed to determine the composition of materials that could be present at the site. This information should be used in making decisions on which parameters should be included in (or deleted from) the analytical scheme for the site.

Analytical Requirements Under a Site-Specific Administrative Consent Order (ACO)

The preceding sections have dealt with specific requirements for selecting water-quality parameters under several regulatory programs. These requirements have been developed to provide a broad-based analytical strategy in order to detect and measure chemical species, particularly contaminants that might be present at a site. In some instances, a regulatory agency will require the facility or responsible party to enter into an Administrative Consent Order (ACO). A list of compounds and constituents for analysis under an ACO is developed on a site-specific basis. Because of this, the benefits obtained from historical sampling and analytical events can be significant.

Analytes That Are Site-Related

As indicated by the size of the RCRA Appendix IX list, the range of chemicals associated with major manufacturing categories is very broad. The Priority Pollutant list was developed from the chemicals most frequently detected in industrial wastewater effluents. However, those waste streams represent only a fraction of the total number of chemicals that are stored, handled, or discharged by industry. A comprehensive guide to industrial waste chemicals is beyond the scope of this chapter.

Selection of an Analytical Method

Just as important as the selection of the analytical parameters is the selection of the analytical method. The selection of the analytical method is in turn determined by the purpose and objectives of the investigation. For example, if the purpose of the investigation were to determine the presence of a specific organic contaminant within certain concentration bounds, the submission of samples for total organic carbon (a nonselective analysis) would not accomplish the objective. Neither would specifying an analytical method that could not obtain the required detection sensitivity.

After establishing the purpose and objectives of a project, an investigator must select the appropriate analytical methods for the parameters of interest. Quite often, the investigator may not be aware of the differences between methods. In such a case, it is important for the investigator to involve personnel with appropriate chemistry and analytical methods expertise during the planning phase of the investigation.

There may be several analytical methods that are capable of meeting project objectives to choose from for the same parameter. Each method should, ideally, give a similar result. However, due to the variables within each method, the results between various methods can vary dramatically, particularly if the method is operationally defined. For this reason, some methods may be preferred or even mandated, depending on whether the analytical results are to be prepared for, or in conjunction with, a regulatory agency.

Specific Requirements for an Analytical Method

Before ground-water samples are submitted to a laboratory, the specific requirements of the analysis, as dictated by the purpose of the investigation, must be communicated to the laboratory so the investigator does not have to assume that the laboratory understands the requirements of the investigation. Passively allowing the laboratory to conduct an analysis by its standard procedures could lead to production of analytical data that are inappropriate (or useless) for the investigation. Perhaps the most important specific requirement for ground-water investigations is the detection limit that will be reported for the requested analysis.

If ground-water samples are to be taken to show that contamination is not present, the concentration at which contaminants can be detected by current environmental technology must be specified. To say that an analyte is not present is correct only to the quantitative extent that the analysis is capable of detecting the analyte of interest. This minimum detectable level is commonly referred to as a "detection limit." In laypersons' terms, a detection limit is the quantitative point at which the analyte will be detected 99% of the time. Detection limits for aqueous samples are typically reported on a weight-by-volume basis (i.e., μg/l or mg/l), or on a statistical basis (i.e., ppb or ppm).

An expensive ground-water sampling and analysis investigation may result in useless information if the detection limits are not low enough to accomplish the objective and satisfy the purpose of the study. An example of this is the analytical detection limit required for a risk determination. Quite often, the primary objective of a ground-water investigation is to assure that human health and the environment are not at risk based upon exposure to analytes of interest that may be present in the ground water. Accordingly, the detection limits that will be needed to accomplish these objectives are levels less than the specific human health-based criteria and environmental-based criteria for the analytes of interest. Obviously the data are of limited usefulness if the resultant analytical detection limits are higher than the most relevant health-based criteria required.

Other specific information that should be discussed with laboratory personnel prior to the sampling and analysis include sample bottle types and volume requirements, field and laboratory quality control (QC) samples, chain-of-custody, hard copy and electronic reporting (documentation) formats, and sample turnaround time.

Description of Analytical Methods

After determining the purpose and objectives of the investigation and defining the specific analytical data requirements, the analytical method can be selected. Some of the most popular references for analytical methods are Standard Methods for the Examination of Water and Wastewater (APHA, AWWA and WPCF, 1989), Methods for Chemical Analysis of Water and Wastes (U.S. EPA, 1979), and Test Methods for Evaluating Solid Waste (SW846) (U.S. EPA, 1986). The latter reference is also available on CD-ROM or from the U.S. EPA Web site (www.epa.gov/epaoswer/hazwaste/test/txsw846.htm). The following sections will discuss some of the more general methods available for ground-water investigations, some of the more commonly analyzed organic and inorganic parameters, and the potential benefits and problems associated with the various methods.

Screening or Diagnostic Tests

Screening or diagnostic tests are procedures that provide an initial indication of the quality of water with an economy of time and expense. Although they can seldom be used alone because they are screening methods, they can provide valuable information when sampling a large number of samples in a relatively short period of time. These screening or diagnostic test analytical procedures have traditionally been conducted in the laboratory, although over the last few years they have been more routinely conducted in the field.

Specific Organic Compound Analysis

Organic analyses are typically divided into three fractions: the volatile (VOA) fraction, base–neutral–acid (BNA) extractables (also referred to as the semi-volatile fraction), and the pesticide or polychlorinated biphenyl (PCB) fraction. To facilitate discussion of organic parameter analysis, these will be discussed by fraction. Many of the aspects discussed below are common to all organic analyses and should provide a basis for selecting an appropriate analytical method. Where applicable, the appropriate U.S. EPA method reference will be provided.

Volatile Organic Compounds

The organic fraction analyzed most frequently in ground-water investigations is the volatile fraction. This is particularly true over the last decade, with the detection of methyl-tert-butyl-ether (MTBE) and other gasoline-related oxygenates in ground water.

Although many of the VOCs are fairly soluble, the primary fate of VOCs in surface-water systems is loss to the atmosphere. However, many VOCs can be fairly persistent in ground water. Depending on the purpose and requirements of the investigation, different analytical methods can be applied to detect the presence of VOCs.

Because VOCs often present health and safety concerns, it is prudent to use field analytical instruments such as screening devices when sampling for these compounds. This provides a warning to the sampler as well as a preliminary indication of the presence of contamination. An example of such an instrument is the organic vapor analyzer (OVA). An OVA provides an approximation of airborne volatile organics, but is not capable of identifying specific VOCs or their individual concentrations without certain modifications to the instrument since the OVA is calibrated to a specific compound such as isobutylene. The OVA is not ordinarily used as a primary analytical method, but is more appropriately used as a screening tool: (1) to monitor volatile vapor releases when a well head is opened; (2) to ensure that vapors are not present in the samplers' ambient breathing zone; and (3) to provide an estimate of relative contaminant concentrations. While this measurement may provide an indication of the presence of volatile contaminants in ground water, it can be deceiving because the measurement is of airborne levels in the well casing and not of the water itself. Other useful screening techniques using the OVA are routinely performed, including, but not limited to, headspace analysis of split-spoon soil samples during borehole drilling and monitoring well installation.

Another field analytical method for VOCs is headspace analysis by portable gas chromatography. Figure 9.1 is a schematic showing the major components of a gas chromatograph (GC). The graphical representation of the compounds as they elute from the GC column are referred to as "peaks" on a gas chromatogram, as represented in Figure 9.2. Peaks are produced during GC analysis by compounds that are present in a sample. Within the limitations and configuration of the GC, if compounds were not present, a flat baseline (without peaks) would be observed on the chromatogram. Chromatographic peaks elute in the order of their boiling points or melting points, with lighter molecular weight

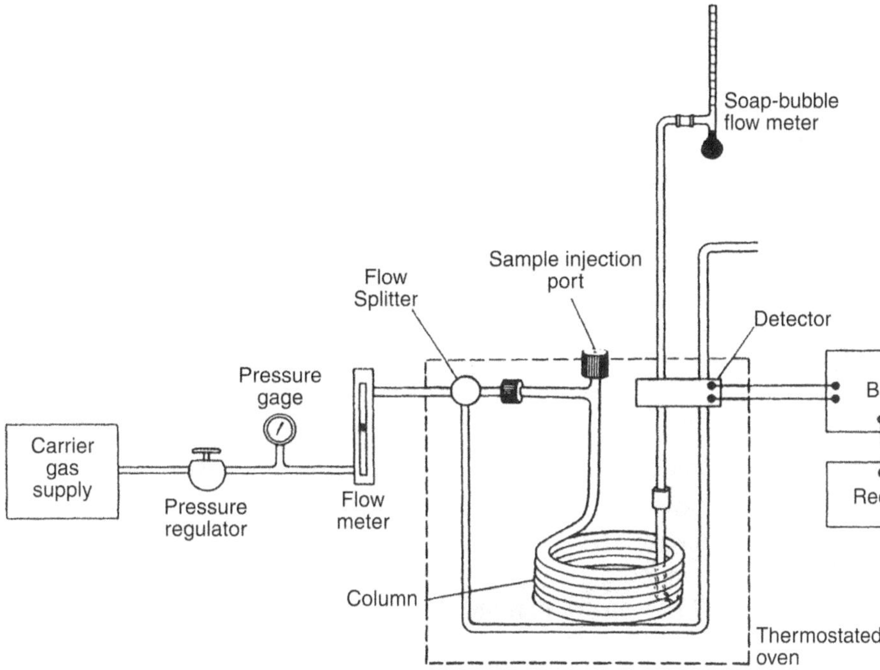

FIGURE 9.1
Schematic diagram of a gas chromatograph. (*Source*: Skoog, D.A., 1985, *Principles of Instrumental Analysis*, 3rd ed. With permission.)

compounds eluting earlier and heavier compounds eluting later on the chromatogram. Compounds can be selectively detected through the use of different types of detectors located at the end of the GC column. This aids in the identification of specific analytes and is particularly important when analyzing complex samples that contain a variety of organic compounds.

Electron-capture detectors (ECDs) and Hall detectors are sensitive to chlorinated compounds (e.g., trichloroethene) but are not very sensitive to straight-chain (normal) or branched alkanes (petroleum hydrocarbons) such as pentane or hexane. Photoionization detectors (PIDs) are selective to unsaturated hydrocarbon compounds such as mononuclear aromatic compounds (e.g., benzene, toluene, and xylene isomers). Flame ionization detectors (FIDs) are used as non-compound-specific detectors for assessing the presence of a variety of organic compounds.

A field-portable GC, which is commonly used in the early stages of environmental site characterization projects, is calibrated with a mixture of standards for those compounds to be analyzed. Once the GC is calibrated, retention time and response information are established for each compound of interest. A retention time is the specific point in time that a compound (peak) elutes on the gas chromatogram.

The most frequently used GC analytical technique used in the laboratory for VOCs is purge-and-trap (e.g., U.S. EPA Methods 601 and 602; see SW846) (Federal Register, 1984). While the technology of purge and trap concentrators and GCs have evolved considerably over the last decade (and continue to evolve), the basic analytical principle remains the same; analytes are effectively transferred via air sparging from the water sample to the sample headspace above the sample. Figure 9.3 presents a schematic of the purge-and-trap system. Identification of target analytes is based on a single peak that matches the retention time of the compound of interest from previously analyzed

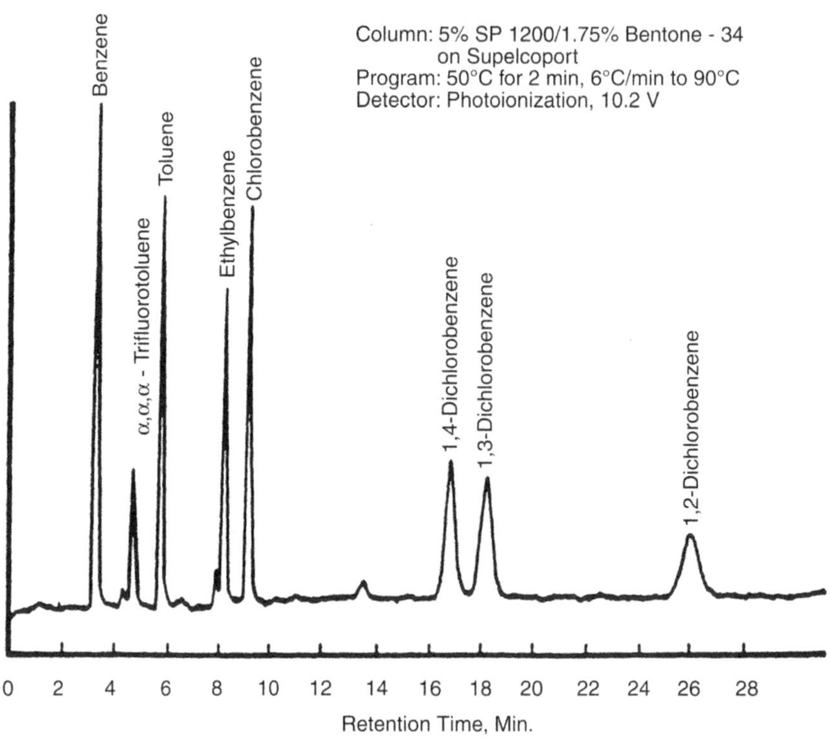

FIGURE 9.2
Gas chromatogram of purgeable aromatics. (*Source*: *Federal Register*, CFR 40 Part 136.)

calibration standards. The method detection limit for most VOCs by purge-and-trap GC is between 0.1 and 1.0 µg/l. The detection limit for some highly water-soluble compounds (e.g., ketones) may be significantly higher. Because of the volatility of this class of organic compounds, samples for this analysis are collected in 40-ml vials with no headspace

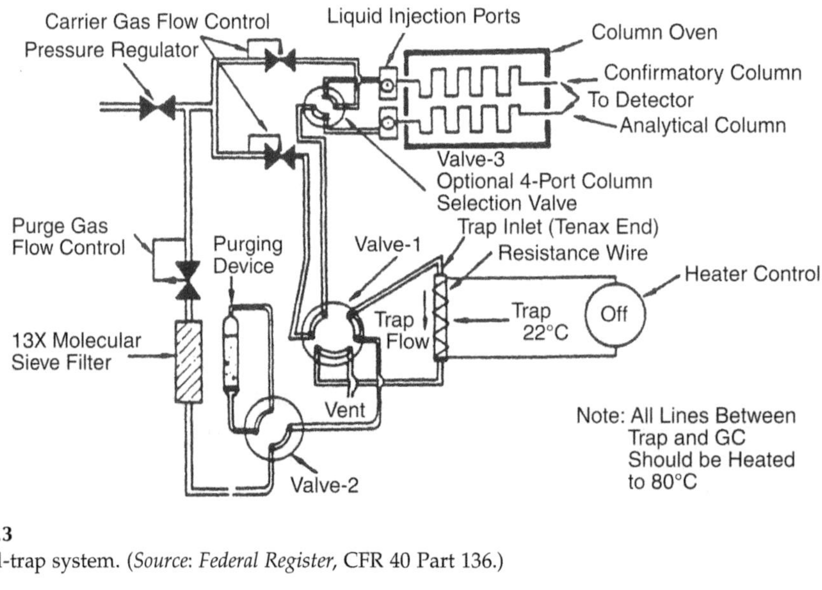

FIGURE 9.3
Purge-and-trap system. (*Source*: *Federal Register*, CFR 40 Part 136.)

(no bubbles). The bubbles will act to liberate the analytes of concern in the same way the analytical method liberates (e.g., sparges) the compounds from the water sample. Holding times are of particular importance due to analyte losses over time. These losses can be attributed to vapor losses through the vial septa, but they have also been shown to be the result of biological degradation.

While GC methods can be used successfully for the analysis of previously characterized ground-water samples, the analytical technique for VOCs that generally provides the most reliable data is a purge-and-trap concentrator, interfaced with a GC, interfaced with a mass spectrometer (MS) (e.g., U.S. EPA Method 624; see SW846) (Federal Register, 1984). This is referred to as a GC–MS (Figure 9.4). Like GCs, GC–MS technology has evolved (and continues to evolve) considerably, but the basic analytical principle remains the same. After organic compounds are separated by the GC column, they are sent through the MS. If the MS detects the presence of a primary mass ion of a targeted analyte, a response will be recorded and processed by the accompanying data system. Typically the MS will listen for the primary mass ion within a certain retention time window, based upon those established during an earlier calibration. With the exception of many isomeric compounds, the identification of target compounds is established confidently because each target compound has a unique mass spectral fingerprint. Because many isomeric compounds have identical mass spectra, isomer specificity enhancement is achieved through GC retention times. Isomeric compounds are compounds that can have several possible

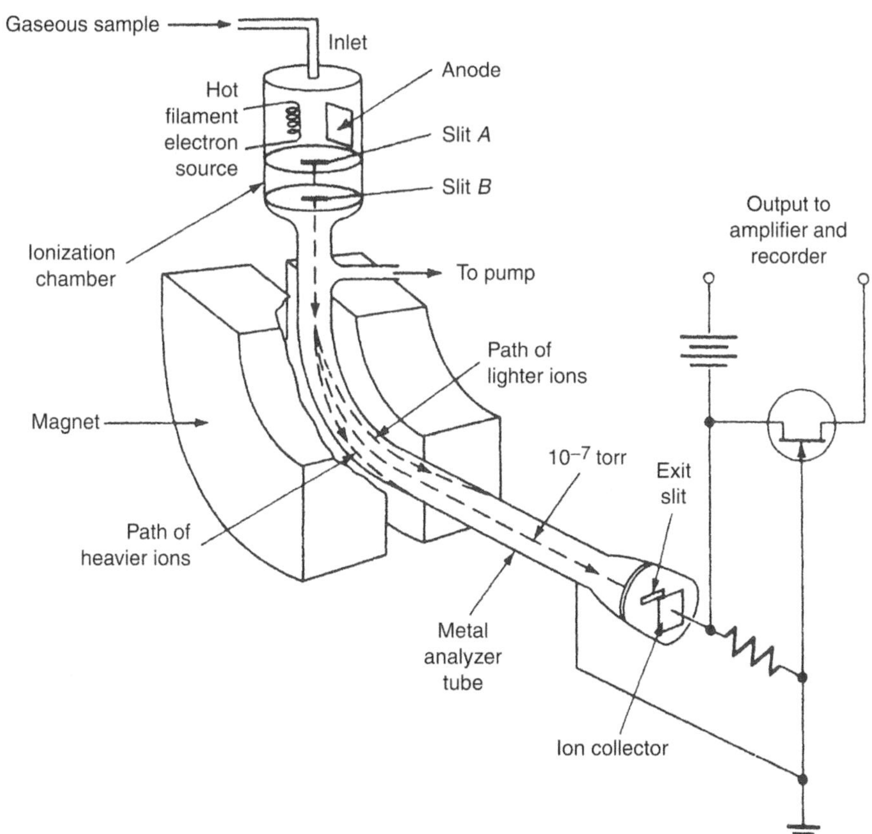

FIGURE 9.4
Schematic diagram of a mass spectrometer. (*Source*: Skoog, D.A., 1985. *Principles of Instrumental Analysis*, 3rd ed. With permission.)

orientations for the same organic compound. For example, 1,2-xylenes and 1,4-xylenes are both xylene (or dimethyl benzene) isomers. Typically, the detection limits for VOCs by GC–MS are between 1.0 and 5.0 μg/l. Like GC detection limits, the detection limit for some highly water-soluble compounds (e.g., ketones) may be significantly higher.

Quite often, large peaks may be present on the chromatogram (detected by the FID), but the MS has not identified any of the organic compounds as being target analytes. Through the use of the accompanying data system, the mass spectra representing these peaks can be compared with a mass-spectral library in order to attempt to ascertain the identity of these nontarget compounds. Compounds detected during these library searches are referred to as tentatively identified compounds (TICs). These identifications should be considered qualitative to semiquantitative at best, although in some instances reasonably good qualitative mass spectral identifications are possible. It is also important to note that investigators can request these TIC mass spectral library searches to be performed years after the analysis is complete as the relevant data are captured on the accompanying data system during the analysis.

From a cost standpoint, ground-water samples collected for the purpose of quantitative volatile organic analysis should first be characterized by GC–MS techniques. This ensures both positive identification and quantification. Analyses for subsequent sampling rounds can then be conducted by less expensive GC techniques. Volatile organic analysis by GC–MS is typically more costly than by GC, although exceptions to this generalization can be found when the analytical laboratory marketplace is extremely competitive.

Semivolatile Organic Compounds

Because the vapor pressures of semivolatile compounds (also referred to as extractable compounds) are lower than those observed for volatile compounds, semivolatile compounds must be removed from ground water samples via solvent extraction. Semivolatile compounds generally have lower solubilities than VOCs, ranging up to tenths of a μg/l. One important variable that governs how semivolatile organic compounds will partition into the solvent is the pH of the sample. The pH of ground-water samples is thus varied during the extraction process to ensure that the target compounds will be extracted. Hence, these compounds are also classified according to the pH at which they were extracted, being either base-, neutral-, or acid-extractable (BNA) compounds.

Ground-water samples for semivolatile organic analyses are typically prepared by taking 1 l of ground water and adjusting the pH at various points during the extraction process. The initial extraction solvent is usually methylene chloride and involves either manual (e.g., separatory funnel) or more extensive automated (e.g., continuous liquid–liquid) extraction techniques. Once the extraction is complete, the extracts are combined and concentrated (evaporated) with a gentle flow of nitrogen or one of several other currently automated solvent concentration techniques. Depending on the type of instrumental analysis being performed, the extract may be exchanged into alternate solvents such as hexane (for pesticides and PCBs by GC), acetonitrile (for polynuclear aromatic hydrocarbons [PAHs] by high-performance liquid chromatography [HPLC] or LC–MS), or toluene (for chlorinated dioxins and furans).

Generally speaking, the most qualitatively reliable analytical method for semivolatile organic compounds is GC–MS for the same reasons previously provided for VOCs. GC–MS semivolatile organic analysis is conducted by injecting microliter amounts of the concentrated methylene chloride extract onto the capillary column. The MS analysis then proceeds in the same manner as for the VOCs, including library search procedures for non-TCL compounds. The typical quantitation limit for most semivolatile compounds by low-resolution GC–MS is 10 μg/l.

There are some specialized GC–MS methods for semivolatiles that can result in substantially lower quantitation limits than 10 μg/l. Methods such as isotope dilution, single-ion monitoring (SIM), and extraordinary concentration techniques can result in parts-per-trillion (ppt) quantitation limits in ground-water samples. Similar quantitation limits can also be achieved by use of a high-resolution MS.

For certain types of semivolatile organics (e.g., PAHs and explosives), the necessity to attain lower detection limits than can be normally attained by GC–MS mandates the use of HPLC and, more recently, liquid chromatography interfaced with a mass spectrometer (LC–MS). HPLC can typically achieve detection limits between 0.1 and 0.3 μg/l. Like a GC, once the HPLC is calibrated, retention time and response information are established for each compound of interest. Identification of target analytes is based on a single peak that matches the retention time of the compound of interest from previously analyzed calibration standards. For PAHs, typically an ultraviolet detector is used with confirmation for positive results by either a fluorescence or diode array. Even with alternate detector and secondary dissimilar column confirmation techniques, as with analysis by GC, false-positive results are possible without a qualitatively definitive mass spectral confirmation.

Pesticides, Herbicides, and PCBs

Chlorinated pesticides, herbicides, and PCBs are classes of semivolatile organic compounds that can be particularly toxic at low concentrations. Because trace concentrations of these chemicals can bio-accumulate and elicit acute and chronic toxic responses in living organisms, the quantitation limits for these compounds must frequently be in the subparts-per-billion range. Most of the routinely analyzed pesticides, herbicides, and PCBs contain at least one chlorine atom (most contain many chlorine atoms), so most are effectively analyzed with the requisite sensitivity by GC with an ECD detector. As stated earlier, an ECD is particularly sensitive to chlorinated compounds. Because these compounds are semivolatile in nature, they must also be extracted and concentrated from ground-water samples by solvent extraction. Although methylene chloride is the solvent of choice for the analysis of semivolatiles, it is inappropriate for GC analysis by ECD, because the ECD is sensitive to chlorine. For the analysis of these compounds, the solvent of choice for GC–ECD analysis is hexane.

Typically, a 1-l ground-water sample is extracted with hexane (or extracted with methylene chloride and then exchanged into hexane). Microliter volumes of hexane are then injected onto the GC column. Quantitation limits for these compounds by GC–ECD are on the order of 10–200 ppt in ground-water samples.

As with the analysis of volatile organics by GC, the primary pitfall of pesticide or herbicide analysis by GC is that identification solely by GC cannot be considered qualitatively confident because the identification of most pesticides and herbicides appears as a single peak on a gas chromatogram. Although these compounds can be further confirmed by analysis on a second dissimilar GC column (i.e., a GC column with a different type of packing material), false-positive results can still be a problem with pesticide or herbicide analysis by GC–ECD.

Some pesticides, such as toxaphene, technical chlordane, and lindane, and all PCBs are mixtures of a base compound with a range of chlorination (congener compounds). The GC analysis of these compounds will result in a unique multi-peak pattern. The positive results for multi-peak pesticides and PCBs can be considered more qualitatively confident (than single-peak pesticides) because the unique multi-peak fingerprint is difficult to randomly generate chromatographically unless the analyte is truly present. Although one column will suffice for multi-peak identifications, the laboratory should be required to confirm the identification on two dissimilar columns. Interpretation of results also has to account for peak shifts or disappearances with the effects of environmental exposure.

Pesticides, herbicides, and PCBs can also be analyzed by GC–MS. These compounds can also be tentatively identified by using the same data system library search procedures as those used for chromatographic peaks from the semivolatile fraction. However, unless extraordinary methods of MS analysis (i.e., SIM, high-resolution MS) are performed, the quantitation limits by GC–MS may not be lower than the health-based criteria for most of these compounds.

Specific Constituent Inorganic Analysis

Inorganic parameters that are typically analyzed in ground-water samples for determination of the presence or absence of contaminants include trace metals and cyanide. The U.S. EPA has established specific holding times, container types, preservatives, and storage requirements for these and other inorganic parameters in ground water (U.S. EPA, 1986).

Trace metals and cyanide can exist in either a nonsoluble solid precipitate (which is immobile in the ground-water system, unless it is in colloidal form and the formation has large enough pore spaces to allow colloidal movement) or in a soluble form. Samples of ground water that are collected by methods that agitate the water column in the well (e.g., bailers or inertial-lift pumps) often contain large quantities of clay, silt, and other solids, and make it difficult to differentiate the forms of metals. In an effort to determine in what form the various inorganic parameters exist, ground-water samples are generally field-filtered through a 0.45-μm filter, though this practice does not always produce the desired result (see Chapter 5). It is important to decide early in the project whether or not samples should be filtered and how the resulting analytical data will subsequently be interpreted. If samples are not filtered, the preservation method for trace metals (addition of nitric acid to a pH of less than 2) may liberate metals sorbed to the surface of formation solids, and produce an erroneous analytical result. Addition of NaOH in a separate bottle to a pH greater than 12, to preserve the sample for cyanide analysis, does not produce the same effect, but may precipitate trace metals out of the sample.

The two methods that are commonly used for metal analysis are atomic absorption spectrophotometry and atomic emission spectroscopy. Recently, mass spectrometers have been put in tandem with atomic emission (e.g., inductively coupled plasma [ICP] or MS). These instrumental methods of analysis require that the sample be prepared so that any metals that are present will be in an ionic form. Metals are converted to an ionic form by sample digestion. Digestion of a ground-water sample is performed by gentle heating with addition of nitric or hydrochloric acid.

Atomic Absorption Spectrophotometry

Atomic absorption (AA) spectrophotometry is based upon a measured difference in electronic signal between instrumental optics induced by the sample, which is present as a gas between these optics (Figure 9.5). There are three types of AA methods: flame AA, graphite furnace AA, and cold vapor AA.

For flame AA analysis, the digested sample is aspirated through a very thin tube, drawn by suction, and introduced into an air-acetylene or nitrous oxide flame. A beam of light from a lamp with a cathode of the metal being analyzed is focused through the

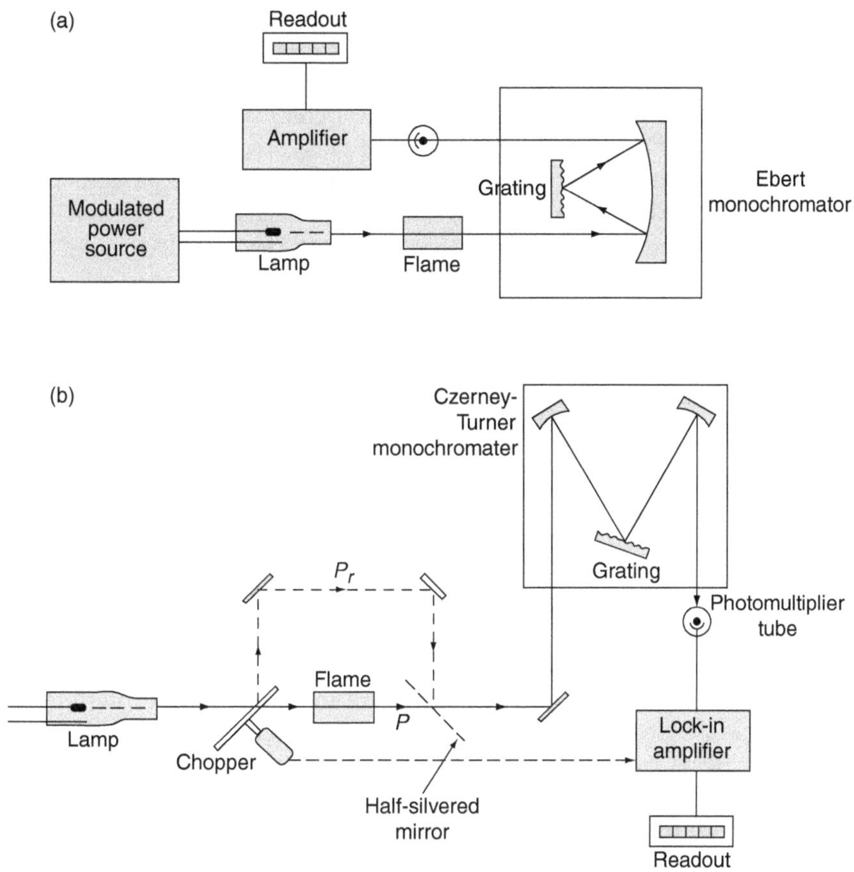

FIGURE 9.5
Typical flame spectrophotometers: single-beam design and double-beam design. (*Source*: Skoog, D.A., 1985, *Principles of Instrumental Analysis*, 3rd ed. With permission.)

flame. Depending on the concentration of the analyte in the digest, the optics will measure an electronic difference in light absorption. Flame AA is used for a wide variety of metals. Some more toxic metals such as arsenic, thallium, lead, and antimony are not sensitive to flame AA methods at concentrations of environmental concern. These elements must be analyzed by graphite furnace AA. Detection limits of 5 pg/l (picogram/l) or less can be achieved for these elements by graphite furnace AA. Typical flame AA detection limits range from 500 to 1000 μg/l for these elements.

The graphite furnace technique involves placing a staged hollow graphite tube in the path of a beam of light set at the wavelength of the analyte of interest. Microliter amounts of the acid digest are placed (or sprayed) into the entrance port of the graphite tube. The temperature of the graphite tube is increased slowly via electrode circuitry. Initially, the liquid is dried within the tube. The temperature is then increased to pyrolysis temperature that will break up various complexes in which the analyte of interest may be tied up. Finally, the temperature is ramped up to the point at which the analyte will be converted to a gaseous form (atomization). When the gas of the analyte of interest passes through the beam of light set at a wavelength unique to that analyte, an electronic difference (absorbance) is measured, which is proportional to the concentration of the analyte.

The cold vapor AA method is used exclusively for the determination of mercury, and can achieve detection limits of 0.1–0.2 μg/l. The theory of cold vapor is similar to that of graphite furnace AA, with one exception. Whereas graphite furnace AA generates the gaseous form of the analyte by a temperature increase, the cold vapor technique generates mercury gas by a chemical reaction. The generation of gaseous elemental mercury is done by the rapid addition of a liquid reagent (stannous chloride) after a complex digestion procedure is carried out in a bottle with a shaved glass stopper (i.e., a BOD bottle). The sample (still in the BOD bottle) is then purged with argon, and any gaseous mercury that is liberated passes quickly across the optics; again, the measured electronic difference is proportional to the concentration of mercury present in the sample.

Atomic Emission Spectroscopy

Just as metals can be measured by an absorbance difference in a gaseous form (some better than others), metals can also be measured by a corresponding emission. The determination of metals by emission spectroscopy can result in much lower detection limits for most metals compared to AA. The two most critical factors in detecting trace-level analytes are the temperature and stability of the flame. To this end, technology has incorporated both these factors by use of ICP and direct-current plasma (DCP) emission spectroscopy.

A plasma is a high-temperature electronic flux that exists at a temperature an order of magnitude higher than conventional flames. Plasmas are extremely stable. The pitfall with ICP (and DCP) is that high concentrations of solids (TDS), salts (sodium, calcium), and common elements (iron, aluminum) can result in severe matrix interferences.

A benefit of the use of ICP, in addition to the lower detection limits, is simultaneous multi-element analysis capability. ICP systems can analyze 15–20 metals in a water sample in a 2-min period. The operation of ICP is similar to that of flame AA. A peristaltic pump draws an acid-digested sample into a chamber, which sprays the sample into the plasma. The optics measure the difference in emissions as intensity at the wavelengths of interest and record the difference in concentration units.

Other Analyses

The preceding discussion focused on the parameters most often required in ground-water investigations in which the emphasis is determining if ground-water contamination is present. There are, of course, many other analytes that may be of interest, depending upon the purpose of the investigation. These include various nitrogen compounds (i.e., ammonia nitrogen, total Kjeldahl nitrogen [TKN], nitrate–nitrogen, and nitrite–nitrogen), sulfur compounds (i.e., sulfate, sulfite, and sulfides), phosphorus (i.e., phosphates), and other soap compounds (i.e., surfactants), cyanide, and total phenols. These organic and inorganic compounds, routinely referred to as "wet chemistries," are analyzed by a variety of colorimetric, potentiometric, titrimetric, gravimetric, and other instrumental techniques (i.e., ion chromatography).

Finally, biological analysis may be of interest in some ground-water investigations, particularly when the ground water is to be used for drinking water. One biological indicator that is commonly used is the analysis of total coliform bacteria. Some coliform bacteria are associated with ground-water contamination by septic systems (i.e., fecal coliform bacteria). Most of the biological parameters are examined by allowing the ground water to incubate with an enriched broth (or administered to agar plates) which the specific microbes of interest can use as a food source. After a designated period of time and temperatures, the culture tubes and agar plates are examined for any activity (i.e., turbidity, the presence of bacteria colonies, etc.).

The procedures for many of these other types of analysis can be found in Standard Methods for the Analysis of Water and Wastewater (APHA, AWWA, and WPCF, 1989).

Quality Assurance/Quality Control

Obviously, selecting the appropriate parameters and methods for analytes of interest are critical steps to properly assessing ground-water quality. However, just as critical is the care taken during sample analysis, the submission of check samples to "test" the sampling process, and a review of the appropriateness of the data after they have been generated. This process is collectively referred to as quality assurance/quality control (QA/QC).

Very often the ability to assess the quality of analytical data is in direct proportion to the degree of confidence that is required in the analytical measurements. For example, taking a pH reading of your garden soil just to satisfy your curiosity does not require stringent QA/QC. However, if you wanted to grow expensive plants that require pH 6–6.5 to survive, more elaborate QA/QC requirements, such as multiple samples, may be required because the measurement will be used to answer a very specific question (i.e., pH between 6 and 6.5). The degree of QA/QC that is implemented should therefore be proportional to the specific requirements with regard to the amount of confidence in the analytical measurements.

The amount of QA/QC that is implemented (bottleware, field notes, duplicate samples, blanks, etc.) is also proportional to the available funds for the investigation. It is intuitive that the costs involved with ground-water investigations will increase as the need for confidence in the data and, hence, the amount of QA/QC in the data increases.

There are various reasons to consider high-confidence analytical data important. Obviously, if expensive decisions are going to be based on analytical data, a high degree of confidence would be desirable. Other reasons to require a high degree of confidence include instances when human health or ecological risks are being assessed or when data are being used for litigation purposes.

Selection of an Analytical Laboratory

A critical quality assurance element is the selection of a qualified analytical laboratory. The selection of a laboratory should be conducted by environmental professionals who are familiar with the types of analysis that are going to be performed, specific methodologies to be used, and various other requirements that are dictated by the purpose of the investigation. The most basic requirement (which should not be assumed) is that the laboratory has the capability (instrumentation, experience, etc.) to perform the specific analyses required for the investigation.

The laboratory should have a good reputation for quality (not to be confused with good service or low prices) and must be willing to cooperate with the investigator, who should set specific requirements for sample analysis. Special requirements such as special holding times, sequence of sample analyses, and frequency of laboratory blanks, duplicates, spikes, calibrations, and data package deliverables are just a few of the requirements that need to be addressed by individuals who are knowledgeable in these areas.

One essential criterion for some projects is that the laboratory be certified under the Contract Laboratory Program (CLP) of the U.S. EPA. A common misconception is that these laboratories are "EPA-certified" laboratories — they are not. This is a contract program that requires an on-site laboratory audit and the successful analysis of a variety of samples submitted by the EPA with known analytical results. This certification is necessary for a laboratory to provide analytical services for projects conducted under the Superfund program.

Another indication of the laboratory's ability to provide quality analytical data is the various state or health department certifications that the laboratory holds. Frequently, laboratories are used outside of the state in which the project is being conducted. However, many states have their own certification programs in order to provide a basis for licensing laboratories. The larger laboratory chains hold licenses in numerous states so they are not geographically confined in terms of accepting samples. Certifications do not exclusively make a good laboratory, but they do indicate that the laboratory is capable (if required) of generating quality work. In order to gain insight into how good (or bad) a laboratory is, an on-site audit should be performed by a competent chemist.

A properly conducted audit will identify those laboratories that can generate high-quality data and those that are not worth further consideration. Additional aspects that should be considered when selecting a laboratory are service considerations, including the availability and flexibility of scheduling sample bottle delivery, sample arrival and analyses, hard copy and electronic reporting formats, and effective and timely communications. For certain ground-water analyses, it is also important to consider the location of the laboratory with regard to the mode of sample shipment, and what impacts, if any, the shipping mode will have on the required sample holding times.

Although the cost of analysis is a consideration, investigators should not select a laboratory solely on a cost basis. Beware of bargain basement prices — more often than not, you get what you pay for. An additional financial consideration that investigators must keep in mind is conducting a due diligence inquiry of the candidate laboratories prior to selection. Many commercial laboratories have been bought or sold or have closed in the middle of a project with little to no notice to their clients, leaving them in a difficult situation.

Preparation of a Quality Assurance Project Plan

For large and small investigations, environmental professionals should consider preparing a Quality Assurance Project Plan (QAPP). In many instances, when the investigation is being conducted under the purview of a state or Federal agency, the preparation of a QAPP is required before work can begin. The purpose of the QAPP is to:

- State data quality objectives as they apply to the investigation.
- State who will perform each task in the investigation (project responsibilities), including the designation of the analytical laboratory.
- Specify what protocols will be used for ground-water sampling.
- Demonstrate sample custody.
- Specify requirements for QC samples.
- Specify analytical methods for each analyte.
- Note holding times.
- Specify sample container and preservative requirements.
- Specify data package requirements.
- Provide data validation and reduction protocols.

- Specify frequency of audits.
- Schedule reports to management.

The ultimate purpose of the QAPP is to describe all the whats, wheres, whens, hows, and whys of the investigation. This is necessary so that all steps in the investigation are understood and nothing is left to the laboratory's interpretation (or imagination). Considering the demonstrated costs associated with disproving bad sample data, a QAPP is a worthwhile effort whether or not a regulatory agency is involved. Nonregulatory QAPPs do not have to be a voluminous, formal document, but rather a concise document that clearly states the needs, requirements, objectives, and logistics of the investigation.

Laboratory QA/QC

The laboratory QA/QC process begins when the laboratory ships sample containers and preservatives to the investigator. The sample containers are usually shipped in insulated shipping containers that will accommodate wet ice for cooling on the return trip. The containers provided by the laboratory will be appropriate for the analytes specified by the investigator. Container labels and chain-of-custody sheets will accompany the sample containers. The topics that follow cover the various aspects of laboratory QA/QC. Figure 9.6 presents the laboratory QA/QC process, beginning with laboratory receipt of samples. The remainder of this chapter briefly discusses each of these elements.

Chain-of-Custody

At the moment the sample bottles are released from the laboratory, a chain-of-custody routinely begins. In other cases, chain-of-custody routinely begins when samples are placed in laboratory bottleware and labeled. When samples are received by the analytical laboratory, chain-of-custody continues with the laboratory sample custodian acknowledging receipt of samples. For certain litigation project needs, an internal laboratory chain-of-custody can be requested in which the internal transfer of samples is documented. Chain-of-custody should be considered a fundamental requirement for all investigations. For shipping samples in coolers by a third-party courier, chain-of-custody cannot be defended without the use of custody seals on the shipping container or cooler.

Sample Storage and Holding Time Requirements

One of the most important aspects of laboratory QA/QC is ensuring sample integrity and strict adherence to holding times. A holding time is the time that has elapsed from the moment the ground-water sample is collected to the moment of sample preparation or analysis. Holding time requirements are generally specified by the regulatory agency and are published in 40 CFR Part 136. Results from samples analyzed beyond established regulatory holding times should be viewed as questionable and potentially cannot be legally defended, depending on the intended use of the data.

Sample Preparation

Depending on the analysis that has been requested, sample preparation may be required before analysis can proceed. The most common types of sample preparation include extractions, digestions, and distillations. Some holding times are specified from the time of sample collection to the time of sample preparation, and others are specified from the time of sample collection to the time of sample analysis.

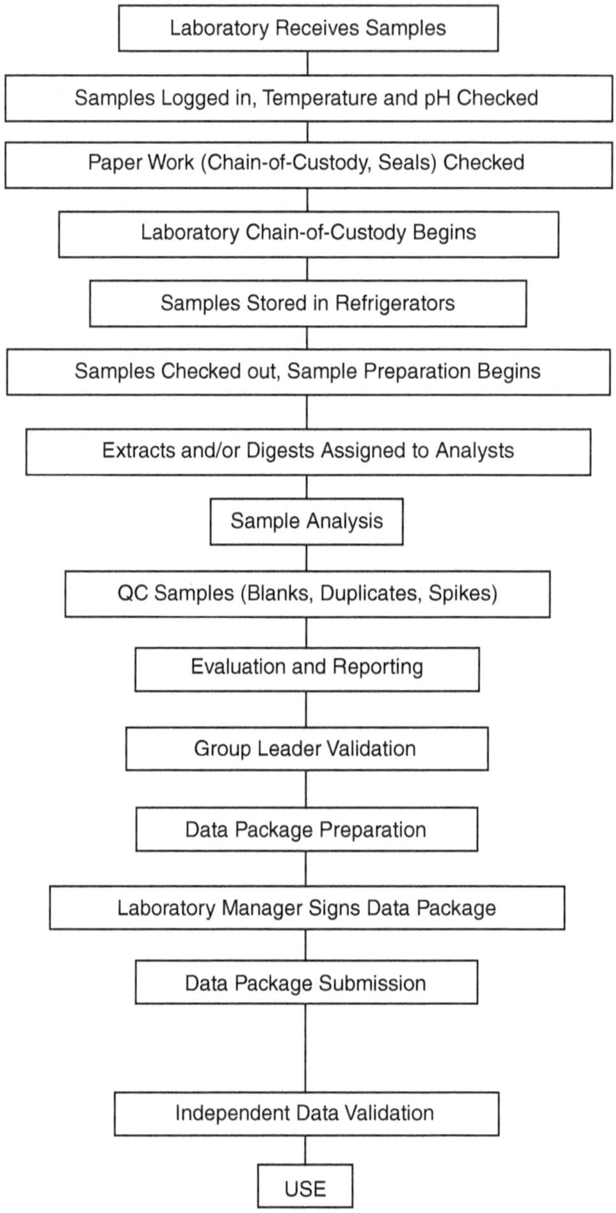

FIGURE 9.6
The laboratory QA/QC process.

Laboratory QC Samples

The analysis of QC samples is another important aspect of laboratory QA/QC and, more specifically, providing a mechanism by which data quality can be assessed. Analysis of certain types of QC samples may require the collection of additional sample volume, which is an important consideration that field samplers must be aware of. The intent of these QC samples is that they must be concurrently prepared and analyzed with the investigative samples with which they are associated. These QC samples, including, but not limited to various types of laboratory blanks, duplicates, and spikes, are described below.

Method Blanks

A method blank is a portion of deionized water that is carried through the entire analytical scheme, including all sample preparation. It is important that the volume used for method blanks be the same as the volume used for the samples. Method blanks must be analyzed (purged) every 12 h for VOC analysis. Method blanks for volatile organic analysis are typically prepared by the laboratory the same day they are analyzed; therefore, they do not monitor contaminants introduced during sample storage as holding blanks do. Method blanks are also referred to as preparation blanks.

Duplicates

Laboratory duplicates are two separate aliquots of the same sample, which have been independently prepared and analyzed for the same parameters to determine the precision of the analytical system. The analytical laboratory should perform a duplicate analysis on a minimum of one sample in 20. Duplicate results are typically compared as relative percent difference (RPD):

$$\text{RPD} = \frac{\text{Sample A} - \text{Sample B}}{\text{Average sample A} + \text{B}} \times 100$$

Spiked Samples

A spike is a sample in which the compound being analyzed is actually added (or spiked) into the sample to determine the accuracy of the analytical system. The question is, can you get back as much as you put in? The results of a spike are expressed in terms of the percent recovery with regard to the amount added. There are two types of spikes that are used by an analytical laboratory: matrix spikes (for both organic and inorganic analyses) and surrogate spikes (organic analysis exclusively).

$$\%\ \text{Recovery} = \frac{\text{Spike result} - \text{unspiked sample result}}{\text{Concentration added}} \times 100$$

Matrix Spikes

To determine laboratory accuracy and precision, a sample is analyzed unspiked (to determine a baseline). A second portion (or aliquot) of sample is then typically spiked with the target compounds or analytes of interest. The results (percent recoveries) of these spikes are a direct measure of analytical accuracy. A third portion of sample (matrix spike duplicate) is spiked in the same manner (for organic compounds). The comparison of respective recoveries between the two spikes is a measure of analytical precision. Depending on anticipated levels and types of parameters, the laboratory can be instructed to add additional matrix spike parameters at appropriate concentrations.

Surrogate Spikes

Surrogate spikes are added to every sample for organic analysis. A surrogate compound is a special compound, sometimes synthetically prepared and other times a compound rarely observed to be present naturally in environmental samples. Surrogate compounds are added to every sample analyzed for organic compounds to test the analytical procedure. As stated earlier, surrogate compounds are typically not found naturally. However, they are similar in structure to several routinely examined target analytes. There are three typically volatile surrogates, six semivolatiles, and two pesticide surrogates, which are added at predesignated concentrations.

Surrogate compound percent recoveries are calculated concurrently with the target compound of interest. Because the sample characteristics will affect the percent recovery, these percent recoveries indicate the accuracy of the analytical method. More importantly, surrogate recoveries are direct measures of how well the method has worked on each individual sample. If poor recoveries are obtained, the sample should be reextracted and/or reanalyzed a second and possibly third time, using less sample. However, using less sample will result in higher quantitation limits.

The "acceptable" recoveries for surrogate compounds vary depending on the individual compound and the type of compound. The U.S. EPA has generated acceptable ranges for both matrix spike and surrogate recoveries to be used as defaults. These ranges were generated from statistical manipulation involving numerous studies based on recoveries reported from many laboratories. The "acceptable" recoveries for some compounds can be as low as 11%. For some purposes, these recoveries are not acceptable, and the laboratory must be instructed to follow specific criteria for reextraction and reanalysis. One problem, which should be noted with multiple reextractions, is the required sample volume and holding time constraints. For this reason, it is always prudent for samplers to collect and submit large volumes of sample.

Instrument Calibration

Aspects of instrument calibration include how the instruments used for analysis will be calibrated, how often the calibration will be checked, what actions will be taken if poor calibration checks are obtained, and who prepared the calibration standards. If instruments are calibrated properly using incorrect concentrations, incorrect data will be generated. The source of primary initial calibrations, calibration check standards, and the resident spiking solutions must be different (usually obtained by different vendors) so that the preparation and accuracy of each standard and spiking solution can be a check on the other.

The laboratory must perform an initial multi-point calibration to determine instrument sensitivity and linearity. The linearity of initial multi-point calibrations is typically judged on the basis of the best linear fit on either a relative standard deviation or correlation coefficient basis. Calibration checks are performed to assess instrument stability relative to the previously performed initial multi-point calibration. Calibration checks are typically judged on the basis of percent difference, percent drift, or percent recovery relative to the response of the initial multi-point calibration.

Sample Analysis

Once the instrument has been calibrated and blanks have been analyzed verifying that the instrument is free of contaminants, the analysis of samples can proceed. An important aspect of sample analysis is the analytical sequence. A highly contaminated sample may severely contaminate the instrument. Carryover or memory effects can generate false-positive results for subsequently analyzed samples, particularly during volatile organic analysis.

Laboratory Validation and Reporting

Once the analytical results have been generated, they must be validated. This is typically done by laboratory section heads or the laboratory manager. The QAPP (if prepared) should specify how calculations will be verified and document the procedures that will be used to assess how the results of laboratory QA/QC samples impact the data.

The laboratory manager should sign the analytical data before results are released and, quite often, the submission of a complete (inclusive of all "raw data") data package may be required. The submission of a laboratory data package is a decision that the investigator must make. The importance of the investigation and the implications of the results should be considered when making this decision.

Documentation and Recordkeeping

To ensure the integrity of the QA/QC process, full documentation should be provided to account for all laboratory activities. The following are some of the areas requiring laboratory documentation:

- Sample log-in procedures, including assignment of laboratory control number, taking pH and temperature, and verifying field paper work.
- Internal laboratory chain-of-custody — every transfer must be entered.
- Verification of refrigerator storage, including daily temperature verification log.
- Time chronicle to verify holding times.
- Extraction and digestion instrument logs to verify analyses and analytical sequence.
- Lab narrative to describe any problems encountered during the analyses.
- Summary forms allowing brief examination of pertinent QA/QC information.
- Raw data — every item of data (i.e., standards, blanks, spikes, duplicates, and samples) relating to analysis; this includes *all* instrument printouts and copies of analysts' notebooks.
- All information pertaining to the case should be stored in a laboratory file, and all analytical data should be stored on magnetic tape or computer disk for future reference for a period of at least 5 years.

Independent Laboratory QA Review

Whenever possible, analytical results and laboratory documentation should be independently validated by a qualified, experienced quality-assurance chemist. The QA process does not end when the laboratory delivers the data to the investigator. Laboratory chemists and managers may not give special attention to your samples to make sure that *your* results are valid. Unless an independent data review is performed to make certain the results are correct, it must be assumed that the laboratory made no mistakes. If errors did occur, this may be a costly assumption.

In addition to assessing the validity of the analytical data, an independent data review can also provide the interpretation of analytical bias. Bias is the tendency for the results to be skewed higher or lower than the actual number. For example, if spike recoveries are consistently 130% for benzene and the laboratory reports benzene in a monitoring well at 20 μg/l, intuitively it can be stated that the actual concentration of benzene in the well may be slightly less than the 20 μg/l reported.

Once the data review has been performed, a report should be prepared that qualifies certain areas of the data before the results can be utilized. Another item to consider is data presentation (data reduction). Considering the number of samples and compounds that may be analyzed during an investigation, the number of analytical results can be quite extensive and cumbersome. Typically, a preferred alternative is to reduce the data by use of computer spreadsheets to just those compounds with positive results. Specific codes that provide an indication of data reliability should be placed next to results as appropriate during the independent data validation. Once the validity has been assessed and the positive results have been tabulated, it is highly recommended that individuals familiar with the hydrogeology of the site examine the data tables. This troubleshooting process can identify anomalies and data gaps.

Summary

The primary thrust of most ground-water investigations is determining ground-water quality. Considering the costs associated with well installation, laboratory analysis, and the implications of discovering human health and ecological threats, it is apparent that all necessary steps should be taken to maximize the results of this effort to assure the quality of the data collected. The quality assurance program starts with the design of the investigation includes the parameters to be analyzed, the analytical methods to be used, and the level of laboratory QA/QC to be followed.

References

APHA, AWWA, and WPCF, *Standard Methods for the Examination of Water and Wastewater,* 17th ed., American Public Health Association, Washington, DC, 1989.

Bellar, T.A. and J.J. Lichtenberg, Determining volatile organics at the microgram-per-liter levels by gas chromatography, *Journal of the American Water Works Association*, December, 39 – 744, 1974.

Federal Register, Friday, October 26, 1984 Guidelines for Establishing Test Procedures under the Clean Water Act, Part VIII, Environmental Protection Agency, 40 CFR Part 136.

Federal Register, Wednesday, November 13, 1985 National Primary Drinking Water Regulations, Volatile Synthetic Organic Chemicals, Vol. 50, No. 219, Part II, Environmental Protection Agency, 40 CFR Parts, pp. 141 – 142.

Skoog, D.A., *Principles of Instrumental Analysis,* 3rd ed., Saunders College Publishing, 1985.

U.S. EPA, Methods for Chemical Analysis of Water and Wastes, EPA-600/4-79-020, U.S. Environmental Protection Agency, Environmental Monitoring and Support Laboratory, Cincinnati, OH, 1979.

U.S. EPA, Test Methods for Evaluating Solid Wastes, SW-846, U.S. Environmental Protection Agency, Office of Solid Waste and Emergency Response, Washington, DC, 1986.

10

Organization and Analysis of Ground-Water Quality Data

Martin N. Sara and Robert Gibbons

CONTENTS

Introduction

Water-quality analyses and interpretative data summaries are important to Phase II site characterization efforts, but these data are even of greater importance under detection and assessment activities associated with facility compliance. Detection monitoring efforts are performed to verify attainment of performance objectives, and assessment monitoring is made of efforts to identify facility noncompliance in terms of nature, location, and extent of contamination. One should not lose sight of the fact that geologic conditions and observed hydraulic heads are typically more important field data than water-quality data to sort out the contamination flow paths and an ultimate remediation solution for a particular site.

Hydrogeologists and others who make use of water-quality analyses must incorporate individual values or large numbers of analyses (data sets) into their interpretations. On the basis of these interpretations, final decisions are made regarding detection and assessment monitoring programs. In the last 15 years few aspects of hydrogeology have expanded more rapidly than interpretation of water-quality data at and around industrial plants and waste management facilities. The expansion of water-quality programs was based on two factors (McNichols and Davis, 1988):

- Improvements in analytical methods have greatly increased our ability to accurately and precisely analyze a vast number of trace elements and organic compounds in water. Automation of analytical processes now allows statistically significant studies of constituents that formerly were beyond the analytical detection capabilities of all but the most sophisticated instrumentation.
- The expansion of water chemistry technology has occurred in response to public and professional concern about health, particularly as related to analyses of radionuclides and trace-level organic hydrocarbon compounds.

As a result, many comprehensive programs for monitoring water quality at waste management facilities have resulted in analyses of thousands of individual parameters. Interpretation of such massive quantities of data must include attempts to determine correlations among the parameters and demonstration of correlations that exist between water-quality parameters and the hydrogeology of the site. Comparison of water quality in upgradient (background) and downgradient wells may also be necessary as part of detection monitoring programs. In the Superfund program, data are being collected by U.S. EPA regional offices, states, other Federal agencies, potentially responsible parties (PRPs), and contractors. The data are used to support the following functions:

- Waste site characterization
- Risk assessment
- Evaluation of cleanup alternatives
- Monitoring of remedial actions
- Monitoring post-cleanup conditions

In general terms, reports of water quality should contain an organized evaluation of the data, including graphics as necessary, to illustrate important environmental relationships. The recommended procedure for assessment of water-quality baseline and detection monitoring is illustrated in Figure 10.1.

The interpretative techniques and correlation procedures described herein do not require extensive application of chemical principles. The procedures range from simple comparisons and inspection of analytical data to very extensive statistical analyses. Typically the first step in evaluating ground-water quality is to review existing hydrogeologic information and try to define subsurface stratigraphy and ground-water flow. Most regulations require comparisons of data between upgradient to downgradient conditions. This is usually only useful in homogeneous aquifers that have very rapid flow (e.g., hundreds of feet per year). As will be fully explained in the following sections, more than one upgradient well is necessary to account for natural subsurface spatial variability present on most sites. When facilities are located over low-hydraulic-conductivity soils and rock that are heterogeneous in composition, additional spatial variability considerations must be addressed in the evaluation of water quality. Upgradient to downgradient comparisons for natural constituents may not be possible for those sites where vertically downward gradients predominate. These situations require sufficient background sampling points to establish the ambient spatial and seasonal variability. Landfills along hillsides often have recharge and discharge conditions that create different chemical evolution pathways and natural differences in upgradient to downgradient ground-water quality (Freeze and Cherry, 1979). In some cases, wells can be located "side-gradient" (along the downgradient directions of ground-water flow) at these sites if enough land is available to eliminate concerns about landfill impacts. The federal regulations recognize that if a site is located on a ridge, for example, where there are no upgradient sites for wells available, then wells can be compared to themselves. This comparison is called a trend analysis or intra-well comparison.

Natural ground-water quality is known to vary both spatially between wells and temporally at a single well. Anthropogenic (or man-made) effects also contribute to the variability observed in water-quality data. To evaluate the potential releases from a facility to ground water, the sources of natural variability, and the additional interrelationships of human activities to ground-water quality must be fully understood. Sources of variability and error in ground-water data are listed in Figure 10.2.

Natural spatial variability of ground-water quality is often due to variations in lithology within both aquifers and confining units (Sen, 1982). Soil and rock heterogeneity may cause the chemical composition of ground water to vary even over short distances. Spatial variability water-quality data may be additionally affected by variations in well

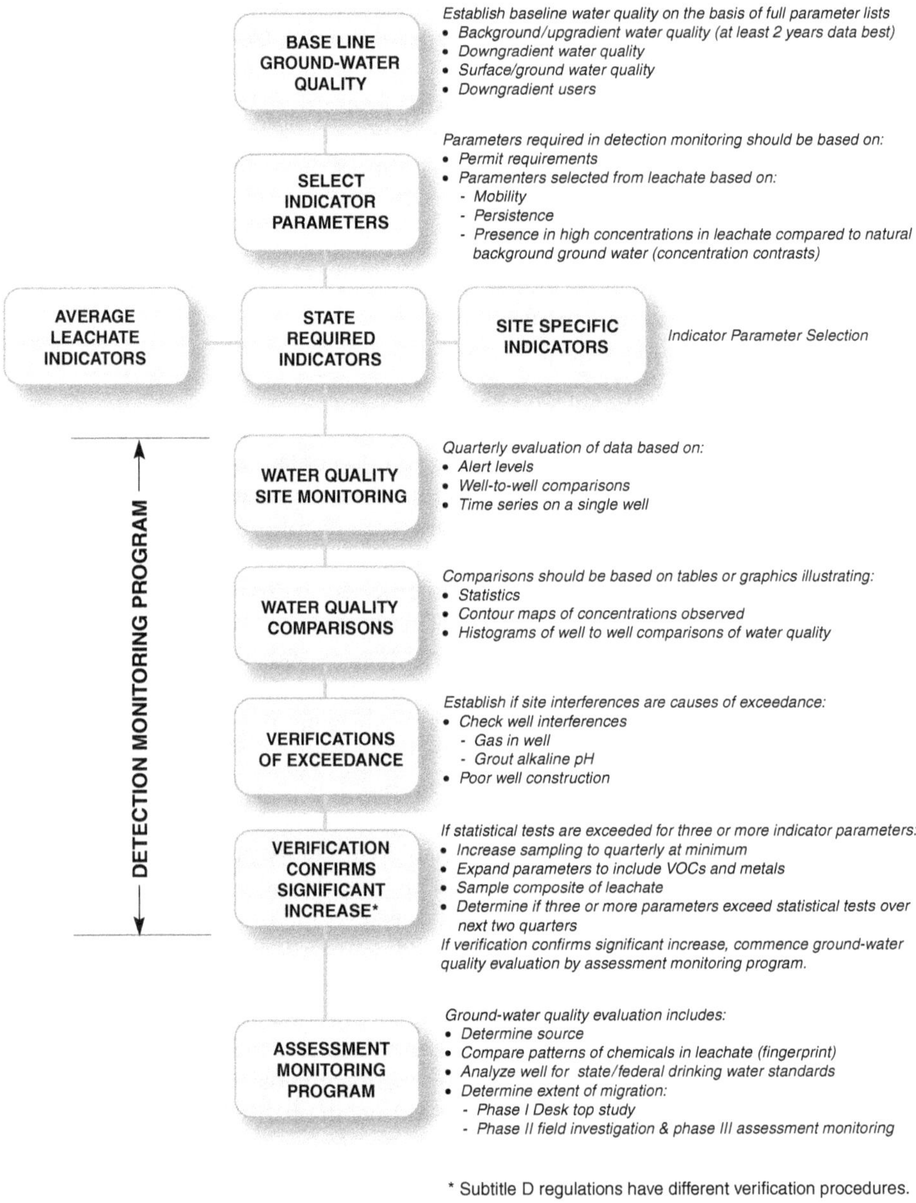

FIGURE 10.1
General water-quality assessment procedure.

installation and development methods as well as the sampling techniques used in the program (Doctor et al., 1985).

Temporal or seasonal effects are usually associated with annual cycles in precipitation recharge events to shallow, unconfined aquifers; these effects are especially pronounced where surface water and aquifer interactions are significant (Harris et al., 1987). Also, seasonal pumping for irrigation and high summer recharge from nonpoint pollution sources may be causes for seasonal fluctuations in background water quality (Doctor et al., 1985). A literature review on seasonality in ground-water data is presented by Montgomery et al. (1987).

SPATIAL	TEMPORAL	WELL CONSTRUCTION	SAMPLING
GEOLOGIC PROPERTIES • Lithologic composition sorting and grain size • Structure of lithologic units • Bedding planes • Fractures (joints and faults) • Soil development • Properties of vadose zone HYDRAULIC CONDITIONS • Location of recharge/ discharge zones • Proximity of water • Presence of aquitards • Pumping OTHER • Other chemical sources • Non-point source inputs	TRENDS SEASONAL • Recharge • Irrigation • Fertilization • Pesticide/herbicide application • Frozen ground PERIODIC • Short-term precipitation • Pumping • River flooding	DRILLING PROCESS • Drilling fluids • Type of borehole • Interaquifer transport of materials WELL DESIGN • Casing and screen material • Diameter • Screen length, depth, slot size • Filter pack material • Annular seal WELL DEVELOPMENT	COLLECTION • Purging methods • Purging rate/duration • Sampling apparatus • Cross-contamination between wells • Field versus laboratory measurements • Sample preparation filtering/container/ preservatives/storage time • Operator error • Incomplete well development ANALYTIC ERROR • Analytic methods • Operator experience • Instrument calibration • Interference from other constituents • Holding times • Clerical/transcription errors

FIGURE 10.2
Sources of variability in ground-water data. (*Source*: From Doctor et al., 1985. With permission.)

The relative importance of these sources of variability is clearly site-specific. Doctor et al. (1985) observed that natural temporal and spatial variability was greater in magnitude than sampling and analytical error, unless gross sample contamination or mishandling of the samples occurs. Goals and procedures used in developing a monitoring program (i.e., baseline or detection) and descriptions of tasks are illustrated in Figure 10.1.

Baseline Water Quality

Characterizing the existing ambient or baseline quality of ground water is an important task for a number of reasons. First, existing drinking water quality standards normally define the baseline ground-water conditions, against which risks to human health and the environment are evaluated. Second, existing ground-water quality in part determines current uses and affects potential future uses of the water. In addition, determining ground-water uses is an important initial step in identifying potential exposure pathways downgradient from the site.

In evaluating the background water quality for an area, the investigator must consider possible background concentrations of the selected indicator chemicals and the background concentrations of other potential constituents of leachate. Existing chemical parameters associated with indicator chemicals (i.e., chloride or iron) or other Resource Conservation and Recovery Act (RCRA) hazardous constituents may be due to natural geologic conditions in the area; prior releases from the old, unlined landfills; or prior or current releases from other upgradient sources. Evaluation of water-quality parameters in ground water is necessary to establish an existing baseline of ground-water quality to which the incremental effects of a potential release can be added.

Measuring ambient concentrations of every RCRA-listed hazardous constituent is not feasible during most baseline studies. To adequately assess background ground-water quality, the investigation should attempt to identify other potential sources in the area (e.g., the Comprehensive Environmental Response, Compensation and Liability Act [CERCLA] sites, RCRA facilities, municipal landfills, agricultural areas or NPDES discharges to surface water) and to identify which constituents are most likely to originate from each source. Some of the background chemicals may also be site-specific indicator parameters, particularly if the facility has experienced a prior release. When determining which chemicals to include on a list of background parameters, the investigator should include all indicator chemicals described as baseline water-quality parameters in the next section.

Where sufficient data from historical monitoring are unavailable, the investigator may install a ground-water-monitoring system or expand an existing system in order to adequately assess the background quality of ground water. The design of a monitoring program should be based on guidance in Nielsen (2006). At a minimum, background water quality should be based upon at least two separate sampling rounds of existing or newly installed monitoring wells.

For facilities that have experienced a prior release, the investigator should also establish the results of any sampling, monitoring, or hydrogeological investigations conducted in connection with the release (if available) and should provide references to any reports prepared in connection with that release.

Selection of Indicator Parameters

The United Nations Statistical Office defines "environmental statistics" as "multidisciplinary in nature, encompassing the natural sciences, sociology, demography and economics. In particular, environmental statistics: (a) cover natural phenomena and human activities that affect the environment and in turn affect human living conditions; (b) refer to the media of the natural environment, i.e., air, water, land or soil and to the man-made environment which includes housing, working conditions and other aspects of human settlements."

Environmental indicators are environmental statistics or aggregations of environmental statistics used in some specific decision-making context to demonstrate environmentally significant trends or relationships. An environmental indicator can be a representative indicator that is selected by some procedure, such as expert opinion or multivariate statistical methods, to reflect the behavior of a larger number of variables, or it can be a composite indicator that aggregates a number of variables into a single quantity (i.e., an index).

The concept of the "indicator parameter" forms the basis for water-quality sampling programs. Because an investigator cannot include all chemical parameters that may be present in a natural or contaminated ground-water system, a selection process must be used to bring the spectrum of chemical parameters down to a workable number. These indicator parameters are selected to provide a representative value that can be used to establish performance of a facility (detection) or quantify rate and extent of contamination (assessment).

Each chemical analysis, with its columns of parameter concentrations reported to two or three significant figures, has an authoritative appearance which can be misleading.

Indicator parameters in general terms must represent the movement of ground water or change in water quality in a clear-cut and understandable descriptive presentation.

Detection Monitoring Indicator Parameters

Detection monitoring programs require that individual chemical parameters be selected to represent the natural quality of the water, as well as the chemical parameters that may be changed or adversely affected through facility operation. These parameters, called "indicators," are selected with consideration of a number of criteria:

- Required by permit, state, or federal regulation or regulatory guidance.
- Are mobile (i.e., likely to reach ground water first and be relatively unretarded with respect to ground-water flow), stable, and persistent.
- Do not exhibit significant natural variability in ground water at the site.
- Are correlated with constituents of the wastes that are known to have been disposed at the site are easy to detect and are not subject to significant interferences due to sampling and analysis.
- Are not redundant (i.e., one parameter may sufficiently represent a wider class of potential contaminants).
- Do not create difficulties during interpretation of analyses (e.g., false-positives or false-negatives, caused by common constituents from the laboratory and field).

Selection of indicator parameters should consider natural levels of constituents in the detection process. Because chemical indicators include naturally occurring chemicals, Table 10.1 provides an example indicator parameter list with ranges of values occurring in natural aquifers, as well as the persistent and mobile parameters typically present in leachates from sanitary landfills.

These indicators represent a restricted selection of parameters measurable in an aquifer and limit the ability of an investigator to assess baseline water quality. However, they are the most likely parameters to undergo change when ground water is affected by a chemical release from a solid-waste management facility.

TABLE 10.1

Example Indicator Parameters for Sanitary Landfills

Indicators of Leachate	Ranges in Natural Aquifers
TOC (filtered)	1–10 ppm
pH	6.5–8.5 units
Specific conductance	100–1000 mm/cu.
Manganese (Mn)	0–0.1 ppm
Iron (Fe)	0.01–10 ppm
Ammonium (NH_4 as N)	0–2 ppm
Chloride (Cl)	2–200 ppm
Sodium (Na)	1–100 ppm
Volatile organics[a]	<40 ppb

[a]via U.S. EPA Method 624.

TABLE 10.2

A Complete Water Quality Parameter List

Ammonia (as N)	The volatile organic compounds (VOCs) established in Method 624
Bicarbonate (HCO_3)	Total organic carbon (TOC)
Calcium	pH
Chloride	Arsenic (As)
Fluorides (F^-)	Barium (Ba)
Iron (Fe)	Cadmium (Cd)
Magnesium (Mg)	Chromium (Cr^{3+})
Manganese (Mn^{2+})	Cyanide (Cn)
Nitrate (as N)	Lead (Pb)
Potassium (K)	Mercury (Hg)
Sodium (Na^+)	Selenium (Se)
Sulfate (SO_4)	Silver (Ag)
Silicon (H_2SiO_4)	Nitrogen, dissolved (N_2)
Chemical oxygen demand (COD)	Oxygen, dissolved (O_2)
Total dissolved solids (TDS)	

Complete Detection Parameter List for Sanitary Landfills

Although individual definitions vary, a "complete" analysis of ground water includes those natural constituents that occur commonly in concentrations of 1.0 ppm or more in ground water. Depending on the hydrogeologic setting, a complete analysis is shown in Table 10.2. In general, the investigator should examine closely the water-quality results if these indicators are above the natural ranges of ground water given above. The concentration of total volatile organics (40 ppb) was established from tolerance intervals on numerous upgradient wells at 17 facilities (Hurd, 1986) and includes cross-contamination interferences from the collection and analysis process.

Analytical Laboratories

The importance of laboratory selection for evaluation of water-quality samples cannot be overstressed. Significant legal and technical decisions, many of which will determine the success of the environmental monitoring program, depend on the quality of the lab's work. The choice of a laboratory may ultimately make the difference between a successful project and one that falls into a pattern of persistent failure, frustration, later recrimination, and resampling.

The general requirement of a laboratory program is to determine the types and concentrations of both inorganic and organic indicator parameters present in samples submitted for analysis. Depending on the project requirements, specific laboratory testing methodologies have been approved within the project scope or are specifically required. For example, under Subtitle C of RCRA, analytical methods contained in Test Methods for Evaluating Solid Waste, Physical Chemical Methods (SW-846) (U.S. EPA, 1988a) are specified.

Under the Federal CERCLA or Superfund Amendments and Reauthorization Act (SARA) program, the Contract Laboratory Program (CLP) was established by the EPA in 1980. The CLP program provides standard analytical services and is designed to obtain consistent and accurate results of demonstrated quality through use of extensive quality assurance/quality control (QA/QC) procedures.

The selection of an analytical laboratory service depends primarily on the client needs and the intended end use of the analytical data. While laboratories performing analytical services must use standard methods and employ method-specified quality control procedures, the choice of laboratory may be based on other factors, as described in the following sections.

Laboratory analyses are critical in determining project direction. Therefore, the reliability of the analytical data is essential. The use of QA/QC must be an integral part of laboratory operations and an important element in each phase of the technical review of data and reports.

Steps in a Lab Evaluation

The first step in the laboratory selection process is for the client or for the consultant to organize a detailed document defining the analytical and quality control (QC) requirements of the program determined by the project scope of work. A typical laboratory would be assigned the responsibility to:

- Evaluate the scope of the project
- Confirm its capacity to comply to the program
- Resolve identified discrepancies in the scope of work requirements
- Propose viable analytical alternatives consistent with the data quality objectives (DQOs) of the program
- Confirm project commitment to within the specified turnaround times

Assessment monitoring programs often require that a Quality Assurance Project Plan (QAPP) be approved by the responsible regional EPA office, state regulatory or other regulatory agency. The QAPP documentation describes:

- The full scope of the project field and laboratory activities
- The analytical methods to be used with their QC requirements
- Project reporting and documentation standards

An experienced laboratory will normally perform a complete and independent assessment of the QAPP and document the laboratory's complete understanding of project responsibilities.

Very large or complex projects may require data collection activity over a broad spectrum of soil and water analyses that may require multiple laboratories. These very large projects can be handled in several ways: (1) contract with additional laboratories as needed to encompass the full scope of the project or (2) contract with a primary or lead laboratory, which then has the direct responsibility to obtain subcontracting laboratory services. This is not a job for amateurs; as additional laboratories are added to the project, complexities mount rapidly that require significant experienced project management efforts.

SOPs and QAPPs

The majority of analytical laboratories have standard procedures for how the laboratory conducts its analytical quality and reporting programs just as consulting firms have standard operating procedures (SOPs) for field-testing procedures. Sample and data pathways should form part of the documents provided for review from the laboratory. Simple listing of analytical procedures tells only part of the necessary documentation; sample preparation and instrumentation procedures should refer to approved methods (as designated in the QAPP or work plan). Procedures for sample handling and storage, sample tracking, bottle and glassware decontamination, document control, and other important project elements are described in the nonanalytical SOPs.

As with any quality assurance program documents, the laboratory SOPs should employ formal document control procedures so that revision numbers and dates are presented on each page. All SOPs should include the staff position performing the task, the specific analytical and quality procedures involved, and the individual responsible for resolving difficulties before taking corrective action when out-of-control events occur. Formal approval by the designated QA manager and laboratory manager should appear on the SOP permanent training documentation and include each staff member's review and understanding of the SOPs. All copies of earlier revisions of SOPs should also be retained within the laboratory documentation system.

The QAPP is the document that brings together the laboratory QA/QC plans and SOPs and specific project requirements. The QAPP should include, at a minimum, the information presented in Table 10.3. Laboratory quality systems must pay particular attention to data quality assessment and corrective action procedures. The document, through reference to the laboratory SOPs and QA/QC program, specifically addresses the laboratory's mechanisms for a program of QC samples analyzed at the appropriate or predetermined frequencies. The QC sampling requirements within the quality assurance program are usually client-, method-, or contract-dependent. The QA plan should specify the mechanisms by which the laboratory identifies these requirements.

Control and reporting of analytical results are important elements of an environmental laboratory's responsibilities. Laboratory data-quality assessment procedures should include:

- General description of all data review levels
- Responsibilities at each level
- Examples of the documentation accompanying the assessment

TABLE 10.3

Laboratory Quality Assurance Program Plan (QAPP) Guidelines

Title page
Table of contents
Laboratory and quality assurance organization
Facilities and equipment
Personnel training and qualifications
Laboratory safety and security
Sample handling and chain-of-custody
Analytical procedures
Holding times and preservatives
Equipment calibration and maintanence
Detection limits
Quality control objectives for accuracy, precision, and completeness
Analysis of quality control samples and documentation
Data reduction and evaluation
Internal laboratory audits and approvals from other agencies
Quality assurance reports to management

- Analytical data-quality criteria used by the reviewers
- Final accountability or "sign-off" on the data report

The control and reporting section should also address the use of data qualifiers (tags) and whether or not it is the laboratory's policy to adjust results based on discovery data or observed blank sample contamination. Because very low levels of organic parameters can cause significant data evaluation problems, the policy and procedure used for adjustment of data by the laboratory must be well known by project staff responsible for data interpretation. As general guidance, data tags are generally preferred over data adjustment.

Custody and Chain-of-Laboratory Security

Environmental laboratories should be restricted to authorized personnel only. Security should extend to sample and data storage areas even for the smallest laboratories. The work plan applicable to the project should contain specific chain-of-custody requirements. The basic components for maintaining sample chain of custody are:

- Samples must be delivered into the possession of an authorized laboratory staff member by the sample handling or transporting organization (such as FedEx or specific sampling teams).
- Samples must be within the authorized staff member's line-of-sight.
- Samples must be locked in a secured storage area with restricted access.

Samples should be kept in locked storage with restricted access when not being processed (refrigerated, as required). The chain-of-custody form is used to document the transfer of these sample fractions (such as splits, extracts, or digestates) as part of the permanent sample-processing record.

Facility and Equipment

A quality assurance program typically contains documentation on equipment maintenance and calibration. An analytical laboratory must maintain such documentation as part of its QA/QC program. Standards used in the analytical process must also be traceable to a certified source such as the U.S. EPA, the National Institute of Standards and Technology (NIST), or commercial sources.

A very important part of the success of an environmental sampling program for state or Federal regulatory programs is the turnaround time of the sample. The turnaround time is defined as the time from field sample collection to receiving QA/QC confirmed analytical results usable for evaluating the performance of the facility. Turnaround times provided by laboratories are typically based on the current sample load and capacity, average turnaround times for data delivery, and history in meeting sample-holding times. Holding time is the maximum allowable time between sample collection and analytical testing. Each chemical parameter has a specific holding time attached to the sample, i.e., 24 h, 2 weeks, or 30 days. For most environmental monitoring projects, data for analytical samples not meeting the required holding times will cause the results to be rejected or, at best, qualified. Exceeding holding times has caused many environmental programs to get into very serious trouble with both permit requirements and stipulated penalties for the project deliverables.

Analytical laboratories are often plagued by persistent low levels of organic parameters such as methylene chloride or acetone. These parameters are common laboratory chemicals used in various organic extraction processes. These organics often show up in analytical results as low background levels. Some laboratories commonly subtract these values from results; other laboratories report the values and let the investigator explain the results to regulatory agencies; others tag the data as background for the lab. Whatever method used by the laboratory, the investigator should expect to see such low levels of common laboratory organic chemicals in analytical results. The laboratory should report in QA/QC plans how they deal with such data.

The laboratory may purchase reagent-grade water or produce its own using a water purification system. A logbook should also be maintained to document checks for water purity, whatever the source. The product water should also be the source for QC method blanks (i.e., samples) in order to verify the absence of organic and inorganic constituents.

Data Accuracy and Availability

Reliability of laboratory-generated environmental data depends on a series of program procedures that include proficiency test samples, mechanisms for handling data inquiries, QA reporting to management, organized ways of handling corrective action, long-term data storage, and access. Initially, analytical results must be reviewed in relationship to the other analytes reported for the project. The purpose of this type of review is to attempt to identify trends, anomalies or interferences that can mislead investigators, or bias the overall use of the data. The technical review process begins with an initial review of the testing program and the overall project requirements. Once samples are analyzed according to project plans and analytical results generated, the laboratory should conduct an initial math check, a QC review, and a laboratory supervisor's technical release of the data. Reviewers consider the relative accuracy and precision of each analyte when interpreting the analytical data. Several alternative methods are available for entering results into a database. Procedures such as double-key entry and internal computer error-checking routines are employed to compare both data entries and generate an exceptions report. Data must be reviewed by qualified staff before changing any analytical or field-generated results. These procedures, along with those described below, are used to establish the reliability of the results before moving to evaluation of the actual project data sets.

Data Inquiries

The mechanisms in place for handling data inquiries are often vital to the success of a project. No matter what the length or the extent of the program, data inquiries will happen on a recurring basis. In general, procedures used in the laboratory should describe how the data are requested from storage, the individuals responsible for resolving the inquiry, and the standard response time.

Expect to see questionable data coming from even the best analytical laboratories. The laboratory should have an SOP in place for responding to client inquiries, both technical and administrative (invoicing, sample shipping logistics, requests for additional copies, etc.).

QA Reports to Management and Corrective Action

When an out-of-control incident is observed on water-quality samples from an environmental monitoring program, it is essential that the event be documented and a form of corrective action be taken. Out-of-control events may be:

- Isolated to individual QC sample recoveries or calibration criteria failures
- Systematic — having widespread effect on the analytical data generation system

When the sample has triggered an out-of-control action, it may, for example, require reextraction or may require qualification with a notice to the data end user that identifies the criteria that were not met and the effect on data acceptability. When sufficient sample volume is not available to reprocess a sample, resampling may be required for an extreme out-of-control event.

Laboratory records should be archived so that individual reports or project files can be easily retrieved. As with any QA program, access to data must be restricted to specified individuals. If data are also stored on magnetic tape or on computer disks, the tapes or disks should be similarly protected with back-up copies stored at a second location. As part of the QA program, the resumes and qualifications of key technical staff must be maintained along with training records for the staff.

MDLs, PQLs, IDLs, and EMLRLs

Site assessment projects generate a great deal of analytical data that may be reported by the laboratory in numerous ways. These reported values often reference some form of detection limits including: method detection limits (MDLs), instrument detection limits (IDLs), practical quantitation limits (PQLs), or reporting limits (RLs). Each of these limits evolves around a detection limit. These detection limits are only a way of statistically expressing how low a particular measuring system can measure. There are a number of ways to evaluate the limit of detection (LOD) of a particular measuring device. For example, one could take an object for which the weight is known accurately, such as a 10-pound weight. The 10-pound object is weighed a series of times using a typical spring-loaded scale. The results of this process will vary depending on the temperature in the room, how the object is placed on the scale, how accurately the results are read, who reads the results, and the quality of the scale (Jarke, 1989). This is called "variability" of the measuring device.

If, for example, your results were 10.2, 10.4, 10.7, 9.1, 9.8, 9.3, 10.0, then the average value is 10.07 pounds and the standard deviation is 0.4461. In such exercises it is a good practice to carry more figures than are really significant until you make your final calculation, and then report only those figures that are significant.

The U.S. EPA's definition of MDL (40 CFR Part 136 Appendix B) describes the detection limit for this scale as 1.1 lb — any value less than 1.1 lb cannot be determined to be different from zero. Even if the scale shows a value, the significance of this value remains questionable. To obtain a lower MDL result than the 1.1 lb, one must go to a scale with a much lower detection limit to get to an accurate or reliable value.

The example of the simple weight scale is similar in many respects to any measuring device, as every measuring device has a detection limit and every device's detection limit is different depending on who, what, how, when, and where it is used. Because all of these components can vary, detection limits are not constants, especially for analytical instruments.

Every instrumental measuring device used in an analytical laboratory has an inherent minimum LOD, as described above. This LOD is usually referred to as the specific IDL. For simple devices, the IDL is based on the smallest unit of measurement that the device is capable of reporting. For example, if a ruler has markings of a sixteenth of an inch, the IDL (if based on one half of the smallest unit of measure) would be one thirty-second of an inch. While the overall concepts of IDL and MDL are quite similar, IDLs for instruments are generally far below the experimentally determined MDLs. The analytical instrument can be optimized for a specific parameter, with fewer and more easily controlled sources of variability within the IDL procedure. MCL determinations include many more sources of variability and therefore have higher experimentally determined MDLs.

In 1980, the U.S. EPA began to administer the RCRA. One of the requirements of this law was that landfills begin to monitor ground water. The agency established MDL in 40 CFR Part 136, Appendix B, to ensure that analytical laboratories were conducting the testing at an acceptable level. This regulation requires that each analytical laboratory must establish MDLs on a routine basis for every analyte, for every analyst, and for every instrument. The goal of the regulation was to demonstrate that the analytical laboratories could obtain results as good as or better than those published with many of the U.S. EPA methods. The U.S. EPA MDL studies are always performed in highly purified water, with only a single known analyte added. The resultant MDLs, therefore, reflect the best performance a laboratory is capable of under the best conditions. Site assessment projects produce environmental samples that do not contain a single known analyte in highly purified water. Rather, samples are delivered to the laboratory containing many types of organic and inorganic parameters, sometimes residing in a significantly concentrated liquid. This produces a matrix effect that can significantly raise MDLs many times over U.S. EPA-reported values. Additional sources of variability presented by real samples can include sampling, site location variability, and interferences that can be caused by compounds in the sample other than the target compound. As one can imagine, the effective MDL for these field samples can be many times larger than those used in establishing laboratory performance.

Although 40 CFR Part 186, Appendix B requirements to establish MDLs are clearly explained, there is little standardization in how the regulation is applied at analytical laboratories. A full spectrum of applications of MCLs is observed applied in analytical laboratory work (Jarke, 1989):

- Laboratories perform MDL studies that meet or exceed the published values but use the published values in their reports.
- Laboratories do not perform MDL studies and assume that if they are using a U.S. EPA-approved method, then the published MDLs can be used in reporting without performing the MDL study.
- Laboratories perform MDL studies and use these as the RLs in their reports.
- Laboratories either do or do not perform MDL studies, but use RLs that are significantly different from the U.S. EPA-published MDLs, such as PQLs or RLs.

Site assessment water-quality evaluations should be based on using analytical laboratories that have performed MDL studies to verify that they can perform a method and provide QA/QC data on how well they are performing that method.

The definition of MDL includes the phrase, "the minimum concentration of a substance that can be measured and reported with 99% confidence that the analyte concentration is greater than zero." Using this definition, if an analytical laboratory reports all results above experimentally determined MDLs, 1% of reported values are false positives. False positives are statistically valid reported values. They appear to be real values, but in reality are not; therefore, many laboratories that perform environmental programs have recognized the need to set meaningful RLs. The CLP, organized by U.S. EPA to control site remediation analytical programs, has also recognized the false MDL rates for analytical data. The methods published for the CLP program use the concept of the PQL.

PQL is considered by the U.S. EPA as the concentration that can be reliably determined within specified limits during routine laboratory operation and is defined as either 5–10 times the MDL or 5–10 times the standard deviation used in calculating the MDL. This definition of an RL still raises technical questions but can be determined experimentally using statistical procedures proposed by Gibbons et al. (1988).

Additional terms have been proposed to address the ability of analytical laboratories to evaluate low levels of chemical parameters. The Environmental Committee of the American Chemical Society (ACS) published a report in 1983 addressing the issue of RLs and detection limits. Figure 10.3 graphically shows this idea. The committee used LOD instead of MDL. A new value, limit of quantitation (LOQ), was defined as 10 times the standard deviation used in the MDL calculation. This value is equal to approximately three times the MDL defined in 40 CFR 136, and is equal to the PQL. The ACS Committee reasoned that data above the LOQ could be reported quantitatively. The region between the LOD and LOQ contained results of 108's uncertain quantitation.

In summary, MDLs should be used in establishing the capability of a laboratory to perform a particular test method in accordance with regulations applicable to the project. The RLs should be established by first determining the intended uses of the data. Reporting any value above the MDL means that some analytical values will still be false positives because they fall in the region of less certain quantification. Each of these detection limit definitions can be summarized using the weight scale example (Jarke, 1989):

- The IDL is the same as the pound scale markings.
- The MDL is determined to be 1.4 lb based on one person (observer) using a single scale.
- The PQL would represent statistically what multiple scales being used by multiple people (observers) could achieve.
- The RL would be a constant value that is above the statistical variation of all people using all similar type scales.

Each type of limit is based on the population observing the operation, from the smallest IDL, where no one is observing, to the single observer (MDL), and finally to the whole population of observers (PQL and RL).

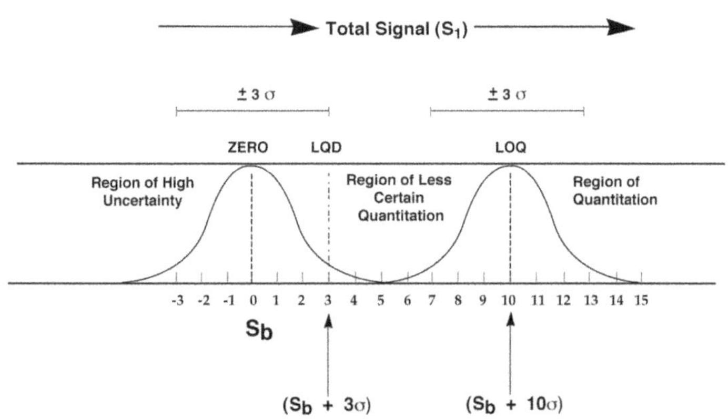

FIGURE 10.3
Relationship of limit of detection (LOD) and limit of quantification (LOQ). (*Source*: Modified from Keith et al., 1983.)

Sample Dilution

In environmental site assessment projects it is often necessary to dilute samples to either eliminate instrument or analyte interferences or to bring down large concentrations to within instrument scale. This reduces the occurrences of "blown columns" during gas chromatographic analysis. Diluting a sample fundamentally affects the MDL first. That is, if the MDL times the dilution factor is still equal to or less than the RL, then the RL remains unchanged. If, however, the effect of diluting the sample results in an MDL above the RL, then a new RL must be established. This may seem to be in conflict with the previous discussion. However, if a laboratory is using MDLs as their RLs, then as the sample is diluted, both the MDL and RLs change because they are equal. If a laboratory is using the concept of an RL that is larger than the MDL, then the dilution factor should only affect the MDL until it reaches the value of the RL and then any further dilution should affect the two simultaneously. The client should only be aware of dilution when it affects the RL.

Low-Level Organic Chemical Results

Evaluation of low levels of organic chemicals in ground water presents one of the more common problems in environmental monitoring programs. The difficulties associated with interpreting low-level analytical results for organic chemicals can be divided into three broad categories:

1. Deficiencies in sampling and analytical methods
2. Background levels for compounds that are commonly present in homes, industrial facilities, transportation facilities, and analytical laboratories
3. Varying significance as well as incomplete data on the significance of organic compounds to public health and the environment

All sampling and analytical methods commonly used for environmental monitoring are subject to variability and error. Replicate samples taken in the field from a single well or samples split in the laboratory will not produce identical analytical results due to:

- Imperfect sampling procedures
- Inability to maintain perfectly constant conditions around a sample point
- Absence of perfect homogeneity in the sample material

Replicate analysis on the same sample by the same method and even by the same analyst will not necessarily produce identical analytical results. At concentrations near the analytical LODs (typically 1.0–10 mg/l for gas chromatography–mass spectrometry [GC–MS] and lower for gas chromatography [GC]) it may be practically impossible to produce two samples that are identical. For ground-water samples, conditions in a well will vary slightly between consecutive sampling events or even during a single sampling process. When a sample is split after sampling, the two splits may not be exposed to the atmosphere in exactly the same way, for exactly the same lengths of time. Furthermore, the slightest amount of suspended solids or turbidity will most likely result in two samples that are not identical. Soil samples can show an extreme lack of sample homogeneity even from samples taken a foot away from a particular coordinate.

The key to evaluation of sampling and analytical data, therefore, is to be cognizant of the types and extent of variability inherent in sampling and analytical methods and to take into account all available QA/QC data when interpreting results.

Background Water-Quality Evaluation

Background refers to chemical parameters introduced into a sample from natural and human-related sources other than those that are the subject of the monitoring program. The problem of background changes in water quality is similar to that of analytical method variability in that it seldom is practical to eliminate it completely. There are many opportunities for a water sample to be exposed to detectable levels of both organic and inorganics at the low detection limits currently available for chemical analysis. As with the problem of method variability, the solution to background sample contamination is to first define to the practical extent the natural variability of the system, then combine these data with documentation of background levels to make reliable interpretations of analytical results.

To give a few examples, some of the most common compounds found as background levels in environmental samples are volatile organics and phthalates. Sources of these compounds include homes, transportation facilities, and analytical laboratories. Some specific examples are included in Table 10.4.

Many laboratories will not even report some of these compounds (e.g., methylene chloride) below certain levels (usually 15–30 ppb) because of assumed laboratory background levels.

Monitoring Site Water Quality

Ground-water data collected during site characterization and detection monitoring is typically restructured or simplified and must be presented in a manner that facilitates verification and interpretation. All analytical data (physical and chemical) are reported through transmittal sheets of laboratory analysis. The data are then compiled into tables and graphic formats that facilitate understanding and correlation of the information. At the very beginning of assessment activities, the investigator should establish common data requirements and standard reporting formats.

TABLE 10.4

Examples of Laboratory and Cross-Contamination Compounds

Compound	Typical Sources
Chloroform	Chlorination of drinking water
Phthalates	Plasticizers used in numerous household and industrial products including pipes, shower curtains, car seats, many bottles and containers, etc.
Methylene chloride	Common in paint strippers, household solvents, septic system cleaners, and spray propellants; used extensively in laboratory procedures
Other solvents	Household cleaners, paints and trichloroethylene, paint strippers, septic system tetrachloroethylene, cleaners and to a limited extent toluene, in laboratory procedures dichloroethane
Trichlorofluoromethane	Common refrigerant (freon) found in freezers, refrigerators, and air conditioners

A list of all data should be provided for each sampling event and updated as new data become available. The data should include the following: well identification number or alphanumeric designation, date of analysis, name of laboratory, units of measurement, LODs, and chemical concentrations. The data are then categorized and organized into the established format to allow quick reference to specific values. Compilation and evaluation of laboratory data into summary reports must be performed without transcription errors. This task is made more achievable by use of standard formatting procedures.

Reporting

Laboratory results for a given analyte generally are presented as a quantified value or as ND (not detected). All chemical data should be presented according to this protocol. Results are reported either as a quantified concentration or as less than (<) the MDL or threshold value (thus, ND results are shown as < on the summary report). To the extent feasible, all laboratory results should be reported in a manner similar to that described above.

Significant Digits

The number of significant digits reported by the laboratory reflects the precision of the analytical method used. Rounding of values is generally inappropriate because it decreases the number of significant digits and alters the apparent precision of the measurements. Therefore, the investigator retains the number of significant digits in the transcription, evaluation, and compilation of data into secondary reports. Variation in the number of significant digits reported for a given analyte may be unavoidable if there is an order of magnitude change in the concentration of a chemical species from one round of sampling to the next or if the precision of the analytical methodology differs from one round to the next.

Outliers

Unusually high, low, or otherwise unexpected values (i.e., outliers) can be attributed to a number of conditions, including:

- Sampling errors or field contamination
- Analytical errors or laboratory contamination
- Recording or transcription errors
- Faulty sample preparation or preservation or shelf-life exceedance
- Extreme, but accurately detected, environmental conditions (e.g., spills, migration from facility)

Gross outliers may be identified by informal visual scanning of the data. This exercise is facilitated by printouts of high and low values. Formal statistical tests are also available for identification of outliers. When feasible, outliers are corrected (e.g., in the case of transcription errors) and documentation and validation of the reasons for outliers are performed (e.g., review of field blank, trip blank, QA duplicate-sample results, and laboratory QA/QC data). Results of the field and laboratory QA/QC, as well as field and laboratory logs of procedures and environmental conditions, are invaluable in assessing the validity of reported but suspect concentrations. Outliers that can reasonably be shown not to reflect true or accurate environmental conditions are eliminated from statistical analyses, but are permanently flagged and continue to be reported within summaries of data.

Units of Measure

Units of measure must be recorded for each parameter in the laboratory reports. Special care must be taken not to confuse "μg/l" measurements with "mg/l" measurements when compiling, transcribing, or reporting the data.

Comparisons of Water Quality

The type of interpretation most commonly required of hydrogeologists is preparation of a report summarizing the water quality in an aquifer, a drainage basin, or some other unit that is under study. The author of such a report is confronted with large amounts of data from a few sources and this information must be interpretable. The finished report must convey water-quality information in ways in which it will be understandable by staff of the regulatory agency and technical management staff of the client.

As an aid in interpreting chemical analyses, several approaches will be discussed that can serve to identify chemical relationships and to predict chemical changes in space and in time. Different types of visual aids, which are often useful in reports, will be described. The basic methods used during interpretation are inspection and simple mathematical or statistical treatment to identify relationships among chemical analyses, procedures for extrapolation of data in space and time, and preparation of graphs, maps, and diagrams to illustrate the relationships.

Inspection and Comparison

A simple inspection of a group of chemical analyses generally will allow distinction of obviously interrelated parameter subgroups. For example, it is easy to group waters that have dissolved solids or chloride concentrations falling within certain ranges. The consideration of dissolved solids, however, should include consideration of the kinds of ions present as well.

Simple visual review of tabulated water-quality data is probably the most frequently used technique by regulatory agencies, to decide if a particular facility is contributing to ground-water contamination. Such analyses commonly exclude consideration of geologic and hydrogeologic conditions at the site. However, placement of water-quality data on maps and cross-sections provides a powerful tool for integration of all chemical and hydrogeologic conditions. These data can be arrayed on maps and cross-sections in a number of ways to enhance interpretation of flow paths and ground-water movement.

Figure 10.4 shows a typical tabular array of water-quality data. Because such a format requires significant efforts to assimilate, it is recommended that alternative formats be employed to display data whenever appropriate for detailed understanding of water-quality information. Water-quality display formats in increasing complexity can be divided into the following categories:

- Tabular presentation
- Contour maps
- Time series displays
- Histograms
- Box plots
- Stiff diagrams

Well Number:	X-106						
ID:	1BN1001	1YN1001	1YN1002	1YN1003	1YN1004	1YN1005/6	
MEA Sample ID:		N1001	1WN1002	9996.11	9961.03		
Collection Date:	1993	1996	1997	1998	1999	2000	2001
Receipt Date:	26-Mar-93	04/17/96	1997	04/21/98	4/27/1999	4/22/2000	04/27/01
Extraction Date:			03/14/97				
Analysis Date:			03/26/97				05/03/01
Time:		1405	00:27:36				
Matrix:	WATER	WATER	WATER	WATER		WATER	WATER
Units:	μg/L	μg/L	μg/L	μg/L	μg/L	μg/L	μg/L
Aluminum	209 U*	120 U*	79.7	48.5 U*	67.6 U*	43.2	710
Antimony	30.0 UJ	2.8	1 U	2.8 U*	1.0 U	2.1 U	1.6 U
Arsenic	3.2	3.0 U	3.3	3.2 U	2.6 U*	2.6	3.1
Barium	49.6	59.1	53.3	63.9	27.2 J	83.0	66.4
Beryllium	1.0 U	1.6 U*	1 UJ	0.10 U	0.15 U*	0.20 U	0.40 U
Cadmium	5.0 UJ	1.0	1 U	0.30 UJ	0.59 U*	0.20 U	0.30 U
Calcium	2,570	2,470	178,000	113,000	859	3,890	365,000
Chromium	6.0 U	28.2	17.9	22.7	15.0	72.1	22.9 J
Cobalt	9.0 U	1.8	1 U	0.50 U	0.40 U	1.4	1.1 U*
Copper	4.0 U	4.0	1 U	1.9	4.6 U*	13.3	6.9 J
Iron	112 U*	307 J	596	434	63 J	206	1,250
Lead	4.8 U*	2.7 U*	1 U	1.5	1.0 UJ	1.3 U	2.5 U*
Magnesium	28,500	587	90,800	53,100	33,800	105,000	193,000
Manganese	52.7	608	138	379	12.5	164	142 J
Mercury	0.20 U	0.20 U	0.2 U	0.10 U	0.05 U	0.10 U	0.10 U
Nickel	13.0 U	15.0	13.9	26.2	21.2	53.7	26.4 J
Potassium	36,400	41,100 J	24,400 J	16,300 J	15,800 J	56,500 J	59,200 J
Selenium	3.0 UJ	4.2	2.2	3.2 UJ	2.5 U*	10.4 J	9.9 J
Silver	5.0 UJ	1.0 U	1 U	5.8 U	0.40 U	0.60 U	0.70 U
Sodium	78,000	105,000	97,200 J	60,700	49,000	141,000	120,000
Thallium	3.0 UJ	3.8	3 U	4.1 U	2.1 U	3.2 U	3.5 UJ
Tin	26.0 U						
Vanadium	4.3	13.6	15.1	12.7	53.1	65.6	68.9
Zinc	5.0 U	11.6 U*	11.8 U*	0.50 UJ	0.40 UJ	0.80 UJ	1.0 UJ
Cyanide	5.0 U	10.0 U	10 UJ	10.0 U	4.7 U	6.7 U*	7.2 U*
Chromium (VI)	10 U	100 U	10 U	50 U	10 U	1,000 U	1,000 U
Sulfide	130,000 J	10,600	1,000 U	2,000	200 U	200 U	1,400

Key:

J = The concentration is approximate due to limitations identified during the quality assurance review

U = Indicates the compound was analyzed but not detected. The associated value is the sample quantitation limit

U* = The compound should be considered "not detected" since it was detected in a blank at a similar concentration level

UJ = Indicates the compound was analyzed but not detected. The associated value is an estimated sample quantitation limit based on a bias identified during the quality assurance review

R = The results were considered unusable during the quality assurance review

Blank = The compound was not analyzed for

(1) Values shown are the highest detected between the investigative sample and its unpreserved, duplicate, reanalysis, or dilution sample.

FIGURE 10.4
Typical water-quality tabular data set for inorganic parameters.

- Scholler diagrams
- Trilinear diagrams
- Correlation coefficients
- Probability plots

Each format can have useful application for understanding variations in water quality and categorization of ground water. Tabular presentations are a necessary evil, the associated tedium of which can be eased by use of summaries and averages. Particular care should be taken in proofreading sets of compiled or merged data, as massive arrays of data almost always contain errors of transcription. Computer-based spreadsheets can decrease time for data reduction; however, any transcription of data must be carefully checked and rechecked for accuracy.

Contour Maps

Presentation of water-quality base maps has been traditionally handled through contouring of data. The technique of mapping of ground-water quality by drawing lines (isocontours) of equal concentration (isograms) of dissolved solids or of single ions has been used in the scientific literature since the early 1930s (Hem, 1970). The applicability of constructing isogram maps depends on several factors, such as:

- Homogeneity of water composition with depth
- Parameter concentration increment between measuring points

Restriction of sampling point density (i.e., insufficient data points) in either a vertical or horizontal direction will limit the usefulness of this technique. However, if the detection or assessment monitoring system at a typical facility is designed using procedures discussed in Nielson (2006) it should provide sufficient data points for construction of isocontour maps. Contour maps can contain either closed isopleths, as shown in Figure 10.5, or open gradient lines, as shown in Figure 10.6. Both these contour maps show isocontours of chlorides. Because chlorides are typically not affected by precipitation or by other reactions that would lower concentrations (decreasing only by dispersion and dilution), this parameter serves as one of the best inorganic parameters to use in contour formats.

Additional parameters such as conductivity, temperature, chemical oxygen demand (COD), or any dissolved parameter with sufficient data density can also be displayed on contour maps. On occasion, lumped organic parameters, such as total volatile organic compounds (VOCs), can also be contoured. Figure 10.7 shows such a presentation. Organic parameters in ground water are difficult to contour effectively due to the typically wide ranges observed in water-quality tests. However, water-quality data from highly concentrated sources, such as product spills or very large-volume, low-concentration organic sources (such as an unlined codisposal facility), may be amenable to such presentations. Questionable data should always be represented by dashed lines on the illustration.

Time-Series Formats

In water-quality evaluations, there is always a continuing interest in observing parameter concentration change over time. To record such data, the standard approach is to make a series of observations at fixed intervals of time — this describes the time-series format. Such time-series formats have the objective of obtaining an understanding of past events by determining the structure of the data or predicting the future by extrapolating from past data. Those responsible for managing data collection systems can appreciate the difficulties of collection of environmental data at regular time intervals.

The variable (the data point) may be directly related to a defined time interval, such as the high and low temperature for the day. Environmental data may also be continuously changing, as would apply to measurements of hydraulic head in a piezometer. These observations are actually samples of instantaneous values, but are expressed as averages

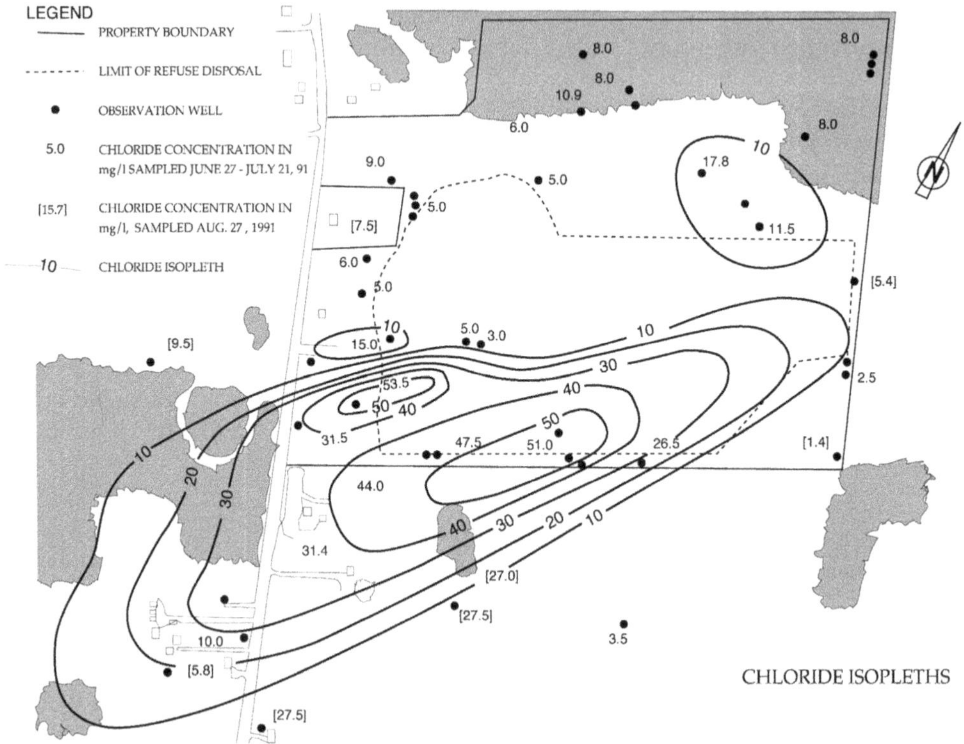

FIGURE 10.5
Closed chloride isopleths (isocontours) at a waste disposal site.

over the measured time interval. Readings taken once per day of a rapidly changing variable establish only a single point on a curve that can vary significantly until the next measurement. Fortunately, water-quality variables obtained from ground water do not vary significantly on a short-term basis due to the typically slow movement of ground water in granular aquifers. Fractured or Karst bedrock may, however, show much faster reaction times both in hydraulic head level changes and variations in water quality. Most detection ground-water monitoring programs sample on a quarterly basis. While a case can be made for somewhat shorter (or longer) sampling periods at some sites, based on ground-water flow rates, these four-times-per-year sampling programs represent a standard period for time-series analysis.

The first step in evaluating a time series is to determine if any structure exists in the data. Structure can be defined as the data behavior at a particular point in time being at least partially predicted by its value at other times. These structure elements in the data can be evaluated by:

- Defining a trend in the data (i.e., do the data increase or decrease with time), using straight lines, higher-order polynomials, or exponential curves.
- Testing for isolated events or unexpected departures from the normal behavior of the data set. This has specific applications for detection-monitoring programs where departures from long-term trends can force environmental programs into assessment actions.

Water quality at a single collection point such as a well or spring should be expected to change with time. Even with the generally slow movement of ground water, long-term

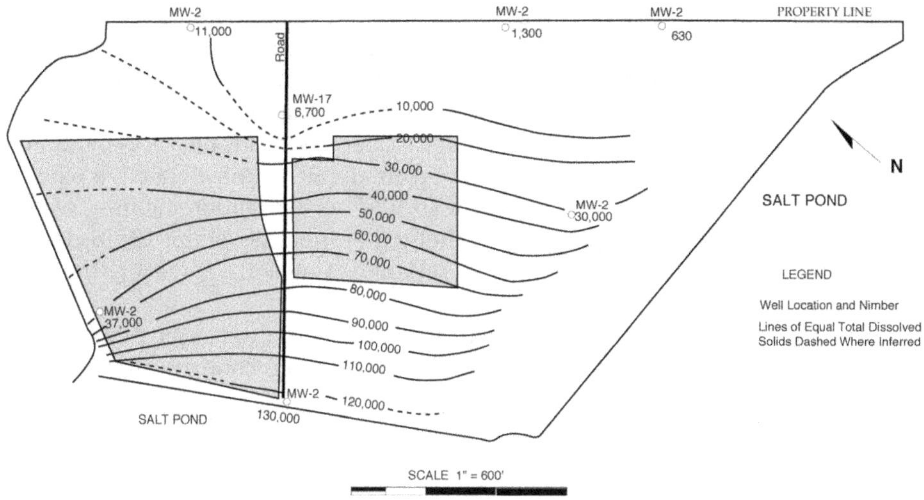

FIGURE 10.6
Open total dissolved solids isopleths (isocontours) at a waste disposal site.

detection or assessment monitoring programs can show gradual changes in water quality. These changes can be best illustrated by time-series presentations. Time-series diagrams can be used to compare individual parameters with time (i.e., compare water quality in a well against itself) or can illustrate changes in multiple parameters with time or changes with time for a common parameter in multiple wells. Figure 10.8 shows a comparison of

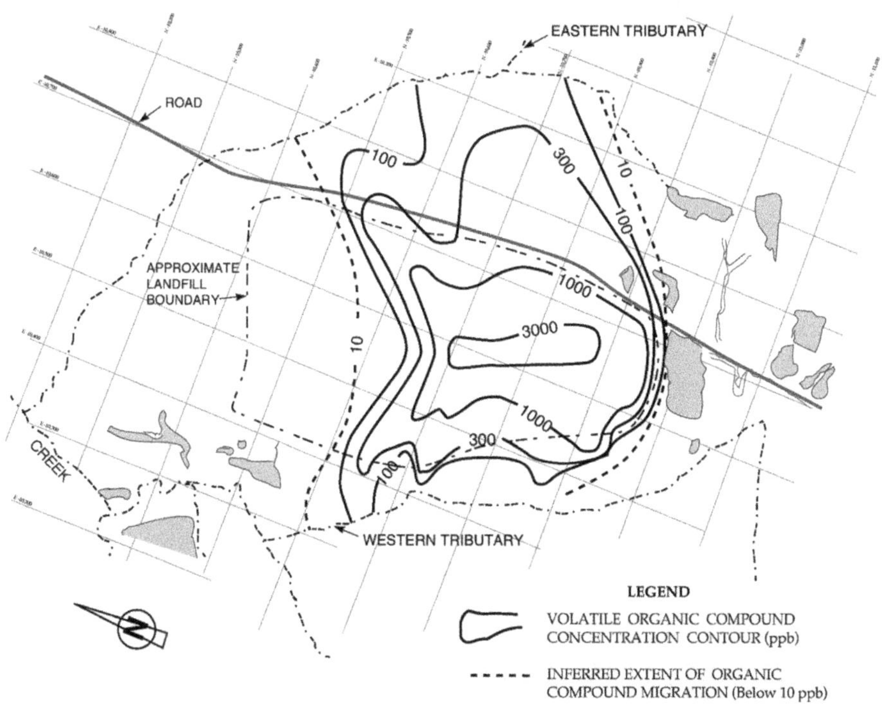

FIGURE 10.7
Isocontour map showing total VOC concentrations.

total dissolved solids in a number of wells. All six wells are compared to each other at any displayed point in time. Figure 10.8 also shows changes in TDS with time for each well. Time-series presentations can be ineffective if too large an amount of data is presented on one plot. Figure 10.9 shows a time-series plot for chloride in eight wells; although only a single parameter is displayed, the variable *Y*-scales used in the presentation make interpretation of trends difficult. Time-series presentations are most effective when single parameters are compared, as shown in Figure 10.10. This illustration includes water-level elevations with chloride concentrations. Whether or not the water-level elevation is related to the chloride concentration is a separate question; however, the data are displayed in an easily understood format.

A similar data set is presented in box-and-whisker plots (or "box plots") in Figure 10.11. Box plots are useful statistical tools for evaluating changes in water quality. Complicated site evaluations may require a series of box plots. For example, all wells screened in a hydrostratigraphic unit may be combined on a single plot or data from a number of well "nests" may be shown on one plot to illustrate vertical trends in water quality. The box plot can be considered as an economical graphical method of presenting the constituent summary statistics. The boxes are constructed using the median (middle value of the data) and the interquartile range (IQR; the range of the middle 50% of the data). These plots separate the results of each well and can clearly show the difference in the data distributions. These plots are generated by ranking the data and may be constructed in a number of different ways (McGill et al., 1978). Some box plots constructed by various software programs use the median and the F-spread. The F-spread, or fourth spread, is a function of the data distribution and measures the variability in the water-quality results, similar to the standard deviation. Hoaglin et al. (1983) provide a full discussion of these order statistics. The median and IQR are analogous to the more commonly used mean and standard deviation of a set of data. The mean and median are measures of central tendency or location, whereas the standard deviation and IQR are measures of variability.

Typically the first step in evaluating ground-water quality for box plot presentations is to review existing hydrogeologic information and to try to define subsurface stratigraphy and ground-water flow. The next logical step is to graph the chemical data as concentration versus time-series plots.

Figure 10.12 shows a chart where the mean values (solid circle), ± 1 standard deviation error bars (vertical line), are plotted for each well next to each box plot. The plots show that the mean for the data is consistently greater than the median. The two standard

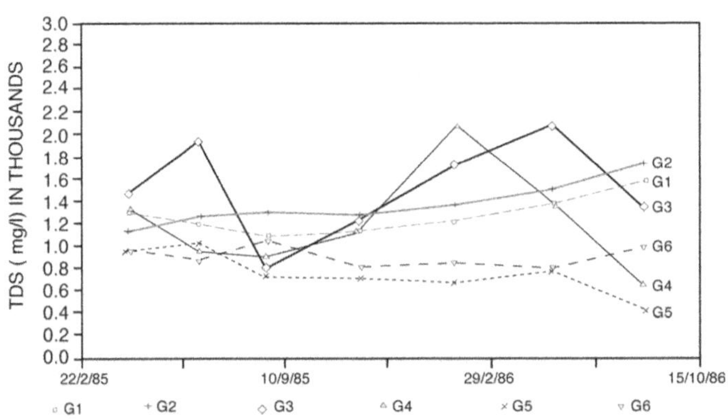

FIGURE 10.8
Time series comparisons for total dissolved solids for six wells.

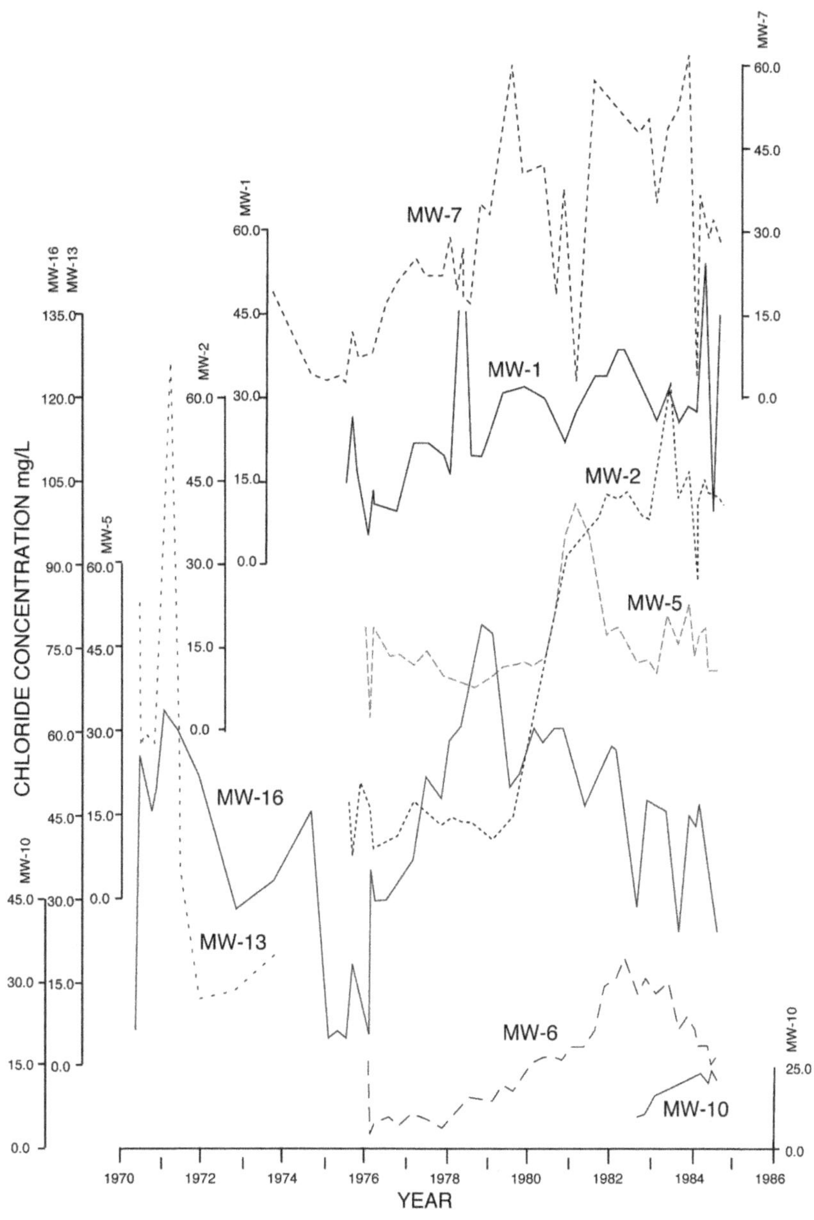

FIGURE 10.9
Time series comparisons for eight wells with sliding scales.

deviations for the data are larger than the IQR. High values otherwise described as "outliers" inflate the estimate of the mean and standard deviation in these statistical plots. The median and IQR are based on ranks and are not particularly sensitive to outlying values. Similar to Figure 10.11, the high variability in the impacted data is revealed by the wide error bars.

The box plots are considered more powerful in illustrating impacted water quality than simple error-bar plots because they contain more information about the actual data distribution. The error-bar plots, however, can be applied to parametric statistics evaluations.

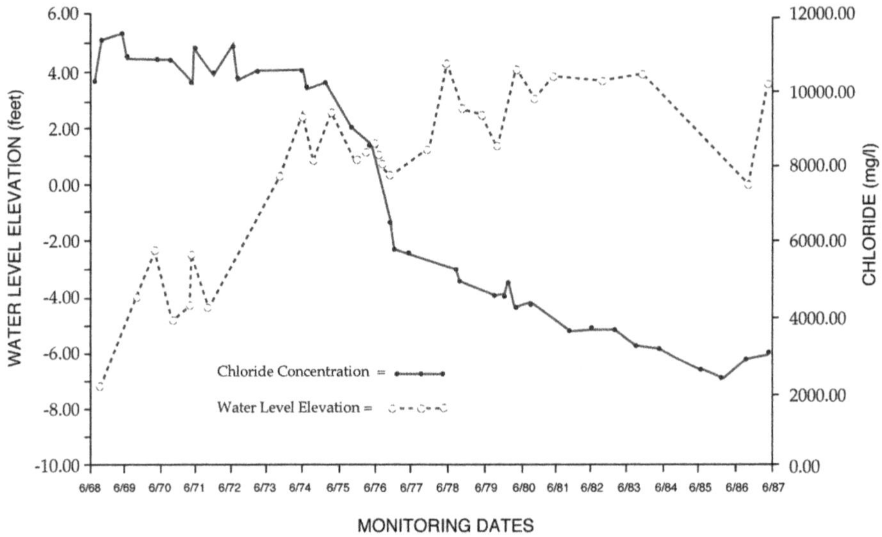

FIGURE 10.10
Time series comparisons for chloride and water levels. (*Source*: Hydro-Search, Inc.)

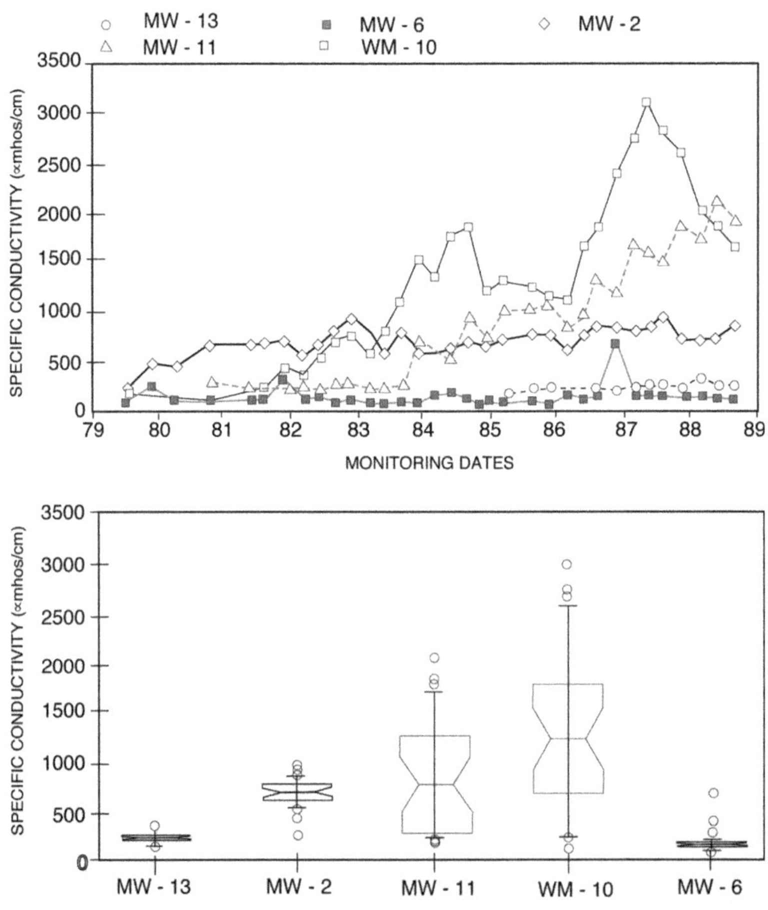

FIGURE 10.11
Time series comparisons with box plots.

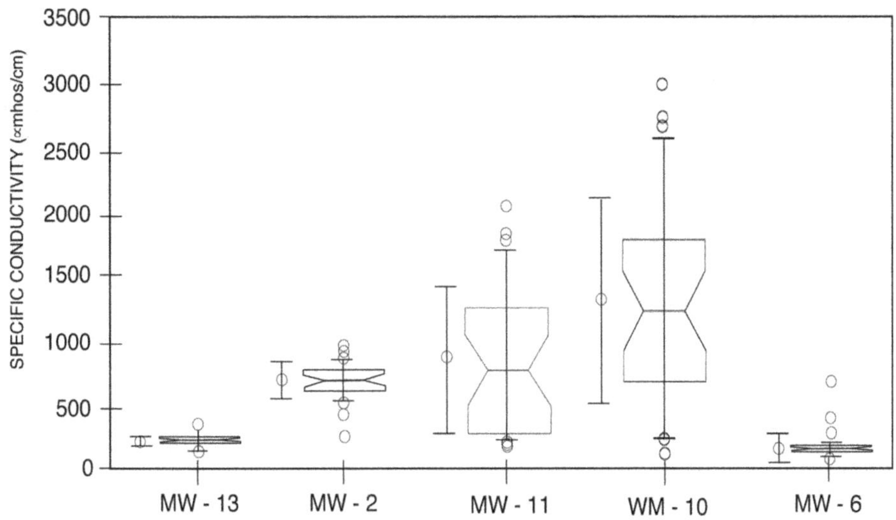

FIGURE 10.12
Error bar plot compared to box and whisker plots.

Histograms

The histogram is a two-dimensional graph in which one axis represents the data and the other is the number of samples that have that value. The *X*- or *Y*-axis of the plot is frequency expressed in terms of the percentage of total samples, rather than as an absolute count. The process of creating a histogram is primarily a counting process. A number of classes or groupings are defined in terms of subranges of the numeric value. These may be set to cover the complete range of the project data or a restrictive range derived from the mean and standard deviation or from knowledge of data ranges from previous project data evaluations. With many computer-based spreadsheet programs offering automated histogram production, these project data can be quickly plotted in a histographic format to evaluate the appearance of the figure.

Even with automated histogram production, the basic usefulness of the display can be enhanced by changing the parameters that influence the appearance of the histogram (Green, 1985):

- The range, which includes the minimum and maximum values
- The number of classes used in the counting
- The size of a class, such as the range of numeric values treated as a unit in the value counting
- Transformations of numeric values including scaling, logarithmic, and exponential

As a general guide, at least one histogram should be produced that covers the complete range of data values to evaluate samples outside the main distribution of data sets. It is recommended that all extreme values be investigated as errors or true anomalies. A common problem with histograms that have broad ranges is that the resultant figure will have poor resolution. The majority of the results in these displays are combined into one or two classes, obscuring the details of the distribution. Exclusion of the outliers results in better resolution of the main data sets.

Plotting of chemical data as a series of comparative histograms (or bar graphs) has been a traditional methodology for representation of variability in water quality. Most of the traditional methods are designed to represent the total concentration of solutes and the proportions assigned to each ionic species (for one analysis or group of analyses). The units in which concentrations are expressed in these traditional diagrams are milliequivalents per liter (meq). Hem (1970) provides descriptions of bar graphs, radiating vector plots, circular diagrams, and stiff diagrams — these methods will not be discussed here. Water-quality data collected during detection or assessment monitoring programs traditionally have not been portrayed in a format of "whole" analysis, that is, with anions and cations given in units of milliequivalents per liter. Rather, results of water-quality analyses are presented in milligrams or micrograms per liter and presented in formats including only a few parameters. These data, especially volatile organics and hazardous metals, have been displayed as histogram fingerprints illustrating variations in water quality. Figure 10.13 shows a series of histograms of hazardous metals obtained from analyses of water in individual wells. Similar histograms have also been used to track plumes of VOCs and to compare relative proportions of organic species in water from individual wells.

Tabular summaries of constituents are another form of comparative histograms. Figure 10.14 shows a summary table used to compare organic parameters observed in leachate with organic parameters observed in off-site wells. Many of the constituents in the

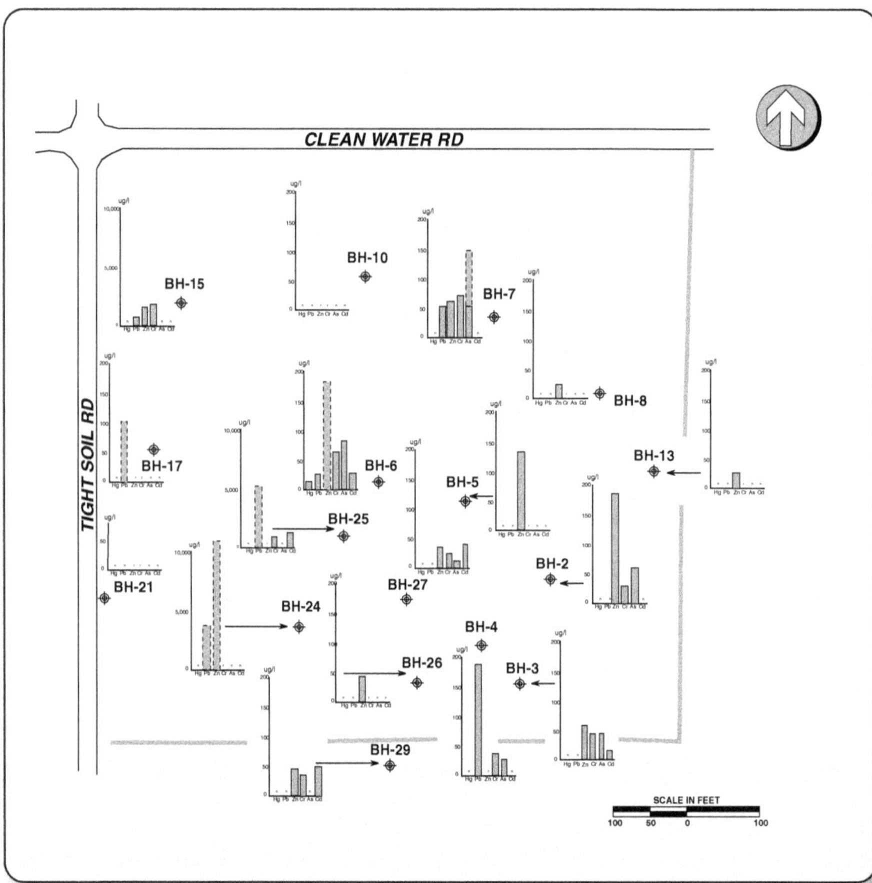

FIGURE 10.13
Histograms of metals data for individual boreholes.

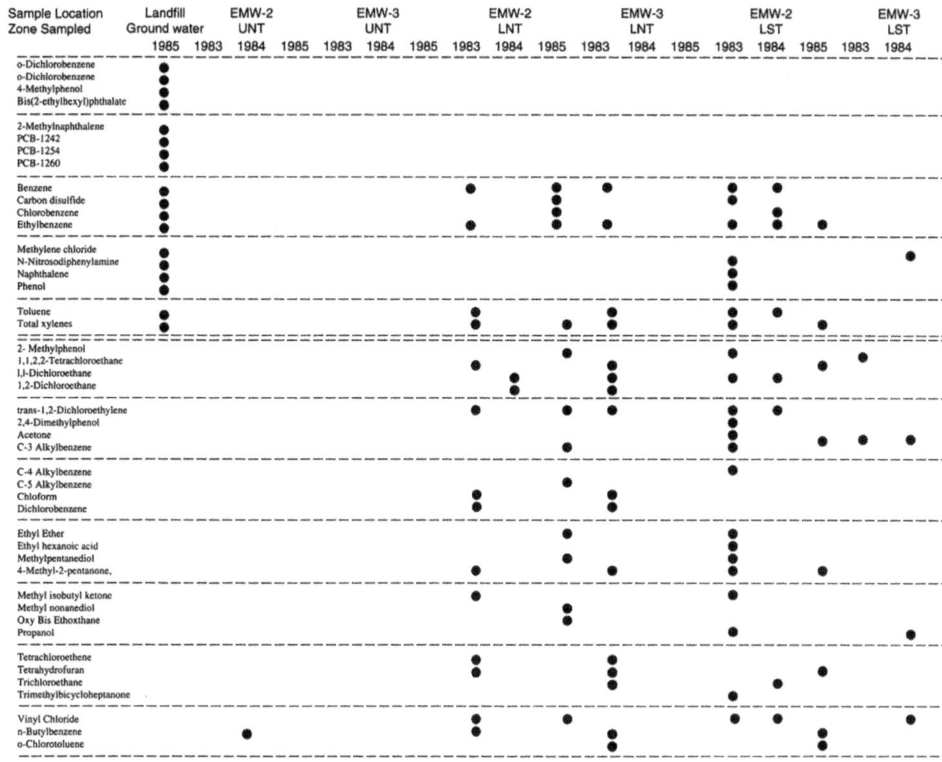

FIGURE 10.14
Histogram of leachate data.

fingerprint of the landfill leachate are different from those in the off-site monitoring well and thus tend to indicate a nonrelationship. Care must be taken to use indicator parameters that will not change with time and therefore provide a misinterpretation of the water-quality fingerprint.

Additional graphical displays of histograms are shown in Figure 10.15. The data shown in Figure 10.15 illustrate over a thousand observations of specific conductivity for two wells. These histograms can be compared to a lognormal distribution (Figure 10.15c) and normal distribution (Figure 10.15d). The histogram construction format for large numbers of observations can be used to investigate the probability distribution of the data. In general terms, the histogram plots values where the higher the bar, the greater the probability that additional measurements will fall in this range. Therefore, the more the sample values are incorporated into the histogram, the closer the graph is to the "true" population distribution. Many statistical tests used in evaluation of water-quality data require knowledge of whether the data come from a normally distributed population. The plotted data distribution illustrated on the histogram can be compared to a normally distributed data set. This provides a qualitative evaluation of the assumption that a normally distributed population is truly represented in the displayed environmental data.

The example project data sets (Figure 10.16) shows that neither of the wells have normally distributed data; both sets of data are skewed to the right. Because the data are not symmetric about the mean, the distribution is considered to be positively skewed. The lognormal distribution is also skewed right as shown. Natural log-scale transformations of positively skewed environmental data can make the data appear more normally distributed. Although histograms represent a good visual tool for evaluation of the

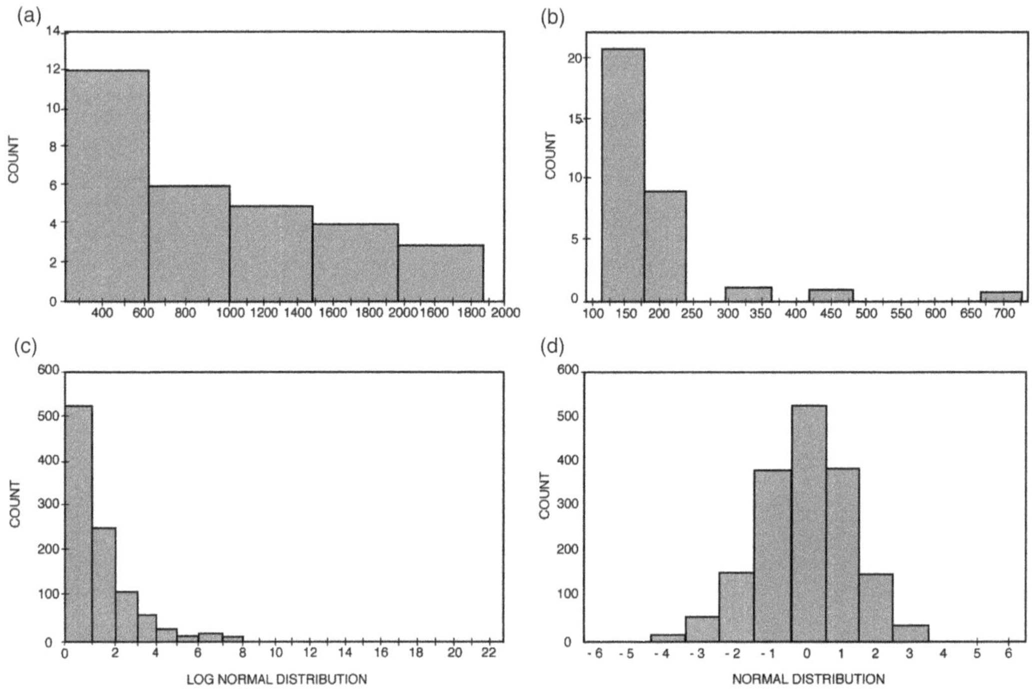

FIGURE 10.15
Histograms used to check data normality.

probability of the environmental data, Benjamin and Cornell (1970) point out that normal probability plots give a better representation of the data and are easy to construct.

Normal probability plots provide an excellent technique to compare environmental data to the normal distribution. Figure 10.16 shows a normal probability plot for the same data as in Figure 10.15a. These are constructed by first ordering the raw data from smallest to largest. Let $x[1] < x[2] < \cdots < X[n]$ denote the ordered data. The $X[i]$ are called the order statistics of the data. The $X[i]$ are then plotted on normal probability paper versus the corresponding plotting position $(1/N+1) \times 100$. If the data are from a normal distribution, the plotted points should lie approximately on a straight line (Fisher and Potter, 1989).

Figure 10.16 (top) illustrates that the data do not plot as a straight line; hence, the assumption of normality is in question. Transforming the data to log scale and then replotting as shown in Figure 10.16 (bottom) does not provide for a straighter line; therefore, we cannot conclude that the lognormal distribution is more appropriate for demonstrating normality. In summary, environmental data from a waste disposal facility can be visually presented in a number of ways that assist inspection of the data sets:

- Time versus concentration plots
- Box and whisker plots
- One standard deviation error-bar charts
- Histograms
- Normal probability plots

The first two graphical tools can clearly illustrate qualitatively the relative water quality between wells (known as inter-well comparisons). The error-bar charts may be valuable

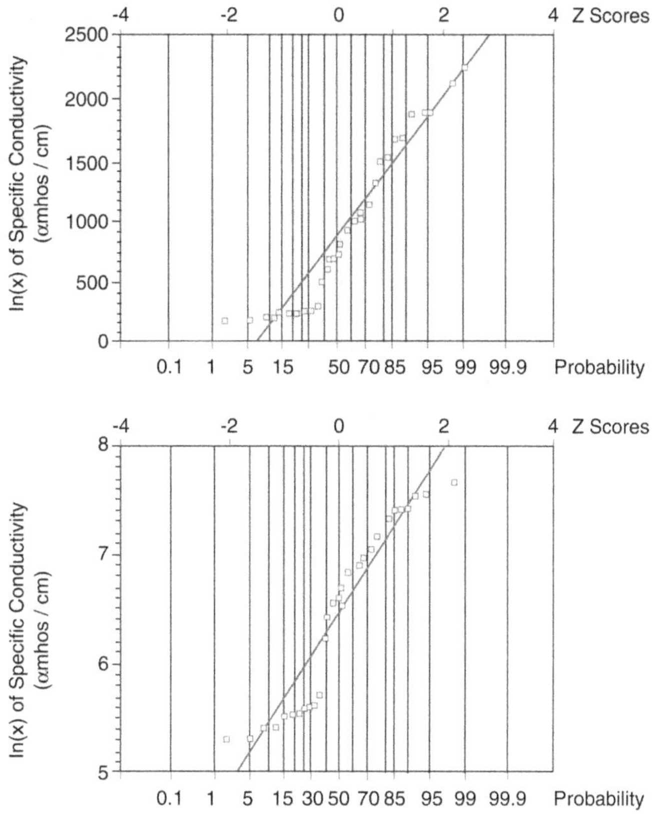

FIGURE 10.16
Probability plots for raw and log-transformed data.

when working with parametric statistics. Histograms may be used to view the probability distribution of the data. When evaluating the assumption of normality, normal probability plots are commonly prepared to observe deviations from normality. The example data illustrate several important points:

- Data outliers tend to inflate the mean and standard deviation of the data.
- The median and IQR are good estimates of the central tendency and variation of data sets, particularly when outliers are present.
- Large data variability is usually associated with high medians (i.e., impacted wells). Natural temporal variability is much lower than the variability observed when contamination is present.
- When histograms and normal probability plots show that ground-water data may not be normally distributed, the median and IQR may be better estimates of the central tendency and variability of the data.

Trilinear Diagrams

If one considers only the major, dissolved ionic constituents, in milliequivalents per liter, and lumps potassium and sodium together and fluoride and nitrate with chloride, the composition of most natural water can be illustrated in terms of three cationic and three anionic species. If the values are expressed as percentages of the total milliequivalents per liter of cations and anions, the composition of the water can be represented conveniently by a trilinear plotting technique.

The simplest trilinear plots utilize two triangles, one for anions and one for cations. Each vertex represents 100% of a particular ion or group of ions. The composition of cations is indicated by a point plotted in the cation triangle and the composition of anions by a point plotted in the anion triangle. The coordinates at each point add up to 100%. Most trilinear diagrams are in the form of two triangles bracketing a diamond-shaped plotting field, as first described by Piper (1944).

The trilinear diagram constitutes a useful tool for interpretation of water analysis. Most of the graphical procedures described here are of value in pointing out features of analyses and arrays of data that require closer study. The graphs themselves do not constitute an adequate means of making such studies, however, unless they can

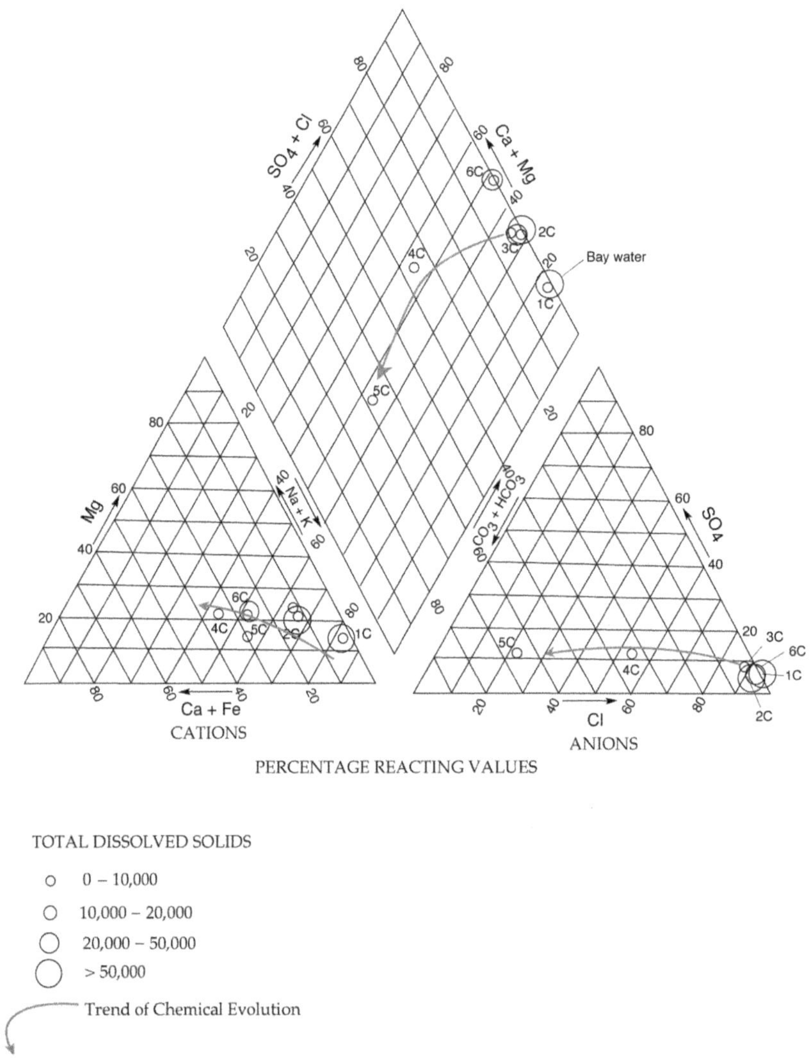

FIGURE 10.17
Trilinear data sets used for comparisons of water quality. (*Source*: Hydro-Search, Inc.)

demonstrate that certain relationships exist among individual samples. The trilinear diagrams sometimes can be used for this purpose.

Figure 10.17 is a trilinear diagram derived from analyses of water from San Francisco Bay and the Newark Aquifer. In any illustration of water-quality data, a diagram should aid interpretation by providing a visual clarification of trends or a comparison of differences in water quality. Trilinear diagrams have become so popular that computer programs have been written to automatically calculate and display the data.

Statistical Treatment of Water-Quality Data

Various procedures, such as averaging, determining frequency distributions, and making simple or multiple correlations, are widely used in interpretation of water-quality analyses. Sophisticated applications of statistical methods, particularly procedures that utilize digital computers, are being applied more and more frequently. Some potential applications of these statistical techniques are covered in the following sections. It is essential that proper consideration be given to chemical principles during application of statistical tests of data sets. Most data sets are evaluated using simple averaging and standard deviations, or just represent the water-quality data changes over time as shown in Figure 10.18; however, more complex statistical treatments of analytical data are necessary when state and federal performance standards are required for waste disposal or remedial cleanup sites.

Federal regulations have been established for statistical determination of compliance for RCRA facilities. Both existing and new hazardous waste facilities are covered by Subtitle C of the RCRA and regulated by 40 CFR Parts 264 and 265. When first issued, Part 264 Subpart F required that Cochran's approximation to the Behrens Fisher Student's *t*-test (CABF) or an alternative statistical procedure approved by U.S. EPA be used to determine whether there is a statistically significant exceedance. This Part 264 Subpart F regulation and, in particular, the CABF procedure, generated significant technical criticism over use of these statistical procedures for use with ground-water quality data and U.S. EPA proposed a new regulation in response to these concerns (U.S. EPA, August 24, 1987). The proposed regulation was revised based on comments EPA received and was then made final (U.S. EPA, 1988b).

The final regulation (October 11, 1988) describes five performance standards that a statistical procedure must meet. The Federal regulations do recommend four types of statistical procedures to evaluate performance of RCRA facilities for releases to ground water. In addition, U.S. EPA has issued (October 1991) amendments to Subtitle D of RCRA to include criteria for municipal solid waste landfills (MSWLFs). The number of samples collected to establish ground-water quality data must be consistent with the appropriate statistical procedures (discussed below). The sampling procedures are defined in the applicable sections for detection monitoring (§258.54[b]), assessment monitoring (§258.55[b]), and corrective action (§258.56[b]):

- Owner or operator must specify in the operating record one of the following statistical methods to be used in evaluating ground-water monitoring data for each hazardous constituent. The statistical test chosen shall be conducted separately for each hazardous constituent in each well.

Well Number:	MW-16S							MW-16D						
Collection Date :	1993	1996	1997	1998	1999	2000	2001	1993	1996	1997	1998	1999	2000	2001
Units:	μg/L	μg/L	μg/L	μg/L	μg/L	μg/L	μg/L	μg/L	μg/L	μg/L	μg/L	μg/L	μg/L	μg/L
Chloromethane	12,000U	1,000U	250 J	2,500U	2,500U	500 U	100 U	10 U	1 U	0.1 J	1 U	1 UJ	1 U	1 U*
Bromomethane	12,000U	1,000U	2,500UJ	2,500U	2,500U	500 U	100 U	10 U	1 U	1 U	1 U	1 U	1 U	1 U
Vinyl chloride	2,200J	1,000U	650 J	2,500U	2,500U	120 J	250	2 U	1 U	0.2 J	1 U	1 U	1 U	1 U
Chloroethane	6,200U	1,000U	2,500U	2,500U	2,500U	500 U	100 U	19 J	4	6 J	3 J	8	7	6 J
Methylene chloride	9,200U*	2,000U	2,500U*	5,000U*	2,500U	500 U	380	7 U*	2.0 U*	1 U*	2 U*	2 U*	2 U*	1 J
Acetone	1,500UR	3,100R	21,000R	22,000U*	11,000J	2,900	2,500J	7 J	R	7 U*	R	R	5 U*	5 U*
Carbon disulfide	6,200U	1,000U	2,500U	2,500U	2,500U	500 U	11 J	5 U	1 U	1 U	1 U	1 U	1 U	1 U
1,1-Dichloroethene	1,200U	1,000U	2,500U	2,500U	2,500U	500 U	100 U	1 U	1 U	1 U	1 U	1 U	1 U	1 U
1,1-Dichloroethane	2,500U	1,000U	320 J	2,500U	2,500U	110 J	260	2 U	1 U	1 U	1 U	1 U	1 U	1 U
cis-1,2-Dichloroethene	860 J	1,000U	620 J	2,500	2,500U	500 U	110	0.7 J	1 U	1 U	1 U	1 U	1 U	1 U
trans-1,2-Dichloroethene	6,200U	1,000U	2,500U	2,500U	2,500U	500 U	17 J	5 U	0.4 J	0.5 J	1 U	1 U	1 U	1 U
Chloroform	1,200U	1,000U	2,500U	2,500U	2,500U	500 U	100 U*	1 U	1 U	1 U	1 U	1 U	1 U	0.4 J
1,2-Dichloroethane	750 U	1,000U	2,500U	2,500U	2,500U	500 U	100 U	0.6 U	1 U	0.3 J	1 U	1 U	1 U	1 U
2-Butanone	350 UR	3,700J	19,000J	19,000U*	11,000J	3,000J	2,100	2 J	R	2 J	R	R	R	10 U
Bromochloromethane	1,000U	1,000U	2,500U	2,500U	2,500U	2,500U	2,500U		1 U	1 U	1 U	1 U	1 U	1 UJ
1,1,1-Trichloroethane	6,200U	1,000U	2,500U	2,500U	2,500U	500 U	100 U	5 U	1 U	1 U	1 U	1 U	1 U	1 U
Carbon tetrachloride	1,200U	1,000U	2,500U	2,500U	2,500U	500 U	100 U	1 U	1 U	1 U	1 U	1 U	1 U	1 UJ
Bromodichloromethane	1,200U	1,000U	2,500U	2,500U	2,500U	500 U	100 U	1 U	1 U	1 U	1 U	1 U	1 U	1 U
1,2-Dichloropropane	2,500U	1,000U	2,500U	2,500U	2,500U	500 U	38 J	2 U	1 U	1 U	1 U	1 U	1 U	1 U
cis-1,3-Dichloropropene	1,200U	1,000U	2,500U	2,500U	2,500U	500 U	100 U	1 U	1 U	1 U	1 U	1 U	1 U	1 U
Trichloroethene	3,800U	1,000U	2,500U	2,500U	2,500U	500 U	100 U	0.8 J	1 U	1 U	1 U	1 U	1 U	1 U
Chlorodibromomethane	2,500U	1,000U	2,500U	2,500U	2,500U	500 U	100 U	2 U	1 U	1 U	1 U	1 U	1 U	1 U
1,1,2-Trichloroethane	620 U	1,000U	2,500U	2,500U	2,500U	500 U	100 U	0.5 U	1 U	1 U	1 U	1 U	1 U	1 U
Benzene	2,500U	1,000U	920 J	2,500U	2,500U	83 J	140	3	0.8 J	1	1	1	1	1
trans-1,3-Dichloropropene	1,200U	1,000U	2,500U	2,500U	2,500U	500 U	100 U	1 U	1 U	1 U	1 U	1 U	1 U	1 U
Bromoform	2,500U	1,000U	2,500U	2,500U	2,500U	500 U	100 UJ	2 U	1 U	1 U	1 U	1 U	1 U	1 U
4-Methyl-2-pentanone	13,000	3,000J	14,000	13,000	8,300J	3,400	3,700	5 U	5 U	5 U	5 U	5 U	5 U	5 UJ
2-Hexanone	720 UR	5,000U	R	R	12,000U	230 J	430	50 UR	R	R	R	5 U	5 U	5 UJ
Tetrachloroethene	2,500U	1,000U	2,500U	2,500U	2,500U	500 U	100 U	2 U	1 U	1 U	1 U	1 U	1 U	1 U
1,1,2,2-Tetrachloroethane	620 U	1,000U	2,500U	2,500U	2,500U	500 U	100 U	0.5 U	1 U	1 U	1 U	1 U	1 U	1 UJ
1,2-Dibromoethane	1,200U	1,000U	2,500U	2,500U	2,500U	500 U	100 U	1 U	1 U	1 U	1 U	1 U	1 U	1 U
Toluene	52,000J	14,000	30,000	44,000	34,000	13,000	12,000	17	1 U	1 U*	1 U	1 U	1 U*	0.1 J
Chlorobenzene	6,200U	1,000U	2,500U	2,500U	2,500U	500 U	100 U	5 U	1 U	1 U	1 U	1 U	1 U	1 U
Ethyl benzene	2,700J	1,700	2,700	2,500	2,900	1,600	1,800	3 J	1 U	1 UJ	1 U	1 U	1 U	1 U
Styrene	1,200U	1,000U	2,500U	2,500U	2,500U	500 U	100 U	1 U	1 U	1 U	1 U	1 U	1 U	1 U
Xylenes (total)	14,000	8,700	12,000	13,000	16,000	9,900	11,000	17	1 U	5 U	1 U	1 U	1 U	1 U
1,3-Dichlorobenzene	6,200U	1,000U	2,500U	2,500U	2,500U	500 U	100 U	5 U	1 U	1 U	1 U	1 U	1 U	1 U
1,4-Dichlorobenzene	6,200U	1,000U	2,500U	2,500U	2,500U	500 U	100 U	5 U	1 U	1 U	1 U	1 U	1 U	1 U
1,2-Dichlorobenzene	12,000U	1,000U	2,500U	2,500U	2,500U	500 U	100 U	10 U	1 U	1 U	1 U	1 U	1 U	1 U
1,2-Dibromo-3-chloropropane	5,000UR	R	2,500U	2,500U	2,500U	500 U	100 UJ	4 UR	R	1 U	1 U	1 U	1 U	1 U
1,2,4-Trichlorobenzene	2,500U	2,500U	2,500U	2,500U	2,500U	500 U	100 UJ				1 U	1 U	1 U	1 U

Key:

J = The concentration is approximate due to limitations identified during the quality assurance review
U = Indicates the compound was analyzed but not detected. The associated value is the sample quantitation limit
U* = The compound should be considered "not detected" since it was detected in a blank at a similar concentration level
UJ = Indicates the compound was analyzed but not detected. The associated value is an estimated sample quantitation limit based on a bias identified during the quality assurance review
R = The results were considered unusable during the quality assurance review
Blank = The compound was not analyzed for

(1) Values shown are the highest detected between the investigative sample and its unpreserved, duplicate, reanalysis, or dilution sample.

FIGURE 10.18
Simple time-series data set for organic parameters.

  - A parametric analysis of variance followed by multiple comparisons procedures to identify statistically significant evidence of contamination.
  - An analysis of variance based on ranks followed by multiple comparisons procedures to identify statistically significant evidence of contamination.
  - A tolerance or prediction interval procedure in which an interval for each constituent is established from the distribution of the background data and the level of each constituent in each compliance well is compared to the upper tolerance or prediction limit.
  - A control chart approach that gives control limits for each constituent.
  - Another statistical test method that meets the performance standards discussed immediately below.

- Any statistical method chosen shall comply with the following performance standards, as appropriate:

  - The statistical method shall be appropriate for the distribution of chemical parameters or hazardous constituents.
  - If an individual well comparison procedure is used to compare an individual compliance well constituent concentration with background constituent concentrations or a ground-water protection standard, the test shall be done at a Type I error level no less than 0.01 for each testing period. If a multiple comparisons procedure is used, the Type I experiment-wise error rate for each testing period shall be no less than 0.05; however, the Type I error of no less than 0.01 for individual well comparisons must be maintained.
  - If a control chart approach is used to evaluate ground-water monitoring data, the specific type of control chart and its associated parameter values shall be protective of human health and the environment.
  - If a tolerance interval or a prediction interval is used to evaluate ground-water monitoring data, the levels of confidence and, for tolerance intervals, the percentage of the population that the interval must contain shall be protective of human health and the environment.
  - The statistical method shall account for data below the limit of detection with one or more statistical procedures that are protective of human health and the environment. Any practical quantitation limit (PQL) that is used in the statistical method shall be the lowest concentration level that can be reliably achieved within specified limits of precision and accuracy during routine laboratory operating conditions that are available to the facility.
  - If necessary, the statistical method shall include procedures to control or correct for seasonal and spatial variability as well as temporal correlation in the data.

- The owner or operator must determine whether or not there is a statistically significant increase over background values for each parameter or constituent required in the particular ground-water monitoring program that applies to the MSWLF unit.

  - In determining whether a statistically significant increase has occurred, the owner or operator must compare the ground-water quality of each parameter or constituent at each monitoring well to the background value of that constituent.

- Within a reasonable period of time after completing sampling and analysis, the owner or operator must determine whether there has been a statistically significant increase over background at each monitoring well.

The statistical test requirements are the same as the RCRA Subtitle C final regulation, as the solid waste rules recommend the same four types of procedures.

The performance standards in these Federal rules allow flexibility in designing statistical procedures to site-specific considerations. Selection of an appropriate statistical test must be made based on the quality of the data available, the hydrogeology of the site, and the theoretical properties of the test. As expressed in previous sections, ground-water quality data can be expected to vary temporally and spatially due to natural effects and the results are also affected by sampling and analytical errors. Due to natural variability observed in ground water, the determination of a significant change in water quality is linked to statistical probability theory.

In order to define if there has been a significant change in water quality, comparison must be made between supposedly "clean" background data and possibly impacted data. Both these ground-water classes are subject to temporal and spatial variability as well as sampling and analytical error. Hence, the problem becomes one of evaluation of variable water quality in time and space with potential statistical inferences. A statistical hypothesis is used to compare water quality:

Null hypothesis, H_0: No contamination exists, therefore the facility is in compliance.

Alternative hypothesis, H_1: Contamination exists, therefore the facility is in violation.

A statistical test is made on the null hypothesis and a conclusion is reached that either the facility is or is not in violation. The null hypothesis starts out with the assumption that there is no real difference between the quality of upgradient and downgradient ground water. The assumption is that they are all from the same population. Thus, the difference between the means of the two samples would be just one possible difference from the theoretical distribution where the mean difference is zero.

The assumption is called the *null hypothesis* because it attempts to nullify the difference between the two sample means by suggesting or forming a hypothesis that it is of no statistical difference. If the statistical difference between the two sample means turns out to be too big to be explained by the kind of variation that would often occur by chance between random samples, then one must reject it (the null hypothesis), as it will not explain our observations. The typical alternative hypothesis would be that the two water-quality population means are not equal. In this context, a violation implies that water quality is significantly different from background. Figure 10.19 illustrates the two types of errors associated with hypothesis testing.

Significant technical discussions surround whether a site has observed a false-positive indicating contamination. A Type I error (false-positive) occurs when a site (or well) is actually in compliance but the statistical test is triggered that decides it is in violation. The probability of a Type I error (or) is defined as the controllable significance level of the test. Usually, this is set at 0.05, giving a 1/20 chance that a false-positive conclusion of contamination will occur.

A Type II error (false-negative) occurs when contamination exists but is not detected. The probability of a false-negative conclusion is more difficult to control, is often difficult to calculate, and is dependent on many factors that may include sample size, the overall

TRUE SITUATION	IN COMPLIANCE	IN VIOLATION
IN COMPLIANCE	Good Decision $1-\alpha$	False Positive Decision α Type I error
IN VIOLATION	False Negative Decision β Type II error	Good Decision $1-\beta$ Power of Test

FIGURE 10.19
Statistical error in hypothesis testing.

magnitude of change in parameter concentration, and choice of statistic tested in the decision process.

Statistical hypothesis testing can be divided into two general categories: (1) parametric, or those which rely on the estimation of parameters of a probability distribution (usually the mean and standard deviation of the normal distribution) and (2) nonparametric, or those which do not fit a normal distribution. Nonparametric methods usually rely on test statistics developed from the ordered ranks of the data. The simplest nonparametric evaluation is the median or middle value of a data set. Both parametric and nonparametric statistical tests are reviewed in the context of ground-water monitoring events in later sections of this chapter.

In general terms, the type of statistical test to use for a facility regulated under federal laws should be consistent with U.S. EPA (1988b) 40 CFR Part 264, Statistical Methods for Evaluating Ground-Water Monitoring from Hazardous Waste Facilities; final rule (*Federal Register*, 53, 196, 39720–39731). Both RCRA solid waste (Subtitle D) and hazardous waste (Subtitle C) sites are keyed into this code.

Data Independence

Independence of data collected in environmental programs must be evaluated by determining if the data show serial correlation. Serial correlation of ground-water sampling data is most likely to occur from very slow ground-water flow. Even with reasonably permeable aquifers (say with hydraulic conductivity $>10^{-3}$ cm/sec) low gradients can slow ground-water velocity to less than 20 ft per year. When ground-water quality measurements are collected too frequently to be independent of each other, one can observe serial correlation in the data. Independence can often be achieved by increasing the time between observations. Several tests have been reported in the literature to evaluate the presence of serial correlation in ground-water quality data. Montgomery et al. (1987) chose the Lag 1 autocorrelation function (ACF). Goodman and Potter (1987) also used this method as well as the nonparametric auto run (AR) test. The application of the ACF test to ground-water quality data is described in detail by Harris et al. (1987). The AR test was applied to hydrologic data by Sen (1979). Most advanced statistics texts and computer packages include these tests.

An example of data that show serial correlation is provided in Figure 10.8. These results indicate that serial correlation may exist in ground-water quality data even though the sampling was at intervals of 3 months.

The reality of most sampling programs dictates the sampling period required by regulatory standards or permit requirements. It is probably sufficient that one is aware of the potential difficulties associated with serial correlation of the data so that independence of the observations can be checked to help in the selection of the statistical test used in the evaluation of the data.

Data Normality

The normal distribution is perhaps the single most important and widely used probability model in applied statistics. This is because many real systems fluctuate normally about a central mean, that is, measurement error of a random variable is symmetric about a true mean and has a greater probability of being small (close to the mean) than large in the tail of the distribution.

The U.S. EPA's RCRA statistical regulations (40 CFR 264 Subpart F) do not require tests for normality or other distributional assumptions unless:

1. A data transformation is made
2. Nonparametric statistical tests are applied

Data transformations are commonly used to normalize skewed data for parametric tests. Many environmental systems are modeled using the lognormal distribution because: (1) it has a lower bound of zero and (2) it is positively skewed, allowing high values to be included (Benjamin and Cornell, 1970).

The hypothesis of normality can be evaluated through any number of statistical goodness-of-fit tests. These tests are used to mathematically compare the shape of the normal distribution to the data set. Care should be taken to only apply these tests to independent, stationary data sets.

In the ground-water quality literature, Montgomery et al. (1987) tested the normality of ground-water quality data using graphical methods, the chi-square test, and the skewness test. Harris et al. (1987) recommend the skewness test for general use with ground-water quality data.

Evaluation of Ground-Water Contamination

A main objective of a ground-water detection monitoring program is to determine if the facility is affecting ground water. Owner and operators are required in federal rules to place detection monitoring wells in both upgradient (background) and downgradient locations around the facility and to monitor those wells at regular intervals, typically quarterly or twice per year, for a series of indicator parameters. Subtitle D defines in [§258.53] (Ground-Water Sampling and Analysis Requirements) that:

> The ground-water monitoring program must include consistent sampling and analysis procedures that are designed to ensure monitoring results that provide an accurate representation of ground-water quality at the background and downgradient wells. The owner or operator must notify the State Director that the sampling and analysis program documentation has been placed in the operating record and the program must include procedures and techniques specified in §258.53(a)(1) through (5):

- The ground-water monitoring program must include sampling and analytical methods that are appropriate for ground-water sampling and that accurately measure hazardous constituents and other monitoring parameters in ground-water samples. Ground-water samples shall not be field filtered prior to laboratory analysis.
- Ground-water elevations must be measured in each well immediately prior to purging each time ground-water is sampled. Ground-water elevations must be measured within a period of time short enough to avoid temporal variations in ground-water flow that could preclude accurate determination of ground-water flow rate and direction.
- The owner or operator must establish background ground-water quality in a hydraulically upgradient or background well(s) for each of the monitoring parameters or constituents required in the particular ground-water monitoring program that applies to the MSWLF unit.
- The number of samples collected to establish ground-water quality data must be consistent with the appropriate statistical procedures (discussed below). The sampling procedures are defined in the applicable sections for detection monitoring (§258.54[b]), assessment monitoring (§258.55[b]) and corrective action (§258.56[b]).

The logic of this sampling strategy is that upgradient water quality represents the background conditions for that particular region and downgradient water quality represents background water quality plus any influence produced by the facility. Section 258.40 states, "The relevant point of compliance specified by the Director of an approved State shall be no more than 150 meters from the waste management unit boundary and shall be located on land owned by the owner of the MSWLF unit." This sets the stage for defining a boundary zone in which to locate the monitoring wells.

The detection monitoring program as described in subtitle D for solid waste (§258.54) then defines the following:

- Detection monitoring is required at MSWLF units at all ground-water monitoring wells specified in §258.51(a)(1) and (2) (background and downgradient). At a minimum, the constituents from Appendix I must be included in the program.
- The director of an approved state may delete any of the Appendix I constituents if it can be shown that the removed constituents are not reasonably expected to be in or derived from the waste contained in the unit.
- A director of an approved state may establish an alternative list of inorganic indicator parameters for an MSWLF unit, in lieu of some or all of the heavy metals in Appendix I.

Note: Deletion of Appendix I constituents or establishing alternative inorganic parameters is not possible unless the state is approved by the U.S. EPA.

- The monitoring frequency shall be at least semi-annual during the active life of the facility (including closure) and the post-closure period. A minimum of four independent samples from each well (background and downgradient) must be collected and analyzed during the first semi-annual sampling event. At least one sample from each well (background and downgradient) must be collected and analyzed during subsequent semi-annual events. The director of an approved state may specify an alternative frequency during the active life (including closure) and the post-closure care period. The alternative frequency shall be no less than annual.

Note: An alternative monitoring frequency is not an option unless the state is approved by the U.S. EPA.

- If the owner or operator determines that there is a statistically significant increase over background for one or more of the constituents in Appendix I or an approved alternative list, the owner or operator must:
  - Within 14 days of the finding, place a notice in the operating record indicating which constituents have shown statistically significant changes from background levels and notify the state director that the notice was placed in the operating record.
  - Establish an assessment monitoring program within 90 days unless the owner or operator can demonstrate that a source other than an MSWLF unit caused the contamination or that a statistically significant increase resulted from an error in sampling, analysis, statistical evaluation or natural variation in ground-water quality. A report documenting the alternate source or error must be certified by a qualified ground-water scientist or the director of an approved state and placed in the operating record.

The issue of the regulatory concept of the simple upgradient and downgradient model is rarely observed in real-world monitoring programs; however, the use of background wells as representative of upgradient water can be used in statistical comparisons. In many cases, particularly when adequate background data are available prior to the installation of the facility, intrawell comparisons may be the most successful technique to use (i.e., each well compared to its own history). The major advantage of this approach is that it eliminates the spatial component of variability from the comparison. One is left with evaluating the local effects on the well installation, such as construction, maintenance, and nearby interferences to water quality (such as wells located near roads that are salted during winter, local spills, etc.). The statistical methodology is illustrated in Figure 10.20, which graphically portrays the variable bases for statistical comparisons between wells and for intrawell statistical comparisons.

Detection monitoring programs at waste-disposal facilities require not only that a release to the environment has occurred, but also that the release observed is directly due to discharges from the facility. Water-quality standards are commonly used as a basis for judging if a release to the environment has occurred. Yet, even a water-quality standard exceedance must be compared to background water quality in order to conclude that the facility is responsible for the ground-water impact. In reviewing facility data, one can expect a number of water-quality standards (especially nonorganic parameters) to be exceeded in natural ground water. Thus, comparison of downgradient water quality to known background water quality is an important part of any detection monitoring program. It is imperative that detection monitoring programs not rely on only one background well as the basis for comparisons to downgradient wells. Natural spatial variability within the geologic environment can significantly affect the statistical comparisons necessary for detection monitoring programs. Evaluation and knowledge of ground-water flow and geology sufficient to design detection monitoring systems, together with time-series graphs of ground-water quality, may clearly show a release.

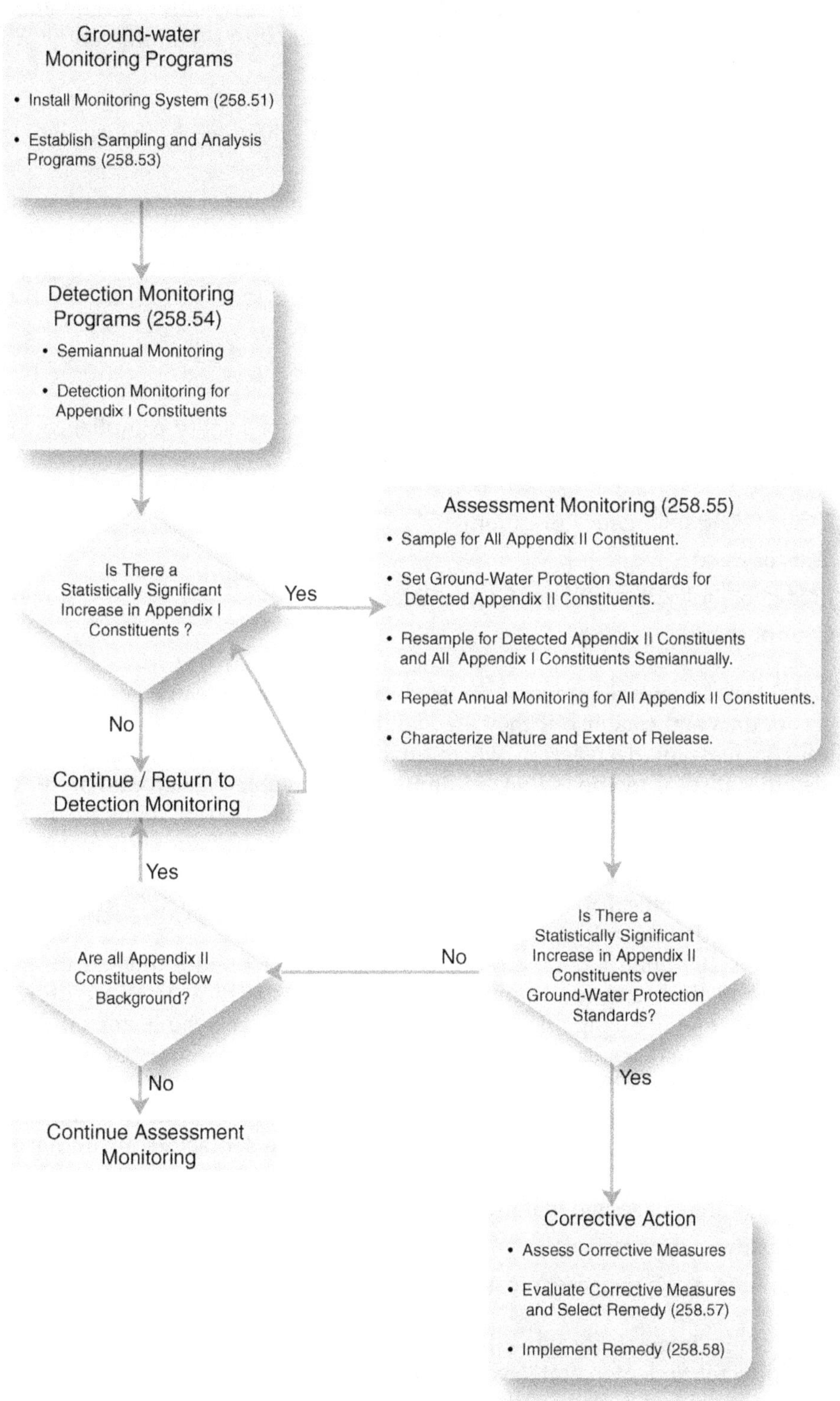

FIGURE 10.20
The variable bases for statistical comparisons between wells and for intra-well statistical comparisons (*Source*: 40 CFR 258.50–258.58).

Statistical tests applicable to water-quality evaluations currently recommended by U.S. EPA with other methods proposed in the water-quality literature can be used as evaluation tools for the following example facilities:

- Existing MSWLFs, new facilities, and existing facilities with historically clean water quality
- RCRA Subtitle C hazardous waste disposal sites, industrial waste disposal sites, land disposal sites for waste water
- Superfund-type evaluations for assessment and aquifer remediation projects for Federal and State cleanup programs

Types of Statistical Tests

The four general categories of statistical methods used for facility compliance comparisons with ground-water quality RCRA regulations are:

- Tests of central tendency (location)
- Tests of trend
- Prediction, tolerance, and confidence intervals
- Control charts

Statistical tests of central tendency are used to compare the mean or median of two or more sets of data and establish if they are significantly different. Tests of trend evaluate significant increase or decrease in water quality over time. Prediction and tolerance intervals are statistical methods that set limits for acceptable background water quality based on historic data sets. These interval tests can also be used to define the number of background measurements required to fully establish background water quality. Confidence intervals also set limits for average background water quality. Control charts are widely used graphical methods for industrial engineering quality control and are similar to the prediction, tolerance, and confidence intervals.

To evaluate water-quality data, a series of questions must be formulated to select the appropriate statistical tests to evaluate the data. These statistical tests are designed to evaluate whether or not a significant difference exists between the historical mean and median of background water quality and the mean and median of each downgradient well. The ability of these tests to detect ground-water contamination quickly (i.e., when applied quarterly with detection within one or two quarters after a release occurs) depends on the choice of the statistical test and other data set factors including:

- Length of the unaffected water-quality record
- Variability in the data sets
- Magnitude of the increase in concentration due to the release

Regulatory requirements for immediate detection of releases to the environment can be extremely difficult to demonstrate even with the monitoring well screen located directly within the ground-water flow path from the facility. Also of interest in a detection monitoring program is to establish if specific water-quality standards have been exceeded. If a sample value exceeds a regulatory standard such as maximum contaminant level (MCL) under the Clean Water Act (accounting for sampling and analytic variability), then a determination of a violation may be made by state or federal regulators. In this situation, a violation means only that a mandated concentration level has been exceeded, not that certain actions must be taken. The problem of regulatory violation of a standard is acute when state or federal standards are at or approach the level of detection of the contaminant, as is the case with some VOCs. Possible decision approaches may include:

- A regulatory mandated "hard" limit where no data should exceed the water-quality standard with consideration given to sampling and laboratory error.
- The more flexible historic mean concentration at a well where the water-quality standard should not exceed this regulatory limit.
- The moving window approach where the last-year's mean concentration should not exceed the limit.
- The statistical limits where 95% of the population must be below the standard.

These provide a number of alternative decision paths for water-quality evaluation or for making decisions on analytical sample values close to a water-quality standard.

Comparison of downgradient water quality to the standard is conceptually straightforward. The indicator parameter is plotted on a simple time-series graph and compared to the concentration called out in the water-quality standard. Background concentrations should be plotted to evaluate if the background water-quality levels exceed the particular standard. If background parameter levels do exceed the relevant standard, the downgradient well parameter concentrations must be evaluated statistically against the background rather than comparing the well data to the water-quality standard. In those cases where background does not exceed the standard, downgradient concentration can be compared to the standard. Parameter trends (especially for inorganic indicator variables) can serve as a useful management tool for implementing corrective actions before primary drinking water standards are exceeded. Once water-quality standards are clearly exceeded and verified through resampling, a release to the environment is confirmed. In the second and third approaches, confidence limits on the mean (where the standard must be below the lower confidence level) can provide the evaluation tool. The last approach can effectively use tolerance interval tests for evaluation of exceedances.

After evaluation of the issues or questions to be answered statistically, the next step is to choose a specific test that answers the question. The test must not only have an appropriate experimental design (i.e., answer the right question), but also the implicit assumptions of the test must not be grossly violated. As previously discussed, ground-water quality data may grossly violate the assumption of normality, even after appropriate data transformation.

A detection monitoring data evaluation must be based on the variable regulatory issues that can change from state to state and from site to site. Some facilities may have specific permit requirements for statistical tests or specified parameter lists that can be significantly different from site to site within a single state. In the following sections, potential releases to ground water can be evaluated using three general types of statistical methods:

1. Tests of central tendency (location)
2. Tests of trend
3. Prediction, tolerance, and confidence intervals

In each section, emphasis is placed on the situations where the type of test is appropriate. The types of water-quality questions these tests can answer are discussed.

Tests of Central Tendency (Location)

The statistical mean and median of water-quality data sets are the most common estimates of central tendency. Tests that compare the mean or median of two or more sets of data are tests of central tendency or tests of location.

The U.S. EPA had previously required that Cochran's approximation to the Student's *t*-test be applied between pooled background water-quality data and each downgradient compliance well. Significant criticism of this procedure (see McBean and Rovers, 1984; Silver, 1986; U.S. EPA, 1988b; and Miller and Kohout, undated) resulted in the Agency change to a parametric one-way analysis of variance (ANOVA) or the nonparametric analog called the Kruskal–Wallis test (U.S. EPA, 1988b). Unfortunately, the ANOVA test also suffers from a high false-positive error rate when many multiple comparisons must be done for sites with, for example, more than five or six compliance wells (see Fisher and Potter, 1989).

Tests of Trend

Tests of trend are commonly used in detection monitoring programs to evaluate whether water-quality parameter values are increasing or decreasing with time. Trend analysis is also useful for evaluating changes in background water quality. Trends in data could be observed as a gradual increase (usually modeled as a linear function) or a step function or even cyclical on a seasonal basis.

Trend evaluations have traditionally been performed by inspection of graphed time concentration plots. Time-series plots can also be used in conjunction with box plots to evaluate trends and seasonal fluctuations. A number of statistical methods can be applied to data sets to evaluate for trends and seasonality. Example procedures such as the Mann–Kendall test for trend evaluates the relative magnitudes of the concentration data with time (Goodman, 1987). The length of time recommended to obtain adequate long-term trends is 2 yr of data (Doctor et al., 1986); for seasonal trends, a much longer period data set may be necessary. Goodman (1987), using a modified Mann–Kendall test, found that at least 10 yr of quarterly data were required for obtaining adequate power to detect seasonal trends. Although few facilities have such a long period of data, the long (post) closure requirements of 10–30 years in state and federal regulations will make such evaluations for seasonal trends possible.

Statistical trend tests alone cannot be used to determine compliance with ground-water quality regulations. These tests can only answer the question, "Does a positive or negative trend exist?" The presence of a minor trend should not be construed to mean there has been a release from the facility. Therefore, if a test of trend is used to support the hypothesis of a release, the results must be linked to exceedance of water-quality standards and to likelihood of the release based on review of potential cross-contamination and interferences.

Tests of trend have been commonly used in evaluating the expected effectiveness of remedial action. However, tests of trend should not be used to predict when a target concentration will be reached since aquifer restoration is usually not a linear but rather an asymptotic process.

A common use of trend tests is to evaluate if background water quality is significantly (gradually) changing over time. Hence, the background water quality represents a moving window that will be compared to downgradient water quality. In this case, the background trend should be removed prior to further analysis (Harris et al., 1987). An apparent trend at a downgradient well cannot be confirmed as evidence of contamination, unless it can be shown that the same trend does not exist in background or upgradient wells.

The nonparametric analogs to the linear regression *F*-test are Kendall's tau statistic and Spearman's (rho) rank correlation coefficient. Usually Kendall's tau is chosen for water-quality data because the test statistic approaches normality at smaller sample sizes than Spearman's rho (Montgomery et al., 1987).

Linear regression is considered a powerful technique of trend, but analysts tend to delete outlying values without physical justification to get a good fit to the data. Also, some users will wrongly try to make predictions of when concentration will return to normal or when a standard will be exceeded. Reviewers should make sure that deletion of data is physically justified. Also, any predictions made with the regression line should be interpreted as no more than a best guess.

Fisher and Potter (1989) reviewed statistical tests for applicability for use in detecting facility ground-water contamination events. They found that tests of central tendency, both parametric and nonparametric, have severe limitations. At least for the cases reviewed, natural spatial variability did not permit ANOVA results to discern between natural variations in the mean and those due to potential contamination. They also observed that ground-water quality data often violated the parametric assumptions of normality for both raw and log-transformed data sets. Even nonparametric tests of central tendency (such as Kruskal–Wallis) are not recommended for detecting contamination but rather should be used for evaluating spatial variability (Fisher and Potter, 1989). Statistical tests based on trend can be used in conjunction with other data evaluation techniques to support the conclusion of observed contamination. Prediction interval tests were recommended by Fisher and Potter (1989) as the most theoretically sound approach to setting background levels, and in the author's opinion interval statistical tests represent the most applicable methods for evaluating detection monitoring programs. As such, the remaining discussion of statistics will concentrate on interval tests.

Recommended Statistical Methods

Ground-water detection monitoring typically involves a series of monitoring wells hydraulically upgradient and downgradient of the facility to compare concentrations of chemical constituents between the upgradient and downgradient locations, assuming that any difference in ground-water quality is caused by leachate released from the facility. However, this assumption is often false because widespread spatial variability in ground-water chemistry exists. In the worst case (often the most typical circumstance), regulations (U.S. EPA, 1988b, 1991) require only one upgradient well and a minimum of three monitoring wells located downgradient from the facility. When a single upgradient well is used to characterize natural variability in background spatial variability, attempts to locate contamination are confounded (i.e., differences between the upgradient and downgradient wells could be due to natural differences between any two locations regardless of their relation to the waste disposal facility). Even with two upgradient wells, characterization of natural background variability may not be possible. That is, two upgradient wells may not display the same amount of variability observed in downgradient wells, which often number between 10 and 100.

Additionally, regulations require each downgradient monitoring well and constituent to be separately tested because releases from a waste disposal facility into ground-water are "plume-shaped," which may influence only a single downgradient well. Pooling data over downgradient wells might mask a release that only affected a single well. In addition, chemical constituents travel at different rates in ground water; the leading edge of the plume may contain only a small number of highly mobile chemical constituents. In many parts of the country, ground water flows quite slowly, in some cases less than a foot per year. Hydrogeologically independent observations from a given monitoring well may be available only quarterly, semiannually, or annually. Pooling data may be impractical because it may result in mixing contaminated and uncontaminated measurements, masking an early-stage release. Therefore, each new datum must be evaluated individually. The two most critical problems are that (a) numerous statistical evaluations must be performed on each monitoring event (typically 100–1000) and (b) environmental data are often censored (i.e., the analyte may or may not be detected when it is present at a level below the capability of the analytical instrument). These two problems complicate analysis of ground-water-monitoring data.

As will be shown in the following sections, solution of these problems leads to the construction of the so-called "prediction limits" adapted to the case of simultaneous statistical inference and sequential testing. Simultaneous statistical inference refers to the construction of limits or bounds that apply simultaneously to all comparisons made on a given monitoring event. In this context, the number of comparisons, which is denoted as k, is the set of all downgradient monitoring wells for all constituents for which statistical evaluation must be performed. Using these methods we can therefore control the overall site-wide false-positive rate (i.e., concluding that there has been an impact when there has not been) at a nominal level (e.g., 5%). As the number of comparisons on any given monitoring event becomes large, however, the associated false-negative rate (i.e., the failure to detect contamination when it is present) also becomes large. To minimize the false-negative rate, we use a sequential testing strategy in which an initial exceedance is then verified by one or more independent verification resamples. In this way, a smaller prediction limit can be used repeatedly, achieving the same site-wide false-positive rate but greatly minimizing the false-negative rate. These ideas are more fully developed in the following sections.

Statistical Prediction Intervals

Single Location and Constituent

If the problem were to set a $(1-\alpha)100\%$ limit on the next single measurement for one location and one normally distributed constituent, a β-expectation tolerance limit (i.e., a prediction limit) (Guttman, 1970; Hahn, 1970) could be computed from n independent background measurements as

$$\bar{x} + t_{(n-1,1-\alpha)} s \sqrt{1+\frac{1}{n}} \tag{10.1}$$

where concern is that the concentration is elevated above background, x and s are the background sample mean and standard deviation, respectively, and t is the $100(1-\alpha)$ percentile of Student's t-distribution on $n-1$ degrees of freedom. If upgradient versus downgradient comparisons are to be performed, then a minimum of two background locations (e.g., wells) should be repeatedly sampled at a time interval sufficient to ensure independence (e.g., quarterly or semiannually). The background time period must include at least 1 yr to ensure that the same seasonal variation present in downgradient locations is rejected in the upgradient background. The reader should note that with multiple upgradient locations, s^2, the traditional estimator of σ^2 is biased (i.e., it is too small) because measurements are nested within upgradient monitoring locations. Alternative estimators for σ^2 based on variance components models have been proposed and should be used where appropriate (Gibbons, 1987a, b, 1994).

Multiple Locations

In practice, multiple comparisons are performed, one for each downgradient monitoring location and constituent. Using the Bonferroni inequality (Miller, 1966), a conservative prediction bound (i.e., the probability of at least one false rejection is at most α) for all kq comparisons (i.e., k locations each tested for q constituents) is

$$\bar{x} + t_{[n-1,(1-\alpha)/kq]} s \sqrt{1 + \frac{1}{n}} \tag{10.2}$$

In the present context, the comparisons are dependent because (1) constituents may be correlated and (2) all downgradient locations are compared to a common background. In this case, the Bonferroni adjustment may be unnecessarily conservative. Some improvement may be gained by adapting the approach of Dunnett and Sobel (1955) originally developed to compare multiple treatment groups to a common control group. The resulting correlation between the comparison of locations i and j to a common background is

$$\frac{\bar{x}_{\mathrm{B}} - \bar{x}_{\mathrm{M}}}{\sqrt{(S_{\mathrm{B}}^2/N_{\mathrm{B}}) + (S_{\mathrm{M}}^2/N_{\mathrm{M}})}} = \frac{7.62 - 7.40}{0.20\sqrt{1/4 + 1}} = \frac{0.22}{0.22} = 1.0 \tag{10.3}$$

where n_0 is the number of background measurements, n_j the number of measurements in monitoring location j, and n_i the number of measurements in monitoring location i. In the measurement of ground water the correlation is constant with value $\rho = 1/(n+1)$, since the number of background measurements $n_0 = n$ and $n_i = n_j = 1$ for all i and j (i.e., we are comparing a single new value in each monitoring location to n background measurements). Dunnett (1955) has shown how required values from the multivariate t-distribution can be reduced to evaluation of the equally correlated multivariate normal distribution for which the required probabilities are easily obtained. These critical points have been tabulated by a number of authors (e.g., Gupta and Panchpakesan, 1979; Gibbons, 1994). As shown below, increasing statistical power can be achieved by generalizing the single-stage Dunnett procedure described here to the case of multistage sampling using verification resampling. Alternative stage-wise comparison procedures have also been considered (Hochberg and Tamhane, 1987) in the context of multiple comparisons to a common control.

Verification Resampling

As the number of future comparisons increases, the prediction limit increases and false-negative rates can become unacceptably large. Gibbons (1987a, b) and Davis and McNichols (1987) noted this problem and suggested sequential testing of new ground-water monitoring measurements such that the presence of an initial exceedance in a downgradient location requires obtaining one or more independent resamples for that constituent. Failure is indicated only if both initial sample and verification resamples exceed the prediction limit. In this way, fewer samples are required and both false-positive and false-negative rates are controlled at minimum levels. Davis and McNichols (1987) derived simultaneous normal prediction limits for the next r of m measurements at each of k monitoring locations, where in the previous example, $r = 1$ and $m = 2$. Their result is a further generalization to Dunnett's test. The derivation is complicated and is reviewed in Gibbons (1994) and Gibbons and Coleman (2001). Complete tables are provided in Gibbons (1994).

Multiple Constituents

Little is known about the correlation between monitoring constituents, except that the interrelationship is highly variable and that there are too few background measurements to precisely characterize the correlation matrix or to use the matrix to construct accurate multivariate prediction limits. For this reason, the Bonferroni inequality has been used to derive conservative prediction bands. This practice will produce prediction limits larger than required when positive association is present (which, in the authors' experience, appears to be common). For example, in the previous illustration, if we were to monitor 10 constituents, $\alpha = 0.05 = 10 = 0.005$ and the limit

$$\bar{x} + 3.36s$$

would be applied to each location and constituent with an overall site-wide confidence level of 95%. Alternatively, there has been some work on multivariate prediction bounds (Guttman, 1970; Bock, 1975), which might apply to those cases where background sample sizes were sufficiently large to obtain a reasonable estimate of the inter-constituent covariance matrix. Unfortunately, the presence of nondetects (i.e., left-censored distributions) violates the joint normality assumption of the multivariate procedure.

Finally, it should be noted that in all cases, the smallest number of constituents that are indicative of a potential release from the facility should be used. Using fewer constituents will decrease the total number of comparisons and provide more conservative (i.e., smaller) prediction bounds. Also note that some constituents exhibit greater spatial variability than others (e.g., geochemical parameters such as chloride versus metals such as barium) and may be less useful at some facilities with heterogeneous geologic formations.

The Problem of Nondetects

In practice, environmental measurements consist of a mixture of detected and nondetected constituents ranging in detection frequency from 0 to 100%. When the detection frequency is high (e.g., >85%) several studies (Gilliom and Helsel, 1986; Hass and Scheff, 1990; Gibbons, 1994) have shown that most estimates of mean and variance of a left-censored normal or lognormal distribution yield reasonable results. This is not true when detection frequencies are between 50 and 85% (Gibbons, 1994). In this case, available methods include maximum likelihood estimators (MLEs) (Cohen, 1959, 1961), restricted maximum likelihood estimators (Persson and Rootzen, 1977), an estimator based on the Delta distribution which is a lognormal distribution with probability mass at zero (Aitchison, 1955), best linear unbiased estimators (Gupta, 1952; Sarhan and Greenberg, 1962), alternative linear estimators (Gupta, 1952), regression-type estimators (Hashimoto and Trussell, 1983; Gilliom and Helsel, 1986), and substitution of expected values of normal-order statistics (Gleit, 1985). In addition, the U.S. EPA has often advocated simple substitution of one-half the MDL. Methods that adequately recover the mean and variance of the underlying distribution from the censored data often inadequately recover the tail probabilities used in computing prediction limits (Gibbons, 1994). In a simulation study (Gibbons, 1994) the MLE was the best overall estimator but the estimator based on the Delta distribution was the best at preserving confidence levels for prediction limits in the presence of censoring.

Nonparametric Prediction Limits

When detection frequency is less than 50%, none of the methods discussed in the previous section work well and an alternative strategy must be employed. In practice, an excellent alternative is to compute a nonparametric prediction limit, which is the maximum of n background measurements. The nonparametric limit is attractive because it makes no distributional assumptions and is defined even if only one of the n background measurements is quantifiable. In some cases, however, the number of background measurements is insufficient to provide a reasonable overall confidence level, therefore the nonparametric prediction limit may not always be an available alternative. Confidence levels for the nonparametric limits are a function of n, kq, and the number of verification resamples similar to the parametric case. For example, let $X_{(\max;\, n)}$ represent the maximum value obtained out of a sample of size n and $Y_{(\min;\, m)}$ represent the minimum value out of a sample of size m. In the present context, $X_{(\max;\, n)}$ is the maximum background concentration and $Y_{(\min;\, m)}$ is the minimum of the initial sample and verification resamples for a constituent in a downgradient monitoring location. The objective is to compare $Y_{(\min;\, m)}$ with $X_{(\max;\, n)}$. The confidence level for the simultaneous upper prediction limit defined as $X_{(\max;\, n)}$ is

$$\Pr Y_{1(min;\, m)} \cdot X_{(max;\, n)};\ Y_{2(min;\, m)} \cdot X_{(max;\, n, \ldots,} Y_{k(min;\, m)} \cdot X_{(\max;\, n)} = 1 - \alpha \tag{10.4}$$

To achieve a desired confidence level (say $1 - \alpha = 0.95$ for a fixed number of background measurements), m must be adjusted; the more the resamples, the greater the confidence. This probability can be evaluated using a variant of the multivariate hypergeometric distribution (Hall et al., 1975; Chou and Owen, 1986) function as

$$1 - \alpha = \frac{n}{km + n} \sum_{j_1=1}^{m} \sum_{j_2=1}^{m} \cdots \sum_{j_k=1} \frac{\binom{m}{j_1}\binom{m}{j_2}\cdots\binom{m}{j_k}}{\binom{km + n - 1}{\sum_{i=1}^{k} ji + n - 1}} \tag{10.5}$$

Based on this result, approximate confidence levels have been derived (Gibbons, 1990) for nonparametric prediction limits defined as the maximum of n background samples in which it is required to pass 1 of m samples (i.e., the initial sample or at least one verification resample) at each of k monitoring locations. To incorporate multiple constituents, the confidence level is adjusted to

$$1 - \frac{\alpha}{q} \tag{10.6}$$

Exact confidence levels for the previous case and approximate confidence levels for the case in which it is required to pass the first or all of m resamples are now also available (Gibbons, 1991). Exact confidence levels for this latter case were recently derived (Willits, 1993; Davis and McNichols, 1994a, b) and extensive tables have been prepared (Gibbons, 1994). The case in which the prediction limit is the second largest measurement has also been considered (Gibbons, 1994; Davis and McNichols, 1994b, 1988).

Intra-Well Comparisons

Upgradient versus downgradient comparisons are often inappropriate (e.g., spatial variability may be present) and some form of intra-well comparisons (i.e., each location compared to its own history) must be performed. Note that intra-well comparisons are only appropriate when (1) predisposal data are available or (2) it can be demonstrated that the facility has not affected that well in the past. In this case, there are two good statistical methods available; combined Shewhart-CUSUM control charts (Lucas, 1982) and intra-well prediction limits (Davis, 1994; Gibbons, 1994). The advantage of the combined Shewhart-CUSUM control chart is that the method is sensitive to both immediate and gradual releases, whereas prediction limits are only sensitive to absolute increases over background. In the intra-well setting, comparisons are independent since each well is compared to its own history. Gibbons (1994, Table 8.3) provides appropriate factors for computing intra-well prediction limits for up to $kq = 500$ future comparisons under a variety of resampling strategies. These factors apply to normally distributed constituents or constituents that can be suitably transformed to approximate normality. In the nonparametric case, selecting a single future sample and setting the confidence level to $(1 - \alpha)/kq$ is also possible; however, overall confidence levels may be poor due to small numbers of background measurements typically available in individual monitoring wells (i.e., generally 8 or fewer). If seasonality is present, adjustments may be required. However, the number of available measurements within a given season is typically one per year, therefore most facilities will have insufficient data to estimate the seasonal effect if present.

Illustration

Consider the data in Table 10.5 for total organic carbon (TOC) measurements from a single well over 2 yr of quarterly monitoring.

Inspection of the data reveals no obvious trends, and these data have a mean $\bar{x} = 11.0$ and standard deviation $s = 0.61$. The upper 95% point of Student's t-distribution on seven degrees of freedom is $t_{[7.1-0.05]} = 1.895$, therefore the upper 95% confidence normal prediction limit in Equation (10.1) is given by

$$11.0 + 1.895(0.61)\sqrt{1 + \frac{1}{8}} = 12.22 \text{ mg/l}$$

which is larger than any of the observed values. This limit provides 95% confidence of including the next single observation from a normal distribution for which eight previous measurements have been obtained with observed mean of 11.0 mg/l and standard deviation of 0.61 mg/l. Assuming that spatial variability does not exist (in many cases a demonstrably false assumption), and that values from this single well are representative of values from each of 10 downgradient wells in the absence of contamination, then the corresponding Bonferroni-adjusted 95% confidence normal prediction limit in Equation (10.2) for the next 10 new downgradient measurements is

$$11.0 + 3.50(0.61)\sqrt{1 + \frac{1}{8}} = 13.26 \text{ mg/l}$$

In contrast, if the dependence introduced by comparing all 10 downgradient wells to the same background were incorporated as in Equation (10.3), the result of

$$11.0 + 3.31(0.61)\sqrt{1 + \frac{1}{8}} = 13.14 \text{ mg/l}$$

TABLE 10.5

Eight Quarterly TOC Measurements

Year	Quarter	TOC (mg/l)
1992	1	10.0
1992	2	11.5
1992	3	11.0
1992	4	10.6
1993	1	10.9
1993	2	12.0
1993	3	11.3
1993	4	10.7

is obtained (see Table 1.4 in Gibbons, 1994). Note that the limit is slightly lower because the multiplier incorporates the dependence introduced by repeated comparison to a common background (i.e., the number of independent comparisons is less than 10 given that they are correlated). Although the Bonferroni-based limit is too conservative, there is little difference in the limits. Extending this result to include the effects of a verification resample as in Equation (10.4) further decreases the limit to

$$11.0 + 2.03(0.61)\sqrt{1 + \frac{1}{8}} = 12.31 \text{ mg/l}$$

If each of 10 constituents in each of the 10 downgradient wells had been monitored, $\alpha = 0.05 = 10 = 0.005$ and the limit would become

$$11.0 + 3.36(0.61)\sqrt{1 + \frac{1}{8}} = 13.17 \text{ mg/l}$$

(see Table 1.5 in Gibbons, 1994). Note that the verification resample allows application of essentially the same limit derived for 10 wells and 1 constituent (13.14 mg/l) to a problem of 10 wells and 10 constituents (13.17 mg/l). Now, consider the nonparametric alternative of taking the maximum of the initial eight background measurements and applying it to the next future monitoring measurements. In this example, the nonparametric prediction limit is 12.00 mg/l. For a single future measurement, confidence is 88% without a resample and 98% with a resample (see Table 2.5 and Table 2.6 in Gibbons, 1994).

For a single measurement in each of 10 monitoring wells, confidence is 44% without a resample and 84% with a resample (see Table 2.5 and Table 2.6 in Gibbons, 1994). With 10 constituents and 10 monitoring wells, an overall 95% confidence level would be obtained with $n = 60$ background samples for one verification resample (see Table 2.6 in Gibbons, 1994) or $n = 20$ samples for passage of one of two verification resamples (see Table 2.7 in Gibbons, 1994). Note that if either the initial sample or both of two resamples must be passed then $n = 90$ background measurements must be obtained (see Table 2.13 in Gibbons, 1994). Other illustrations and further statistical details are available (Davis, 1994; Gibbons, 1994; Davis and McNichols, 1994b).

Some Methods to Be Avoided

Analysis of Variance — ANOVA

In both U.S. EPA Subtitle C and D regulations and associated guidance (U.S. EPA, 1988b, 1989, 1991, 1992), ANOVA is suggested as the statistical method of choice. The U.S. EPA's specific recommendation is a one-way fixed-effect model where the upgradient wells are pooled as one level and each downgradient well represents an additional level in the design. A minimum of four samples is obtained from each well within a semiannual period. In the presence of a significant *F*-statistic, post hoc comparisons (i.e., Fisher's LSD method) between each downgradient well and the pooled upgradient background are performed. Either parametric or nonparametric ANOVA models (i.e., Kruskal–Wallis test) are acceptable. Unfortunately, application of either parametric or nonparametric ANOVA procedures to detection monitoring is inadvisable for the following reasons:

1. Univariate ANOVA procedures do not adjust for multiple comparisons due to multiple constituents. This can be devastating to the site-wide false-positive rate. As such, a site with 10 indicator constituents will have as much as a 40% probability of failing for at least one constituent on every monitoring event by chance alone.

2. ANOVA is more sensitive to spatial variability than to contamination. Spatial variability produces systematic differences between locations that are large relative to within-location variation (i.e., small consistent differences due to spatial variation achieve statistical significance). In contrast, contamination increases variability within the impacted locations, therefore a much larger between-location difference is required to achieve statistical significance. In fact, application of ANOVA methods to predisposal ground-water-monitoring data often results in statistically significant differences between upgradient and downgradient wells, even when no waste is present (Gibbons, 1994), as illustrated in the example below.
3. Nonparametric ANOVA is often presented as if it would protect the user from all of the weakness of its parametric counterpart; however, the only assumption relaxed is that of normality. The nonparametric ANOVA still assumes independence, homogeneity of variance, and that each measurement is identically distributed. Violation of any of these assumptions can corrupt the power of detection, or the false-positive rate.
4. ANOVA requires pooling of downgradient data. Specifically, the U.S. EPA suggests that four samples per semiannual monitoring event be collected (i.e., eight samples per year). However, ANOVA cannot rapidly detect a release since only a subset of the required four semiannual samples will initially be affected by a site impact. This heterogeneity will decrease the mean concentration and increase the variance for the affected location, limiting the ability of the statistical test to detect actual contamination. To illustrate, consider the data in Table 10.6 obtained from a facility in which disposal of waste has not yet taken place (Gibbons, 1994).

Applying both parametric and nonparametric ANOVAs to these predisposal data yielded an effect that approached significance for COD ($p < 0.072$ parametric and $p < 0.066$ nonparametric) and a significant difference for alkalinity (ALK) ($p < 0.002$ parametric and $p < 0.009$ nonparametric). Individually compared (using Fisher's LSD), significantly increased COD levels were found for well MW05 ($p < 0.026$) and significantly increased ALK was found for wells MW06 ($p < 0.026$) and P14 ($p < 0.003$) relative to upgradient wells. These results represent false positives due to spatial variability since no waste has been deposited at this site (i.e., a "greenfield" site).

Most remarkable is the absence of significant results for TOC, notwithstanding the fact that some values are as much as 20 times higher than others. These extreme values increase the within-well variance estimate, rendering the ANOVA powerless to detect differences regardless of magnitude. Elevated TOC data are inconsistent with chance expectations (based on analysis using prediction limits) and should be investigated. In this case, elevated TOC data are likely caused by contamination from insects getting into the wells since this greenfield facility is located in the middle of the Mojave desert.

Cochran's Approximation to the Behrens Fisher t-Test

For years the U.S. EPA RCRA regulation (U.S. EPA, 1982) was based on application of the Cochran's approximation to the Behrens Fisher (CABF) *t*-test. The test was incorrectly implemented by requiring that four quarterly upgradient samples from a single well and single samples from a minimum of three downgradient wells each be divided into four aliquots and treated as if there were $4n$ independent measurements. The result was that most hazardous waste disposal facilities regulated under RCRA were declared "leaking." As an illustration, consider the data in Table 9.7. Note that the aliquots are almost perfectly correlated and add virtually no independent information, yet they are assumed by the statistic to be completely independent. The CABF *t*-test is computed as

TABLE 10.6

Raw Data for All Detection Monitoring Wells and Constituents (mg/l) (This Facility Has no Waste in It)

Well	Event	TOC	TKN	COD	ALK
MW01	1	5.2000	0.8000	44.0000	58.0000
MW01	2	6.8500	0.9000	13.0000	49.0000
MW01	3	4.1500	0.5000	13.0000	40.0000
MW01	4	15.1500	0.5000	40.0000	42.0000
MW02	1	1.6000	1.6000	11.0000	59.0000
MW02	2	6.2500	0.3000	10.0000	82.0000
MW02	3	1.4500	0.7000	10.0000	54.0000
MW02	4	1.0000	0.2000	13.0000	51.0000
MW03	1	1.0000	1.8000	28.0000	39.0000
MW03	2	1.9500	0.4000	10.0000	70.0000
MW03	3	1.5000	0.3000	11.0000	42.0000
MW03	4	4.8000	0.5000	26.0000	42.0000
MW04	1	4.1500	1.5000	41.0000	54.0000
MW04	2	1.0000	0.3000	10.0000	40.0000
MW04	3	1.9500	0.3000	24.0000	32.0000
MW04	4	1.2500	0.4000	45.0000	28.0000
MW05	1	2.1500	0.6000	39.0000	51.0000
MW05	2	1.0000	0.4000	26.0000	55.0000
MW05	3	19.6000	0.3000	31.0000	60.0000
MW05	4	1.0000	0.2000	48.0000	52.0000
MW06	1	1.4000	0.8000	22.0000	118.0000
MW06	2	1.0000	0.2000	23.0000	66.0000
MW06	3	1.5000	0.5000	25.0000	59.0000
MW06	4	20.5500	0.4000	28.0000	63.0000
P14	1	2.0500	0.2000	10.0000	79.0000
P14	2	1.0500	0.3000	10.0000	96.0000
P14	3	5.1000	0.5000	10.0000	89.0000

TABLE 10.7

Illustration of pH Data Used in Computing the CABF *t*-Test

	Replicate				
Date	**1**	**2**	**3**	**4**	**Average**
Background					
November 1981	7.77	7.76	7.78	7.78	7.77
February 1982	7.74	7.80	7.82	7.85	7.80
May 1982	7.40	7.40	7.40	7.40	7.40
August 1982	7.50	7.50	7.50	7.50	7.50
$\bar{x}_B$		7.62			7.62
SD_B		0.18			0.20
N_B		16			4
Monitoring					
September 1983	7.39	7.40	7.38	7.42	7.40
$\bar{x}_B$		7.40			7.40
SD_B		0.02			
N_B		4			1

$$t = \frac{\bar{x}_B - \bar{x}_M}{\sqrt{(S_B^2/N_B) + (S_M^2/N_M)}} = \frac{7.62 - 7.40}{\sqrt{(0.032/16) + (0.004/4)}} = \frac{0.22}{0.05} = 4.82 \quad (10.7)$$

The associated probability of this test statistic is 1 in 10,000, indicating that the chance that the new monitoring measurement came from the same population as the background measurements is remote. Note that, in fact, the mean concentration of the four aliquots for the new monitoring measurement is identical to one of the four mean values for background, suggesting intuitively that probability is closer to one in four rather than one in 10,000. Averaging the aliquots yields the statistic

$$t = \frac{\bar{x}_B - \bar{x}_M}{S_B\sqrt{(1/N_B) + 1}} = \frac{7.62 - 7.40}{0.20\sqrt{(1/4) + 1}} = \frac{0.22}{0.22} = 1.0 \quad (10.8)$$

which has an associated probability of one in two. Had the sample size been increased to $N_B = 20$, the probability would have decreased to one in three. The U.S. EPA eliminated this method from the regulation (U.S. EPA, 1988b).

Summary

Protection of our natural resources is critical. However, statistical tools used to make environmental impact decisions are limited and often confusing. The problem is not only interesting regarding development of public policy, but it also contains features of statistical interest such as multiple comparisons, sequential testing, and censored distributions. Highlighting the weaknesses of currently mandated regulations may lead to further critical examination of public policy in the field of ground-water monitoring as well as heightened interest in statistical analysis.

Reporting Water-Quality Data to Agencies

State and federal regulations require some form of reporting to confirm that the monitoring system is working as required by the codes. Some regulations require the reporting of tabular sets of data on forms or through a formatted electronic media. In general terms, all data should be fully reviewed before transmittal to regulatory agencies. A simple set of guidelines can ease potential errors and embarrassment when submitting water-quality data on your facility:

- Read the permit or waste discharge requirements and follow them.
- Format data as required in a manner that communicates the data most effectively (so everyone reaches the same conclusions).
- If the state requires reporting of exceedances, format the response in a neutral manner:
  - Talk about the specific exceedance issues.
  - Relate progress made on defining causes of the exceedance(s).
  - Propose schedules for establishing the cause of the exceedance or schedules for the remedial actions required.
  - Provide a summary statement on the level of concern.

- Maintain consistency and continuity between quarterly reports:
  - Indicator parameter exceedance changes from quarter to quarter.
  - New personnel should review past data.
  - Always cross-check reports from quarter to quarter.
- Explain what will be done with the data.
- Maintain technical standards and textural reporting consistency between sites; always maintain a consistent standard format for reporting water-quality data.

References

Aitchison, L., On the distribution of a positive, random variable having a discrete probability mass at the origin, *Journal of the American Statistical Association*, 50, pp. 901–908, 1955.

Benjamin, J.R. and C.A. Cornell, *Probability Statistics and Decision for Civil Engineers*, McGraw-Hill Publishing Co, New York, NY, 1970.

Bock, R.D., *Multivariate Statistical Methods in Behavioral Research*, McGraw-Hill Publishing Co, New York, NY, 1975.

Chou, Y.M. and D.B. Owen, One-sided distribution-free simultaneous prediction limits for future samples, *Journal of Quality Technology*, 18, pp. 96–98, 1986.

Cohen, A.C., Simplified estimators for the normal distribution when samples are singly censored or truncated, *Technometrics*, 1, pp. 217–237, 1959.

Cohen, A.C., Tables for maximum likelihood estimates: singly truncated and singly censored samples, *Technometrics*, 3, pp. 535–541, 1961.

Davis, C.B., Environmental regulatory statistics, Chapter 25 in, *Handbook of Statistics: Environmental Statistics*, Patil, G.P. and Rao, C.R. Eds., Vol. 12, Elsevier Press, New York, NY, 1994.

Davis, C.B. and R.J. McNichols, One-Sided Intervals for at Least *p* of *m* Observations From a Normal Population on Each of *r* Future Occasions, *Technometrics*, 29, pp. 359–370, 1988.

Davis, C.B. and R.J. McNichols, Ground-water monitoring statistics update: I: progress since 1988, *Ground-Water Monitoring and Remediation*, 14(4), pp. 148–158, 1994a.

Davis, C.B. and R.J. McNichols, Ground-water monitoring statistics update: II: non-parametric prediction limits, *Ground-Water Monitoring and Remediation*, 14(4), pp. 159–169, 1994b.

Doctor, P.G., R.O. Gilbert, R.A. Saar, and G. Duffield, Draft Statistical Comparisons of Ground-Water Monitoring Data: Ground-Water Plans and Statistical Procedures to Detect Leaking at Hazardous Waste Facilities, Document PNL-5754, Battelle Pacific Northwest Laboratories, Richland, WA, 1986.

Doctor, P.G., R.O. Gilbert, R.A. Saar, and G. Duffield, An Analysis of Sources of Variation in Ground-Water Monitoring Data of Hazardous Waste Sites, Milestone 1, Revised Draft EPA Contract No. 68-01-6871, Battelle Pacific Northwest Laboratories. Richland, WA, 1985.

Dunnett, C.W., A multiple comparisons procedure for comparing several treatments with a control, *Journal of the American Statistical Association*, 50, pp. 1096–1121, 1955.

Dunnett, C.W. and M. Sobel, Approximations to the probability integral and certain percentage points of a multivariate analogue of student's *t*-distribution, *Biometrika*, 42, pp. 258–260, 1955.

Fisher, D.A. and K. Potter, Methods for Determining Compliance with Ground Water Quality Regulations at Waste Disposal Facilities, Wisconsin Department of Natural Resources, Madison, WI, 120 pp, 1989.

Freeze, R.A. and J.A. Cherry, *Groundwater*, Prentice-Hall, Inc, Englewood Cliffs, NJ, 1979, 604 pp.

Gibbons, R.D., Statistical prediction intervals for the evaluation of ground-water quality, *Ground Water*, 25(4), pp. 455–465, 1987a.

Gibbons, R.D., Statistical models for the analysis of volatile organic compounds in waste disposal facilities, *Ground Water*, 25(5), pp. 572–580, 1987b.

Gibbons, R.D., A general statistical procedure for ground-water detection monitoring at waste disposal facilities, *Ground Water*, 28(2), pp. 235–243, 1990.

Gibbons, R.D., Some additional nonparametric prediction limits for ground-water monitoring at waste disposal facilities, *Ground Water*, 29(5), pp. 729–736, 1991.

Gibbons, R.D., *Statistical Methods for Ground-Water Monitoring*, John Wiley and Sons, New York, NY, 1994.

Gibbons, R.D. and D.E. Coleman, *Statistical Methods for Detection and Quantification of Environmental Contamination*, John Wiley and Sons, New York, NY, 2001.

Gibbons, R.D., F.H. Jarke, and K.P. Stoub, Method Detection Limits, Proceedings of the Fifth Annual U.S. EPA Waste Testing and Quality Assurance Symposium, pp. 292–319, 1988.

Gilliom, R.J. and D.R. Helsel, Estimation of distributional parameters for censored trace-level water-quality data: 1. estimation techniques, *Water Resources Research*, 22, pp. 135–146, 1986.

Gleit, A., Estimation for small normal data sets with detection limits, *Environmental Science and Technology*, 19, pp. 1201–1206, 1985.

Goodman, I., Graphical and Statistical Methods to Assess the Effect of Landfills on Ground-Water Quality, M.S. Thesis, University of Wisconsin Madison, 1987.

Goodman, I. and K. Potter, Graphical and Statistical Methods to Assess the Effects of Landfills on Ground Water Quality, Report to Wisconsin Department of Natural Resources, Bureau of Solid and Hazardous Waste, 1987.

Green, W.R., *Computer-Aided Data Analysis: A Practical Guide*, John Wiley and Sons, New York, NY, 268 pp., 1985.

Gupta, A.K., Estimation of the mean and standard deviation of a normal population from a censored sample, *Biometrika*, 39, pp. 260–273, 1952.

Gupta, S.S. and P. Panchpakesan, *Multiple Decision Procedures*, John Wiley and Sons, Inc, New York, NY, 1979.

Guttman, I., *Statistical Tolerance Regions: Classical and Bayesian*, Hafner Publishing, Darien, CT, 1970.

Haas, C.N. and P.A. Scheff, Estimation of averages in truncated samples, *Environmental Science and Technology*, 24, pp. 912–919, 1990.

Hahn, G.J., Statistical intervals for a normal population, part 1: examples & applications, *Journal of Quality Technology*, 2(3), (July), pp. 15–125, 1970.

Hall, I.J., R.R. Prarie, and C.K. Motlagh, Non-parametric prediction intervals, *Journal of Quality Technology*, 7, pp. 109–114, 1975.

Harris, J., J.C. Loftis, and R.H. Montgomery, Statistical methods for characterizing ground-water quality, *Ground Water*, 25(2), pp. 185–193, 1987.

Hashimoto, L.K. and R.R. Trussell, Evaluating Water Quality Data Near the Detection Limit, Proceedings of the American Water Works Association Advanced Technology Conference, American Water Works Association, Denver, CO, 1983.

Hem, J.D., Study and Interpretation of the Chemical Characteristics of Natural Water, U.S. Geological Survey Water-Supply Paper 1473, 363 pp, 1970.

Hoaglin, D.C., F. Mosteller, and J.W. Tukey, *Understanding Robust and Exploratory Data Analysis*, John Wiley & Sons, New York, NY, 1983.

Hochberg, Y. and A.C. Tamhane, *Multiple Comparison Procedures*, John Wiley and Sons, Inc, New York, NY, 1987.

Hurd, M., Personal communication, U.S. Environmental Protection Agency, 1986.

Jarke, F., Is it possible to understand MDLs, PQLs, IDLs, and EMLRLs? *Lab Notes*, 2, 1989.

Lucas, J.M., Combined shewhart-CUSUM quality control schemes, *Journal of Quality Technology*, 14, pp. 51–59, 1982.

McBean, E. and F.A. Rovers, Alternatives for handling detection limit data in impact assessments, *Ground-Water Monitoring Review*, 4(2), pp. 42–44, 1984.

McGill, R., J.W. Tukey, and W.A. Larsen, Variations of box plots, *The American Statistician*, 32(1), pp. 12–16, 1978.

McNichols, R.J. and C.B. Davis, Statistical issues and problems in ground-water detection monitoring at hazardous waste facilities, *Ground-Water Monitoring Review*, 8(4), pp. 135–150, 1988.

Miller, M.D. and F.C. Kohout, RCRA Ground-Water Monitoring Statistical Comparisons: A Better Version of Student's *t*-Test, Mobil Research and Development Corporation, Paulsboro, NJ, undated.

Miller, R.G., *Simultaneous Statistical Inference*, McGraw-Hill Publishing Co, New York, NY, 1966.
Montgomery, R.H., J.C. Loftis, and J. Harris, Statistical characteristics of ground-water quality variables, *Ground Water*, 25(2), pp. 176–184, 1987.
Nielsen, D.M. (Editor), *The Practical Handbook of Environmental Site Characterization and Ground-water Monitoring*, Second Edition, CRC Press/Taylor and Francis Group, Boca Raton, FL, 1318 pp., 2006.
Persson, T. and H. Rootzen, Simple and highly efficient estimators for a type I censored normal sample, *Biometrika*, 64, pp. 123–128, 1977.
Piper, A.M., A Graphic procedure in the geochemical interpretation of water analysis; transactions of the American Geophysical Union, 25, pp. 914–923, 1944.
Sarhan, A.E. and B.G. Greenberg, Eds., *Contributions to Order Statistics*, John Wiley and Sons, Inc, New York, NY, 1962.
Sen, Z., Application of the autorun test to hydrologic data, *Journal of Hydrology*, 42, pp. 1–7, 1979.
Sen, Z., Discussion of statistical considerations and sampling techniques for ground-water quality monitoring by, J.D. Nelson and R.C. Ward, *Ground Water*, 20(4), pp. 494–495, 1982.
Silver, C.A., Statistical Approaches to Ground-Water Monitoring, Open File Report #7, University of Alabama, Environmental Institute for Waste Management Studies, 1986.
U.S. EPA, Hazardous Waste Management System: Permitting Requirements for Land Disposal Facilities; Federal Register, 47(143), July 26, pp. 32274–32373, 1982.
U.S. EPA, Test Methods for Evaluating Solid Waste, Physical Chemical Methods, SW-846, U.S. Environmental Protection Agency, Washington, DC, 1988a.
U.S. EPA, Statistical Methods for Evaluating Ground Water Monitoring Data From Hazardous Waste Facilities: Final Rule, 40 CFR Part 264, Federal Register, Vol. 53, No. 196, October 11, pp. 39720–39731, 1988b.
U.S. EPA, Interim Final Guidance on Statistical Analysis of Ground-Water Monitoring Data at RCRA Facilities, April, 1989.
U.S. EPA, Solid Waste Disposal Facility Criteria: Final Rule, Federal Register, Vol. 56, No. 196, October 9, pp. 50978–51119, 1991.
U.S. EPA, Addendum to Interim Final Guidance on Statistical Analysis of Ground-Water Monitoring Data at RCRA Facilities, July, 1992.
Willits, N., Personal Communication, University of California-Davis, 1993.

Index

H

I

L

M

N

O

P

T

U

V

W

www.ingramcontent.com/pod-product-compliance
Lightning Source LLC
LaVergne TN
LVHW080643011026
849160LV00014B/1301

* 9 7 8 1 4 2 0 0 4 2 7 8 8 *